Stochastic Processes in Physics and Engineering

Mathematics and Its Applications

Stochastic Processes in Physics and Engineering

edited by

Sergio Albeverio

Fachbereich Mathematik, Ruhr-Universität Bochum, F.R.G.

Philip Blanchard

Fakultät für Physik, Universität Bielefeld, F.R.G.

Michiel Hazewinkel

Centrum voor Wiskunde en Informatica, Amsterdam, The Netherlands

and

Ludwig Streit

Fakultät für Physik, Universität Bielefeld, F.R.G.

D. Reidel Publishing Company

A MEMBER OF THE KLUWER ACADEMIC PUBLISHERS GROUP

Dordrecht / Boston / Lancaster / Tokyo

Library of Congress Cataloging in Publication Data

CIP

Stochastic processes in physics and engineering / edited by S. Albeverio . . . [et al.].
p. cm. — (Mathematics and its applications)
Results of a Meeting, BIBOS IV, held April 1986, in Bielefeld.
Includes index.
ISBN-13: 978-94-010-7803-0 e-ISBN-13: 978-94-009-2893-0
DOI: 10.1007/ 978-94-009-2893-0
1. Stochastic processes—Congresses. 2. Mathematical physics—Congresses. 3. Engineering mathematics—Congresses. I. Albeverio, Sergio. II. BIBOS IV (1986: Bielefeld, Germany) III. Series: Mathematics and its applications (D. Reidel Publishing Company)
QC20.7.S8S77 1987
519.2—dc 19 87–31865
CIP

Published by D. Reidel Publishing Company,
P.O. Box 17, 3300 AA Dordrecht, Holland.

Sold and distributed in the U.S.A. and Canada
by Kluwer Academic Publishers,
101 Philip Drive, Norwell, MA 02061, U.S.A.

In all other countries, sold and distributed
by Kluwer Academic Publishers Group,
P.O. Box 322, 3300 AH Dordrecht, Holland.

SERIES EDITOR'S PREFACE

Approach your problems from the right end and begin with the answers. Then one day, perhaps you will find the final question.

'The Hermit Clad in Crane Feathers' in R. van Gulik's *The Chinese Maze Murders*.

It isn't that they can't see the solution. It is that they can't see the problem.

G.K. Chesterton. *The Scandal of Father Brown* 'The point of a Pin'.

Growing specialization and diversification have brought a host of monographs and textbooks on increasingly specialized topics. However, the "tree" of knowledge of mathematics and related fields does not grow only by putting forth new branches. It also happens, quite often in fact, that branches which were thought to be completely disparate are suddenly seen to be related.

Further, the kind and level of sophistication of mathematics applied in various sciences has changed drastically in recent years: measure theory is used (non-trivially) in regional and theoretical economics; algebraic geometry interacts with physics; the Minkowsky lemma, coding theory and the structure of water meet one another in packing and covering theory; quantum fields, crystal defects and mathematical programming profit from homotopy theory; Lie algebras are relevant to filtering; and prediction and electrical engineering can use Stein spaces. And in addition to this there are such new emerging subdisciplines as "experimental mathematics", "CFD", "completely integrable systems", "chaos, synergetics and large-scale order", which are almost impossible to fit into the existing classification schemes. They draw upon widely different sections of mathematics. This programme, Mathematics and Its Applications, is devoted to new emerging (sub)disciplines and to such (new) interrelations as exempla gratia:

- a central concept which plays an important role in several different mathematical and/or scientific specialized areas;
- new applications of the results and ideas from one area of scientific endeavour into another;
- influences which the results, problems and concepts of one field of enquiry have and have had on the development of another.

The Mathematics and Its Applications programme tries to make available a careful selection of books which fit the philosophy outlined above. With such books, which are stimulating rather than definitive, intriguing rather than encyclopaedic, we hope to contribute something towards better communication among the practitioners in diversified fields.

Stochastics is, as a well-developed specialism in mathematics, a relatively young discipline and probably the fastest growing one. It is extraordinarily important in applications. There are, of course, many phenomena in nature and engineering which have a random component. But the applications of stochastic thinking are not limited to such cases. Stochastic models also can be, and are, used to model other phenomena which do not have an obvious random component. For example, very complicated deterministic phenomena (turbulence, chaos) can sometimes fruitfully be handled in this way, or unknown aspects of a given class of models can be incorporated by means of stochastic ideas. Further, there are the interrelations between potential theory and probability and, of course, the evolution equations for the density of suitable continuous-time Markovian processes are partial differential equations, and such equations may well arise in different contexts. And in such cases a stochastic interpretation can be very fruitful. Quite generally there seems to be a tendency in mathematics as a whole for ideas from stochastics to penetrate into other areas. For example, stopping times are a powerful tool in certain parts of analysis.

As already remarked, stochastics has very many 'applications': in physics, in chemistry, in geology, in astronomy, in all kinds of engineering, ... These application areas have all evolved their own corresponding intuitions. In this book, the ways physicists think about these things and the way engineers do are brought together. That was the idea of the meeting on which it is based. There is more on this theme in the preface to the book. And it is clear from the meeting itself that such an interaction can be remarkably useful and stimulating. This book will help to further enhance this beneficial interaction.

The unreasonable effectiveness of mathematics in science ...

Eugene Wigner

Well, if you know of a better 'ole, go to it.

Bruce Bairnsfather

What is now proved was once only imagined.

William Blake

As long as algebra and geometry proceeded along separate paths, their advance was slow and their applications limited.

But when these sciences joined company they drew from each other fresh vitality and thenceforward marched on at a rapid pace towards perfection.

Joseph Louis Lagrange.

Bussum, September 1987

Michiel Hazewinkel

Table of Contents

PREFACE

BIBOS, of course, stands for a rather large research project stretching over a number of years involving the universities of Bochum and Bielefeld. The 'S' stands for stochastics. The project is funded by the Stiftung Volkswagenwerk. The present book is the result of the meeting BIBOS IV which took place in April 1986 at ZiF in Bielefeld. The basic philosophy at the back of this meeting was as follows:

> 'Stochastics in physics has a strong intuitive background. Also in order to deal with the various stochastic ideas which arise in physics an enormous amount of technique and expertise has been developed. It is quite likely that much of all this could also profitably be used in the kind of problems met in engineering and industrial mathematics, in chemistry and geology, as well as in economics, planning and management. The inverse is also likely true. For instance the engineering community has much more of a habit in thinking about problems involving exterior forces (often called inputs or controls in certain settings) and observations (outputs), i.e. about situations in which only indirect data are available about the true physical state of the process involved. Thus the basic philosophy of BIBOS IV is the serve as a meeting ground and forum where engineers, planners, analysts, physicists and mathematicians meet and try to solve each other problems and describe and discuss each others solutions and techniques'.

Let us elaborate a bit on this theme at the hand of the various contributions in this volume.

Simulated annealing (Bernasconi) is a global optimization technique and is a perfect example of how introducing ideas from a completely different, and at first sight totally unrelated, field may yield large and unexpected benefits. Annealing is the process of heating a solid and cooling it slowly to remove strain and crystal imperfections. During this process the free energy of the solid is minimized. The initial heating is necessary to avoid becoming trapped in a local minimum. Virtually every function can be viewed as the free energy of some system and thus imitating on a computer whatever goes on physically during annealing yields optimization algorithms. The result is a powerful new technique called simulated annealing. Still very young; the first book on the topic (Aarts, van Laarhoven, Simulated annealing: theory and practice, Reidel, 1987) just appeared. It is clearly a very powerful tool in some cases. It is almost totally unknown for what kind of optimization problem it will give good results. "Potholes" e.g. are bad (cf. the Bernasconi article). Perhaps here another look at the original physics will prove illuminating.

As already noted the engineering world (signal processing, control) has paid a great deal of attention to situations where the state of a certain process is not directly observable but must be infered from - usually noise corrupted - indirect (derived) observations. Here we take this problem in the large sense of including parameter identification. This problem area, not surprisingly, involves all kinds of deep mathematics and sophisticated computer science and is well represented by five articles (Blankenship, Fresewinkel, Hazewinkel, Koski, Picard) all more or less from the engineering side. Now of course also in physics the true state is rarely directly accessible and must be infered. It would seem likely that the large engineering literature on filtering and identification could be very useful in the natural sciences. On the other hand some of the series expansions in the Blankenship and Hazewinkel articles are very similar to Dyson-Feynman expansions and it is far from unlikely that some of the ideas of theoretical physics to deal with divergent series will be most useful here.

Wavelets of course come out of physics. It is already clear that a wavelet-based Fourier-type analysis (Grossmann) will become most important in such computer science and engineering subjects as pattern recognition and signal and image processing.

Optimal control, stochastic (optimal) control and adaptive control mainly came out of engineering but the applications definitely are not limited to engineering and physics. There are for instance all kinds of applications to economics (Gomez, Kemna-Vorst, Papavassilopoulos to some extent). A more traditional engineering adaptive control problem is discussed by Hut-Olsder. Quite generally adaptive type control should be important in very many fields. So far it seems not to be employed systematically outside of the control electrical engineering and signal processing worlds. The idea is simple: built in extra feedback loops which serve to adapt the control loops to unknown or changing parameters of the system. There are a number of adaptive schemes available and implemented. Many work quite well. The underlying theory is still almost totally missing. Thus there are but a few convergence and stability proofs, and we have no idea to what extent the various available schemes are in any sense near optimal.

By and large the engineering community seems to be more conservative that the physics one, and slower to adopt such bits of weird mathematics as nonstandard analysis, superspace and fractals and also slower to see that the simplest approximation to 'large' may well be 'infinite'. However nonstandard techniques definitely have found their way into optimal stochastic control (Cutland), and in this field yield standard, i.e. implementable, results.

One traditional difficulty with infinite dimensional things is that usually there are many potential infinite dimensional analogues for well studied finite dimensional concepts. For example the Laplacian and harmonic analysis. Here the stochastic approach via Hida's powerful white noise analysis ideas suggests the right analogues for infinite dimensional harmonic analysis, an infinite dimensional Laplacian and an infinite dimensional rotation group (Hida-Saito).

Many applications of stochastic ideas rest in one form or another on a some law of large numbers. This makes any and all studies on limit distributions important also from the applied point of view (Schneider).

Socalled $1/f$ noise occurs frequently in engineering situations (cooled electric circuits e.g., or in road modeling ($1/f^2$ for large f seems more frequent there)) and has been the subject of many conferences. Beyond phenomenological results, descriptions etc. not much could be done until recently. The start was a suggestion, independently by several people, that deterministic chaos could well have some of the characteristics of $1/f$-noise. Now with Takahashi's impressive paper on semi-Bernouilli systems the road seems to be open for a great deal of progress in this area. It is rather pleasant to observe that the engineering community thus will have to master yet another area of sophisticated mathematics (viz. Fredholm determinants and such; and no doubt, as always, they will find yet other uses for these concepts).

Fractals, like deterministic chaos, also are but hesitantly finding a niche in the engineering worlds. Possibly because they are primarily a descriptive device with no physical dynamics behind them and (hence) no control parameters. The word 'are' in the previous sentence is not quite correct anymore. It is one of the most fascinating merits of the Pietronero-Evertsz-Siekesma paper in this volume that they do manage to give an underlying 'genetic' dynamics to fractal structures. This really opens up all kinds of highly promising control applications. For of course approximately fractal structures turn up just about anywhere.

The applications of stochastics are by no means limited to phenomena with a clear stochastic component in them. Many phenomena which are purely deterministic can up to a point be modeled well or even better by means of stochastic ideas. Turbulence for instance can be approached this way (Chow). Here, in the general area of modeling, the engineering communities seem to put rather less stress on the need to interprete the various internal variables of the model in terms of physically existing entities. This leads sometimes to remarkably useful, very flexible and mathematically attractive (exotic) models such as a completely 1-0 model for the heating of a gas flowing through a pipe. Sometimes building a model based on whatever is known about the physics is preferable; sometimes a more 'black-box' type approach yields more useful results.

We already mentioned simulated annealing above. One gift, so to speak, from statistical mechanics to other parts of mathematics. Interacting particle systems constitute another idea from this general field. And these in turn have lots of applications, e.g. in several ways to biology (stepping-stone models for distributed biological populations (Shiga) and models for interacting biological cell systems including cancer growth models (Tautu).

Finally, in this volume, there is the metaplectic group and its 'most singular representation' the oscillator representation or Segal-Shale-Weil representation. It comes up in the (reduction) theory of quadratic forms, i.e. in number theory (Weil), it arises naturally when studying symmetries of boson quantum fields (Segal, Shale), it is definitely one of the most important representations from the purely representation theoretic and harmonic analysis points of view (R. Howe's theory of dual reductive pairs) and it is narrowly related to linear filtering and the Kalman-Bucy filters and hence the matrix Riccati equation (cf. the Hazewinkel article in this volume). That would appear to various enough for any one mathematical object. But no, it is also of fundamental and for reaching importance in laser optics and computerized tomography (Schempp).

Thus we feel that the contributions in this volume form a nice interesting cross-section of stochastics (taken to include some 'probabilistic-like' aspects of quantum mechanics) as it manifests itself across a rather wide area of science and engineering. May these contributions be as stimulating to many readers as they were to the participants.

Organizing a meeting like this is work. It is a real pleasure to thank the most efficient staff of ZiF in Bielefeld for their courteous and plentiful help. It is also a pleasure to thank the Stiftung Volkswagenwerk for the BIBOS funds which went into this meeting.

S. Albeverio
Ph. Blanchard
M. Hazewinkel
L. Streit

STATISTICAL PHYSICS AND OPTIMIZATION: BINARY SEQUENCES WITH SMALL AUTOCORRELATIONS

J. Bernasconi
Brown Boveri Research Center
CH-5405 Baden
Switzerland

ABSTRACT. Recent developments in the statistical mechanics of disordered and frustrated systems have led to the design of new, unconventional algorithms for hard optimization problems. These so-called "simulated annealing algorithms" are briefly introduced and then applied to find binary sequences with low off-peak autocorrelations which play an important role in several communication engineering applications. The problem is equivalent to the problem of finding low-energy configurations for a specific spin model with long-range 4-spin interactions whose statistical mechanics is analyzed in some detail. For long sequences, the energy minima found with the simulated annealing procedure differ by about a factor of 2 from the conjectured "true" minimum, indicating that the lowest-energy configurations are extremely isolated in configuration space.

1. INTRODUCTION

In the engineering of complex systems one is often confronted with hard optimization problems. Examples are problems of the travelling salesman type, graph partitioning problems, placement- and wiring-problems in the design of integrated circuits, scheduling and routing problems, the optimization of codes, etc [see e.g. references 1 to 6].

Quite generally, such problems consist in the optimization (minimization or maximization) of a cost function (objective function)

S. Albeverio et al. (eds.), Stochastic Processes in Physics and Engineering, 1–16.

which depends on a set of variables that describe the possible configurations (states) of the system. In the case of combinatorial optimization, the number of possible configurations is finite (e.g. 2^N or N! if N denotes the number of variables). As this number increases exponentially with N, however, an exhaustive search becomes impossible for many practical applications. On the other hand, in engineering problems one is often not particularly interested in finding a global optimum, but is satisfied with a near-optimal solution if this can be found with an effective approximate optimization strategy. Due to so-called frustration effects, however, complex optimization problems possess a very large number of local optima which considerable increase the difficulty of finding a good optimum.

Conventional heuristic optimization strategies usually consist of iterative local improvements (e.g. steepest descent methods), applied to different randomly generated starting configurations. As such algorithms invariably get stuck in the nearest local optimum, their average performance is thus often rather poor if the optimization problem has a high complexity.

The complications that arise in complex optimization problems are closely related to those encountered in the statistical mechanics of disordered and frustrated physical systems, such as e.g. spin glasses [1,7,8]. The cost function of the optimization problem thereby corresponds to the energy function of a physical system, and the optimal configurations correspond to the low-energy states. The difficulties in finding these states in complex systems are due to the absence of an obvious symmetry and to frustration effects that make it impossible to obtain the true ground state by a local optimization procedure.

Some recent developments in the field of statistical physics have now led to useful insights for the design of new heuristic optimization algorithms [1-8]. An often very effective method is to simulate the evolution of the physical system in a slow cooling process, so that it freezes into a state of very low energy. These so-called "simulated annealing algorithms" will be introduced in section 2. They have been applied to a variety of optimization problems [1-6], most extensively to problems of the travelling salesman type.

In section 3, we introduce the problem of finding low autocorrelation binary sequences [9-11]. Binary sequences with low off-peak autocorrelations play an important role in several communication engineering problems (separation of signals from noise, radar ranging, synchronization, etc.). The problem of finding such sequences is equivalent to the problem of determining the low-energy configurations of a specific one-dimensional spin model with long-range 4-spin interactions. The statistical mechanics of this spin model is analyzed with a simulated annealing procedure, and the results are compared with predictions based on a simple "ergodicity assumption" [9,10], which moreover may be used to derive an estimate for the true ground-state energy. For long sequences, the optimum autocorrelation merit factors found with the simulated annealing procedure are comparable with the best values known today, but differ by about a factor of 2 from the conjectured "true" optimum. This indicates that the lowest-energy configurations are extremely isolated in configuration space.

2. SIMULATED ANNEALING ALGORITHMS FOR OPTIMIZATION PROBLEMS

Let us consider an optimization problem of the form

$$E(S) = \text{minimum} \quad ,$$

where E(S) is the cost function and where S denotes a configuration (or state) of the system. We then identify E(S) with the energy function of a physical system and recall that the probability to find the system in state S at temperature T is given by the Boltzmann distribution

$$P_T(S) = \frac{1}{Z} e^{-E(S)/T} \quad , \tag{1}$$

where Z is the partition function

$$Z = \sum_S e^{-E(S)/T} \quad . \tag{2}$$

If $\rho(E)$ denotes the probability density for the distribution of the energy values, Z can also be expressed as

$$Z = N_s \int dE \, \rho(E) e^{-E/T} \quad , \tag{3}$$

where N_s is the total number of states. The probability density for finding the system in a state with energy between E and E+dE then becomes

$$\rho_T(E) = \frac{N_s}{Z} \rho(E) e^{-E/T} \quad . \tag{4}$$

From the partition function Z we can calculate all thermodynamic quantities of the system, e.g. the average energy at temperature T,

$$\langle E \rangle = - \frac{d\ln Z}{d(1/T)} \quad , \tag{5}$$

or the specific heat,

$$C = \frac{d\langle E \rangle}{dT} = \frac{1}{T^2} [\langle E^2 \rangle - \langle E \rangle^2] \quad . \tag{6}$$

For complex systems, however, an exact analytical determination of the thermodynamic averages is usually not possible. We may then try to obtain the equilibrium distribution $P_T(S)$, or $\rho_T(E)$, numerically. This can be achieved with the "Metropolis Procedure" [12] which simulates, for a fixed temperature T, the temporal evolution of the system. In each step of the procedure, the current state S of the system is changed randomly to create a trial state S', and the energy difference

$$\Delta E = E(S') - E(S)$$

is calculated. The new state S' is then accepted with probability $p(\Delta E)$, where

$$p(\Delta E) = \begin{cases} 1 & \text{if} \quad \Delta E \leq 0 \\ e^{-\Delta E/T} & \text{if} \quad \Delta E > 0 \end{cases} \quad . \tag{7}$$

With this choice of $p(\Delta E)$, the system evolves according to the Boltzmann distribution $P_T(S)$, or $\rho_T(E)$, if the procedure is iterated long enough at a fixed temperature T.

We now observe that $\rho_T(E)$ is concentrated around lower and lower energy values as the temperature T decreases. By combining the Metropolis procedure with a sufficiently slow cooling schedule, we can therefore hope to simulate the freezing of the system into a state of very low energy.

Corresponding "Simulated Annealing Algorithms" have been designed for various hard optimization problems [1-6]. E(S) then represents the cost function to be minimized, and T plays the role of an "effective temperature" or control parameter. The optimization process is started at a high enough temperature, such that the system can evolve freely and is not trapped in local minima near the starting configuration. The Metropolis procedure is then used to simulate the evolution of the system at a sequence of slowly decreasing temperatures until the system "freezes" and no further configuration changes occur. At each temperature, the simulation must be continued long enough, so that the system can reach a steady state. The annealing schedule therefore has to specify the sequence of temperatures and the number of attempted changes of state at a given temperature.

Finding a suitable set of configuration changes $S \to S'$, and constructing an effective annealing schedule may require considerable insight into the specific problem to be solved. An observation of the behavior of thermodynamic averages can also give some information. A large value of the specific heat C, for example, indicates that a freezing process has begun, so that very slow cooling is required.

Simulated annealing algorithms differ from iterative improvement strategies in one important respect. Because of Eq.(7), transitions with $\Delta E > 0$ are always possible at a non-zero temperature. Annealing procedures therefore need not get stuck in a local minimum, in contrast to iterative improvement methods that correspond to an infinitely rapid cooling, or "quenching", of the system.

3. LOW AUTOCORRELATION BINARY SEQUENCES

In this section we are concerned with binary sequences of length N,

$$S = \{s_1, s_2, \ldots, s_N\} \quad , \quad s_i = \pm 1 \quad , \tag{8}$$

and with their aperiodic autocorrelations,

$$R_k = \sum_{i=1}^{N-k} s_i \, s_{i+k} \quad , \quad k = 0,1,\ldots,N-1 \quad . \tag{9}$$

Binary sequences with small off-peak ($k \neq 0$) autocorrelations play an important role in several communication engineering problems [9-11], such as e.g. synchronization, pulsed radar range detection, and smearing filters for noise suppression. The problem of constructing such sequences has a long history (cf. [10] and the references cited therein), and has turned out to be a very hard mathematical problem.

Two criteria have been used for low autocorrelation binary sequences. The first is small values of $\max\{|R_k|, k \neq 0\}$, and the second, introduced by Golay [9,10], is a large "merit factor" F, defined as

$$F = N^2/(2 \sum_{k=1}^{N-1} R_k^2) \quad . \tag{10}$$

Here we shall concentrate on the second one, so that our optimization problem can be stated as

$$F = \text{maximum} \quad ,$$

or

$$E = \frac{N^2}{2F} = \sum_{k=1}^{N-1} R_k^2 = \text{minimum} \quad . \tag{11}$$

By an exhaustive search, the sequences with maximal merit factor F have been found up to N = 32 (see [10]). For higher values of N, however, only limited results have been obtained. The most extensive searches have concentrated on skew-symmetric sequences of odd length N = 2n-1. These sequences satisfy

$$s_{n+\ell} = (-1)^{\ell} s_{n-\ell} \quad , \qquad \ell = 1,\ldots,n-1 \quad , \tag{12}$$

from which it follows that all R_k with k odd vanish,

$$R_{2\ell-1} = 0 \quad , \qquad \ell = 1,\ldots,n-1 \quad . \tag{13}$$

Skew-symmetry reduces the size of the problem by a factor of 2, and Golay [9] has made an exhaustive search to find all optimal skew-symmetric sequences for N up to 59. In Figure 1, the corresponding F-values (for $17 \leq N \leq 59$) are plotted versus 1/N, together with the best results obtained by an extended computer search [11] for low

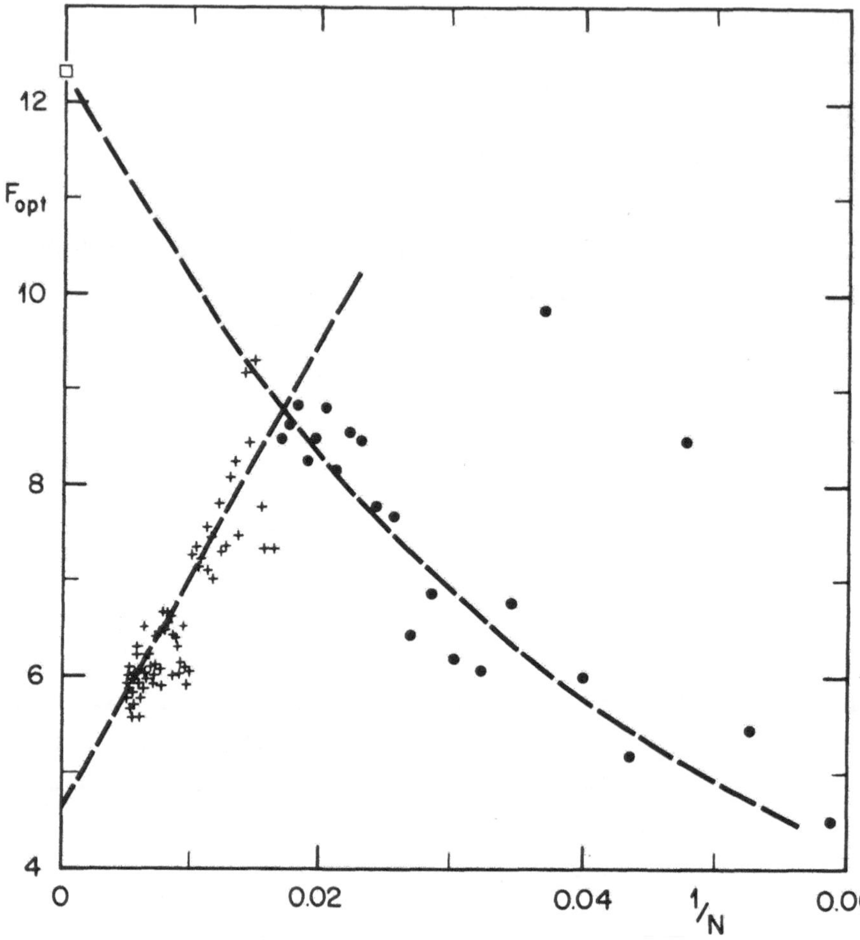

Figure 1: Highest known merit factors for skew-symmetric binary sequences of odd length N, $17 \leq N \leq 199$. The circles refer to an exhaustive search for $N \leq 59$ [9], and the crosses represent the best results obtained by extensive computer searches for higher values of N [11].

autocorrelation skew-symmetric sequences of lengths N = 61,63,..., 199. While the known optima below N = 59 are not inconsistent with an extrapolation to the conjectured asymptotic value [10] of F = 12.32 (see below), the best results obtained by stochastic search procedures decrease to an asymptotic value below F = 5. In this context we also note that the highest F-values known for very long sequences, attained by cyclically shifted Legendre sequences [11,13], are close to F = 6, more than a factor of 2 smaller than the conjectured "true" optimum.

With the hope to obtain some insight into this apparent discrepancy, we shall now attempt to analyze the statistical mechanics of our optimization problem. The cost function,

$$E(S) = \sum_{k=1}^{N-1} R_k^2 = \sum_{k=1}^{N-1} \sum_{i=1}^{N-k} \sum_{j=1}^{N-k} s_i \, s_{i+k} \, s_j \, s_{j+k} \quad , \tag{14}$$

can be regarded as the energy function of a one-dimensional spin system with long-range 4-spin interactions, and we first derive a simple approximate expression for the partition function Z_N of this highly frustrated system. The approximation is based on Golay's "ergodicity postulate" [10] which assumes that the R_k^2 in Eq. (14) can be treated as independent random variables y_k whose discrete distributions are approximated by continuous probability densities $p_k(y)$,

$$p_k(y) = \frac{1}{[2\pi(N-k)y]^{1/2}} \exp\left[-\frac{y}{2(N-k)}\right] \quad . \tag{15}$$

The probability density for the distribution of the energy values, $\rho_N(E)$, is then given by the convolution

$$\rho_N(E) = (p_1 * p_2 \ldots * p_{N-1})(E) \quad , \tag{16}$$

and its Laplace transform,

$$\tilde{\rho}_N(z) = \int_0^\infty dE \; \rho_N(E) e^{-zE} \quad , \tag{17}$$

becomes

$$\tilde{\rho}_N(z) = \prod_{k=1}^{N-1} \tilde{\rho}_k(z) = \prod_{k=1}^{N-1} \frac{1}{(1+2kz)^{1/2}} \quad . \tag{18}$$

For the partition function, see Eq.(3), we thus obtain

$$Z_N = 2^N \tilde{\rho}(\tfrac{1}{T}) = 2^N \prod_{k=1}^{N-1} \frac{1}{(1+\frac{2k}{T})^{1/2}} \quad , \tag{19}$$

and in the limit as $N \to \infty$ this leads to

$$-\frac{2}{N} \ln Z_N \approx \frac{1+2\beta}{2\beta} \ln(1+2\beta)-(1+2\ln 2) \quad , \tag{20}$$

where

$$\beta \equiv N/T \tag{21}$$

is the appropriately normalized inverse temperature. The corresponding approximate expressions for the average energy and the specific heat, Eqs.(5) and (6), respectively, are given by

$$\frac{2}{N^2} \langle E \rangle \approx \frac{1}{2\beta^2} [2\beta-\ln(1+2\beta)] \tag{22}$$

and

$$\frac{2}{N} C \approx 1 + \frac{1}{1+2\beta} - \frac{1}{\beta} \ln(1+2\beta) \quad . \tag{23}$$

In Figures 2 and 3, the predictions of our simple approximate treatment are compared with results obtained by a simulated annealing procedure in which we have used single spin flip configuration

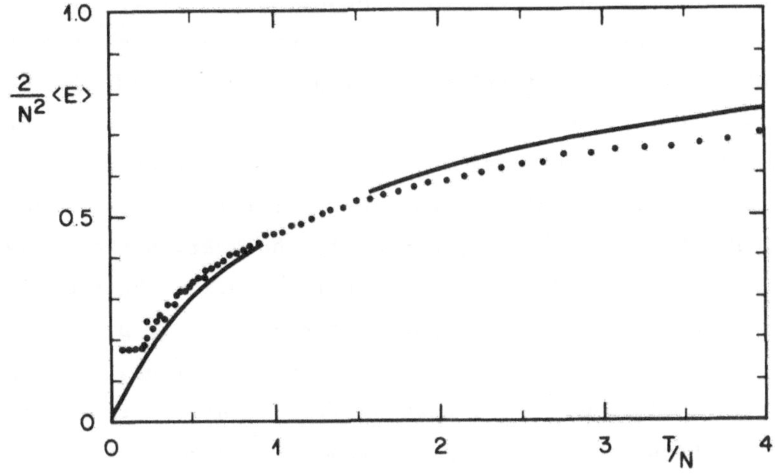

Figure 2: Simulated annealing results (circles) for the temperature dependence of the average "energy" of binary sequences of length N = 31, compared with the approximate asymptotic expression of Eq. (22).

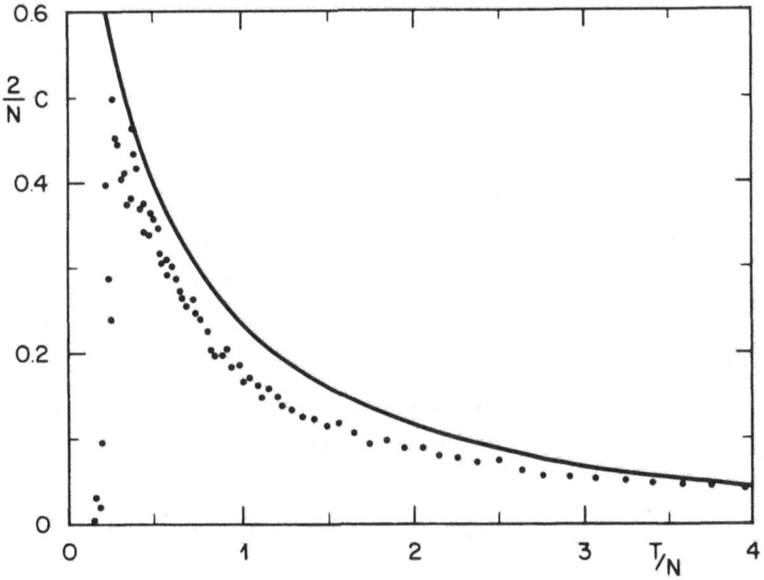

Figure 3: Simulated annealing results (circles) for the "specific heat" of binary sequences of length N = 31, compared with the approximate asymptotic expression of Eq.(23).

changes and an exponential cooling schedule ($T \to 0.95T$). At each temperature, the simulation was continued until 5N flips were accepted or until the number of attempted flips exceeded 20N. Variations of the annealing parameters, as well as the use of multiple spin flip transitions, did not lead to significantly different results.

The numerical data of Figures 2 and 3 refer to binary sequences of length $N = 31$. If scaled appropriately, however, higher N-values give practically identical results. For $T/N \gtrsim 0.25$, the simulations agree reasonably well with the approximate asymptotic predictions of Eqs. (22) and (23). The observed systematic deviations, however, can only partly be attributed to finite N effects. At low temperatures, the approximation based on Golay's "ergodicity hypothesis" breaks down completely, and the numerical data are consistent with the occurrence of an "ergodicity breaking" phase transition at a finite temperature, $T_c/N \approx 0.20\text{-}0.25$.

In Figure 2, the average energy curve reaches a low temperature value of

$$\frac{2}{N^2} \langle E \rangle \equiv \langle \frac{1}{F} \rangle \approx 0.18 \quad , \tag{24}$$

corresponding to an optimal merit factor of $F_{opt} \approx 5.6$. Before turning to more detailed results on the N-dependence of F_{opt} for annealed sequences, we shall first present an elegant rederivation of Golay's estimate [10] for the maximum possible merit factor of very long sequences.

If E_{min} denotes the minimum energy for a sequence of length N, the partition function Z_N obviously satisfies

$$e^{-E_{min}/T} < Z_N \leq 2^N e^{-E_{min}/T} \quad , \tag{25}$$

so that

$$\frac{2}{N^2} E_{min} \beta - 2\ln 2 \leq - \frac{2}{N} \ln Z_N < \frac{2}{N^2} E_{min} \beta \quad , \tag{26}$$

where $\beta \equiv N/T$ as above. As a function of β, $-(2/N)\ln Z_N$ is thus bounded by two straight lines with slope $(2/N^2)E_{min}$. Now our approximate expression of Eq.(20) for $-(2/N)\ln Z_N$ only increases as $\ln\beta$ for $\beta \to \infty$, so that it cannot represent a reasonable approximation for arbitrarily large β. As the function defined by Eq.(20) is concave, however, the slope of its tangent through the origin gives an estimate for the smallest possible value of $(2/N^2)E_{min}$. A straightforward calculation then leads to

$$\frac{2}{N^2} E_{min} \gtrsim \frac{1}{2\beta_o^2} [2\beta_o - \ln(1+2\beta_o)] \quad , \tag{27}$$

where β_o is the solution of

$$(1+\beta_o)\ln(1+2\beta_o) = 2(1+\ln 2)\beta_o \quad , \tag{28}$$

and we obtain

$$\frac{2}{N^2} E_{min} \gtrsim 0.0811 \quad , \tag{29}$$

or

$$F_{max} = \frac{N}{2E_{min}} \lesssim 12.3248 \quad . \tag{30}$$

Eq.(30) is identical with Golay's conjecture [10] for the maximum merit factor of long ($N \to \infty$) binary sequences, which he has derived in a much more cumbersome way. We emphasize, however, that Eq.(30)

is based on an approximate expression for the partition function, so that it can only be regarded as an estimate, and not as an exact upper bound. Nevertheless, the optimal merit factors found by exhaustive searches are consistent with an extrapolation to F ≈ 12 for large N [10], see also Figure 1. On the other hand, the highest merit factors known for very long sequences, where no exhaustive search is possible, do not even exceed F ≈ 6 [9-11,13].

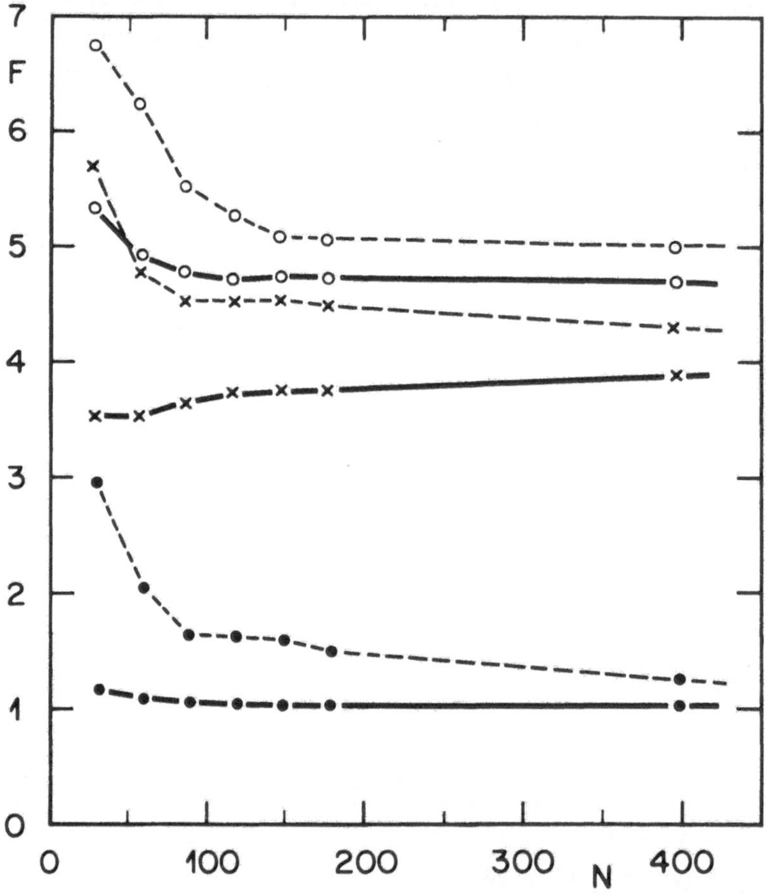

Figure 4: Merit factors F for binary sequences of lengths up to N = 400, obtained by different search procedures. ●●●● : random sequences; xxxx : quenched sequences (local optimization); oooo : annealed sequences. The data points indicate the average (continuous curves) and the best (broken curves) F-value found in a corresponding sample of 200 sequences.

In Figure 4, we compare the merit factors found with different stochastic search procedures. These results give us some information about the distribution of the F-values of long binary sequences. The overall distribution, represented by the F-values of randomly selected sequences, is strongly peaked at $F = 1$, whereas the merit factors of the local optima (quenched sequences) are concentrated around $F \approx 4$. The annealed sequences, finally, which probe the high F tail of the local optima distribution, exhibit F-values between 4.5 and 5 for large N. As indicated by the broken curves, the results can be improved to a certain extent by applying the quenching or annealing process to a large number of different starting configurations. For long sequences, however, the potential for corresponding improvements is very limited, as the size of the samples needed seems to increase exponentially with increasing N. We also note that our asymptotic value of $F \approx 4.7$ for the annealed sequences coincides well with the asymptotic value extrapolated from very extensive stochastic searches for low autocorrelation skew-symmetric binary sequences [11] (see Figure 1). When comparing the results of Figures 1 and 4, we have to keep in mind that skew-symmetry reduces the complexity of the problem by a factor of 2, and that in the search of Beenker et al [11] up to 20'000 and more starting configurations have been used for a given value of N.

We therefore have to conclude that stochastic search procedures will not lead to merit factors larger than $F \approx 5$ for very long sequences. This indicates that the sequences with a very high merit factor ($6 < F \lesssim 12$), which we believe to exist, must represent extremely isolated, "golf-hole like", local energy minima in configuration space.

More detailed results, in particular with respect to the distribution of the annealed minima in configuration space, and with respect to the cooling rate dependence of their energy values, will be presented in a separate publication [14].

ACKNOWLEDGEMENT

The author is indebted to Dr. D. Baeriswyl for numerous stimulating discussions.

REFERENCES

[1] S. Kirkpatrick, C.D. Gelatt Jr., and M.P. Vecchi, Science 220, 671 (1983).
[2] S. Kirkpatrick, J. Stat. Phys. 34, 975 (1984).
[3] E. Bonomi and J.-L. Lutton, SIAM Review 26, 551 (1984).
[4] R.E. Burkhard and F. Rendl, European J. Operational Res. 17, 169 (1984).
[5] P. Siarry and G. Dreyfus, J. Physique Lett. 45, L-39 (1984).
[6] V. Černy, J. Optimization Theory and Appl. 45, 41 (1985).
[7] S. Kirkpatrick and G. Toulouse, J. Physique 46, 1277 (1985).
[8] M. Mézard and G. Parisi, J. Physique Lett. 46, L-771 (1985).
[9] M.J.E. Golay, IEEE Trans. Inform. Theory IT-23, 43 (1977).
[10] M.J.E. Golay, IEEE Trans. Inform. Theory IT-28, 543 (1982).
[11] G.F.M. Beenker, T.A.C.M. Claasen, and P.W.C. Hermens, Philips J. Res. 40, 289 (1985).
[12] N. Metropolis, A. Rosenbluth, M. Rosenbluth, A. Teller, and E. Teller, J. Chem. Phys. 21, 1087 (1953).
[13] M.J.E. Golay, IEEE Trans. Inform. Theory IT-29, 934 (1983).
[14] J. Bernasconi, to be published.

A MACSYMA/EXPERT SYSTEM FOR NONLINEAR FILTERING

G.L. Blankenship,† A. LaVigna,† D.C. MacEnany,†
J.P. Quadrat,†† I. Yan†

Abstract: A prototype "expert" system for the treatment of certain nonlinear filtering problems is described with illustrative examples. The system is written in MACSYMA. It accepts user input in symbolic form; it carries out the basic analysis of the user's problem in symbolic form (e.g., computing the Zakai equation and the estimation Lie algebra for nonlinear filtering problems); and it produces output in the form of automatically generated FORTRAN code for the final numerical reduction of the problem, e.g., FORTRAN code for the conditional statistics of the signal given the observation process. The system is also capable of testing the well-posedness of a limited class of filtering problems by setting up the robust form of the Zakai equation and applying some known theorems for well-posedness of that system. This is done in symbolic form directly in the MACSYMA code. In addition, the system has a module which generates FORTRAN code for the computation of the likelihood ratio arising in connection with the filtering problem. This code implements a sophisticated approximation for the Feynmann-Kac (or Kallianpur-Streibel) representation of the conditional density in the estimation problem. Sample terminal sessions are presented to illustrate its operation. The status of the system and plans for its further development are described.

---

†Electrical Engineering Department, University of Maryland, College Park, Maryland 20742. The research of G.L. Blankenship and I. Yan was supported in part by NSF Grant ECS-83-15965. A. LaVigna and D.C. MacEnany were supported in part by ONR Graduate Fellowships. Address correspondence concerning this paper to G.L. Blankenship.

††INRIA, Domaine de Voluceau, Rocquencourt, B.P. 105, 78153 LE CHESNAY CEDEX, FRANCE.

S. Albeverio et al. (eds.), Stochastic Processes in Physics and Engineering, 17–61.

1. Introduction

Nonlinear filtering theory has undergone a period of intense study over the past five years. Powerful and sophisticated mathematical tools, including constructions from differential geometry, stochastic partial differential equations, functional integration, stochastic mechanics, quantum physics, and asymptotic analysis, among others have been brought to bear on the basic problem of constructing recursive estimates of a signal process from measurements of a related process. The anthology [1] covers many of the key ideas, and the papers [2, 3] are excellent expositions of much of the recent research. Because most of this work has employed sophisticated tools not in the lexicon of most practicing engineers, and because much of the work has been expressed in the theorem - proof format which has become the standard in control theory, it is likely that the impact of this work on engineering design will be marginal at best, at least in the short run.

In this paper we describe a software system based on MACYSMA[1] which embodies several elements of the theory of nonlinear filtering, especially as described in [3] and [4, 5, 6, 7, 8]. The system is easy to use, requiring only user specification of signal and measurement process equations in symbolic form and responses to a few simple questions on whether the model involves vector or scalar signals and whether or not there are small parameters in the model. The system computes the stochastic partial differential equation which describes the evolution of the conditional density of the signal given the history of the observations; if small paramters are present in the model, it does an asymptotic analysis of this equation; it computes the differential geometric structure of the approximating system; producing intermediate expressions which describes this structure; and, as its final output, it writes a FORTRAN code for numerical computation of the conditional mean of the signal. The FORTRAN code is generated "automatically" for each (symbolic) specification of the system model. The system is capable of checking the well-posedness of the conditional density equation using the criteria which have been derived for this purpose, e.g., in [6]. In addition, the system has a module which generates FORTRAN code for the computation of the likelihood ratio arising in connection with the filtering problem. This code implements

---

[1]MACSYMA is a language for symbolic manipulation developed at Project MAC at MIT. MACSYMA is a trademark of *Symbolics, Inc.*

a sophisticated approximation for the Feynmann-Kac (or Kallianpur-Streibel) representation of the conditional density in the estimation problem [7, 8].

Each of the mathematical computations carried out in MACSYMA is accessible to the user; however, he need not access any of them at any time in the execution, if he is interested only in the numerical code output. The advantages of this kind of system are clear in a CAD/CAE setting. Given a new problem, the time involved in analyzing the conditional density equation and then writing the FORTRAN code to solve it is eliminated. The mistakes and the time required to test and debug the FORTRAN code are also eliminated - and this is a major advantage of the system in its present form. Perhaps most importantly, the system allows the engineer to interact with the computer for design at the symbolic manipulation level. In this way he can modify his analysis or design problem by modifying the symbolic functional form of the model (and the design objectives and constraints in other cases.) The FORTRAN subroutines that he might have to modify by hand to accomplish this in conventional design procedures are written automatically for him. The advantages of working at the higher level are clear.

This system is part of a prototype *expert system* based on MACSYMA which has been developed for the analysis and design of stochastic control systems. The research was initiated, and has been carried out in the main, at the *Institute National de Recherches en Informatique et en Automatique* (INRIA) under the direction of Dr. J.P. Quadrat [9]. The system for nonlinear filtering which is described here has been developed in a collaborative effort between INRIA and the University of Maryland.

The total system which now exists has four major components: (i) a modular system of programs written in MACSYMA which "solve" certain stochastic control and nonlinear filtering problems in symbolic form; (ii) a natural language interface for optimal stochastic control; (iii) a "theorem proving module" using PROLOG[2] capable of checking the well-posedness (existence and uniqueness of solutions on a Sobolev space determined by the module) of certain linear and nonlinear PDE's in symbolic form; and (iv) a general-purpose module for generating FORTRAN code

[2]PROLOG is a powerful programming language originally developed at the University of Marseilles, as a practical tool for programming in logic.

from the symbolic manipulation modules of the system. The ultimate, intended function of the system is to accept input from the user in natural language with model equations expressed in symbolic form, to "expertly" (automatically) select a solution technique for his control or signal processing problem, reducing the model equations by symbolic manipulations to a form suited to the chosen technique, checking well-posedness of the model along the way; and to automatically generate a numerical language code (e.g., FORTRAN) realizing the solution algorithm. It is apparent from the sample sessions shown in [9, 10] that the system will ultimately be capable of this function in realistic design exercises.

In artificial intelligence terms the system does automatic program generation. It is an *expert system* [11] in the sense that it embodies portions of the knowledge that a control theorist has, and it is able to use that information to produce numerical code which consititutes a "solution" to specific problems. It is possible to enhance the system to behave more "expertly;" and these enhancements are now being developed. For example, it is possible to expand the capabilities of the program by adding *production rule* "reasoning" procedures to the program modules to take advantage of the special structures associated with control problems. These facilities could be used to select or guide the selection of analytical reduction procedures or numerical algorithms to treat specific problems. The existing natural language interface can be enlarged to take advantage of the specialized vocabulary of (stochastic) control theory. This would facilitate use of the system by novices and designers who are unfamiliar with the detailed mathematics of control theory. In addition, it is possible to allow languages other than FORTRAN to be selected as the medium for the output. For example, producing output code in LISP is very simple since MACSYMA itself is written in a dialect of LISP. These enhancements and others are now under development and are described elsewhere [12, 13, 14].

1.1. Comparison with other work

Several software systems for computer-aided-design of control systems have been developed recently including:

(1) ORACLS by NASA [15] ;

(2) DELIGHT.MIMO by the University of California, Berkeley and Imperial College, London [16] ;

(3) DELIGHT.Marylin by the University of Maryland [17] ; and

(4) LAS by S. Bingulac.

In addition, there has been some excellent work on general and specific algorithms for numerical problems associated with control system design [18]. The cumulative product of these efforts is a strong library of *numerical* systems and packages for the analysis and design of numerically defined control systems.

The system described here and in [9, 10, 12] has a very different function. It is based on symbolic manipulation programs; and consequently, it permits design efforts to take place at the "next level up" from FORTRAN code. It interfaces naturally with AI constructs since it is based on LISP and symbolic data, as opposed to the numerical structures of FORTRAN. This facilitates the development of a natural language interface. It would be difficult to design such an interface in FORTRAN; and it would be equally awkward to write theorem proving modules in FORTRAN.

1.2. Summary

In the next section we explain how the MACSYMA portion of the code is developed for the construction of approximate nonlinear filters for systems with small parameters. We describe modules of the system for computing the likelihood ratio and for checking well-posedness of the conditional density equation. In section 3 we present samples of the operation of the system. In section 4 we describe samples of the MACSYMA code. In section 5 we comment on plans for further development of the filtering module.

2. Developing a MACSYMA generator for approximate nonlinear filters

The nonlinear filtering problem involves some sophisticated mathematical constructions; and its treatment using MACSYMA is a good illustration of the capabilities of this tool in control systems design, and the way in which it permits sophisticated mathematics to be imbedded a CAE design system. The system described here is not complete and its development is continuing. We shall say more later about our intentions to convert the system into a useful design tool.[3] The examples chosen and the methods used to analyze them are intended to illustrate the main ideas in using symbolic manipulation programs in a control theoretic context. We do not claim that they embody the best algorithms for designing approximate nonlinear filters.

In the section 2.1 we describe the approach to the approximation of nonlinear filters developed in [5] and in [3]. In section 2.2 we describe the realization of this approximation procedure in a MACSYMA system. In section 2.3 we describe an alternative approximation technique based on the use of Gaussian qudrature approximations to the (stochastic) function space integral representation of the likelihood ratio associated with the nonlinear filtering problem. This approach is based on the approximation formulas derived in [7, 8]. In section 2.4 we describe a MACSYMA module for checking the well-posedness of nonlinear filtering problems by testing existence and uniqueness criteria for the robust form of the Zakai equation. The conditions are those derived in [6].

[3]We would like to thank P.S. Krishnaprasad, C.-H. Liu, and S.I. Marcus for their contributions to the ideas in this section. In particular, the use of MACSYMA in the asymptotic analysis of nonlinear filtering problems was first explored in C.-H. Liu's dissertation [4].

2.1. Algebraic and asymptotic simplification of nonlinear filters

The problem of estimating one diffusion Markov process $\{ x_t \}$ in terms of observations of another $Y_t = \sigma\{ y_s , s \leq t \}$ based on the model (Ito calculus)

$$dx_t = f(x_t)dt + g(x_t)dw_t \tag{1}$$

$$dy_t = h(x_t)dt + dv_t , \quad 0 \leq t \leq T < \infty$$

is fundamental in many engineering applications. Here f , g , and h are smooth functions, w_t , v_t are independent standard Wiener processes, and x_0 is a random variable independent of (w_t , v_t) for all t . The problem of recursively estimating x_t given the σ-algebra Y_t generated by the observations, may be formulated in terms of the (Stratonovich) stochastic partial differential equation for the unnormalized conditional density $u(t,x)$

$$du(t,x) = (L^* u)(t,x)dt + h(x)u(t,x)dy_t$$

$$L^* u = \frac{1}{2} \partial_{xx}(g^2(x) u) - \partial_x(f(x) u) - \frac{1}{2} h^2(x) u \tag{2}$$

$$u(0,x) = p_o(x), \text{ the density of } x_0$$

That is, if (2) has a nice solution, then the conditional density of x_t given Y_t is

$$p(t,x) = u(t,x)/[\int u(t,z)dz].$$

A considerable effort has been devoted to the search for (recursive) finite dimensional "representations" of estimators in terms of various "sufficient statistics" of either the solution of (2) or other conditional statistics (such as the conditional mean). This effort, which has produced few such estimators, has nevertheless increased our understanding of the system (2) and of the tools available for its treatment [1]. In [5] a systematic procedure was developed for constructing approximations to the solutions of (2) in terms of certain Lie algebras of differential operators associated with the equation. See also [3, 19, 20]. The procedure is based on the presence of a small parameter in the model (1), and it uses the methods of asymptotic analysis to derive the approximations. Suppose, for example, that the model equations contain a small nonlinear drift and are otherwise linear; i.e.,

$$f(x,t) = ax + \epsilon \tilde{f}(x), \quad g(x,t) = b \tag{3}$$

$$h(x,t) = cx.$$

In effect, the system is a small nonlinear perturbation of the Kalman-Bucy filtering problem. The associated unnormalized conditional density satisfies

$$du^{\epsilon}(x,t) = \{ \frac{1}{2} b^2 \partial_{xx} u^{\epsilon}(x,t) - \partial_x [ax \; u^{\epsilon}(x,t)] - \frac{1}{2} c^2 x^2 u^{\epsilon}(x,t) \} dt \tag{4}$$

$$+ \epsilon \{ \partial_x [\tilde{f}(x) u^{\epsilon}(x,t)] \} \, dt + cxu^{\epsilon}(x,t) \, dy_t .$$

If we formally set $\epsilon = 0$ in this equation, then we obtain the equation for the Kalman-Bucy conditional density $u_0(x,t)$, which is, of course, Gaussian. By formally expanding the conditional density $u^{\epsilon}(x,t)$ in a power series in ϵ, we obtain a sequence of stochastic partial differential equations which can be solved in terms of the Gaussian conditional transition density, operating as the Green's function for each problem, and the function $\partial_x [\tilde{f}(x) u_0(x,t)]$ as a forcing function in the first order approximation. If, for example, $\tilde{f}(x)$ is a polynominal, as in [5], then the first order in ϵ approximation to the conditional density involves various moments of a Gaussian density, and it can be written down explicitly. In this case higher order approximations (in ϵ) can also be computed explicitly. The computations are, however, very involved, as the formulas in [3, 5, 19, 20] demonstrate. It would be a trial to carry out the calculations by hand for multidimensional systems. In [4, 5] MACSYMA was used to do some of the analytical calculations.

There are two basic ways to compute the expressions for the approximate conditional density. One can proceed directly and solve the sequence of stochastic partial differential equations for the higher order approximations to the conditional density by using the explicit Green's function and computing the integrals (e.g., moments) analytically (using MACSYMA). It is difficult to ascertain *a priori* how much computational effort this procedure will require; e.g., how many moments or integrals must be evaluated.

By using the techniques of Lie theory to construct the solutions to the sequence of approximate density equations, it is possible to determine the computational effort *a priori*. Lie theory also provides a natural structure to the computations which permits systematic construction of approximate finite dimensional filters in many situations; a fact first articulated by Brockett [21] and Mitter [22] and then developed in specific contexts by many others [19, 20, 23, 24, 25].

The idea is as follows: Consider the simple perturbation of the Kalman filtering problem treated in [5].

$$dx_t = ax_t\,dt + dw_t$$
$$dy_t^{\epsilon} = [\, x_t + \epsilon(x_t)^k \,]dt + dv_t \tag{5}$$
$$y_0^{\epsilon} = 0, \quad x_0 \text{ has Gaussian density } p_o(x),\ k \geq 2.$$

The conditional density of x_t^{ϵ} given Y_t^{ϵ} satisfies

$$du^{\epsilon}(t,x) = [L^* - \frac{1}{2}(x + \epsilon x^k)^2]u^{\epsilon}(t,x)dt + (x + \epsilon x^k)\, u^{\epsilon}(t,x)\, dy_t^{\epsilon} \tag{6}$$
$$u^{\epsilon}(0,x) = p_o(x)$$
$$L^* u = \frac{1}{2}\partial_{xx} u - a\,\partial_x(xu).$$

Assume that u^{ϵ} has an asymptotic expansion

$$u^{\epsilon} = u_0 + \epsilon u_1 + \epsilon^2 u_2 + \cdots \tag{7}$$

Substituting this into (6) and equating coefficients of like powers of ϵ gives

$$du_0 = L^* u_0\, dt + x\, u_0\, dy_t^{\epsilon} \tag{8}$$
$$u_0 = p_o(x)$$
$$du_k = L^* u_k\, dt + x\, u_k\, dy^{\epsilon}_t + x^k u_{k-1}^{\epsilon_t}$$
$$u_k(0,x) = 0, \quad k = 1,2,\ldots$$

Defining the vector

$$U_n(t,x) = \begin{bmatrix} u_0 \\ u_1 \\ \cdots \\ u_n \end{bmatrix} \tag{9}$$

we have

$$dU_n(t,x) = [(L^* - \frac{1}{2}x^2)E_1 - x^{k+1}E_2 - \frac{1}{2}x^{2k}E_3]\,U_n(t,x)\,dt \qquad (10)$$

$$+ [xE_1 + x^k E_2]\,U_n(t,x)\,dy_t^\epsilon$$

where the $(n+1)$ x $(n+1)$ matrix E_i has zero entries everywhere except for $1's$ on the subdiagonal $(i,1)(i+1,2),...,(n+1,n+2-i)$ (i.e., on a particular subdiagonal). Note that E_1 is the identity matrix.

The equation for u_0 is the Kalman conditional density equation (given y_s^ϵ, $s \leq t$). Therefore, it has a closed form analytical solution - a Gaussian density. This density defines the Green's function for all the subsequent equations, $k = 1,2,...$, in (8). Since u_1 involves this Green's function and $x^k u_0\, dy_t^\epsilon$ as a forcing function, u_1 can also be computed explicitly. It follows that all the terms $u_1, u_2, \cdots$ can be computed explicitly; however, the complexity of the calculation increases very rapidly. By computing $U_n(t,x)$ we produce an approximation to the conditional density which is accurate to $O(\epsilon^{n+1})$.

All these calculations can be done automatically in symbolic form by MACSYMA. To do this, we use Lie theory. First, we rewrite (10) abstractly as

$$\dot{U}_n(t) = AU_n(t) + BU_n \dot{y}_t^\epsilon$$

$$A = [(L^* - \frac{1}{2}x^2)E_1 - x^{k+1}E_2 - \frac{1}{2}x^{2k}E_3] \qquad (11)$$

$$B = [x E_1 + x^k E_2]$$

Note that A is a second order differential operator and B is multiplication by a matrix valued function. The Lie algebra generated by A and B consists of all linear combinations of A and B and their consecutive commutator products, e.g.,

$$[A,B] = AB - BA \qquad (12)$$

and subsequent repeated products (taken as differential operators). This is always a finite dimensional *solvable* (e.g., "lower triangular" structure) Lie algebra[4] for the class of filtering problems

[4]A Lie algebra is *solvable* if the derived series of ideals $L^{(0)} = L$; $L^{(n+1)} = [L^{(n)}, L^{(n)}]$, $n \geq 0$ is the trivial ideal $\{0\}$ for some n.

with polynominal coefficients described above.

Consequently, we can use the Wei-Norman theory [26] to solve (11) globally in the form

$$U_n(t,x) = [\, e^{g_1(t)A_1} e^{g_2(t)A_2} \cdots e^{g_d(t)A_d} \,](x) \tag{13}$$

where $(A_1, A_2, \ldots, A_d)$ is a basis for the Lie algebra generated by A and B, and the $g_i(t)$ are defined by a lower triangular system of nonlinear (stochastic) ordinary differential equations. These equations must be solved numerically (in general). The $g_i(t)'s$ may be thought of as a finite dimensional family of (approximate) conditional statistics for the original nonlinear filtering problem. It is useful to elaborate on the derivation of these equations, since this is the most important and challenging computation done in the system, and since it provides a good illustration of the capabilties of MACSYMA in control theoretic analysis. The necessary steps are summarized nicely in [3]. See also [4, 5, 27].

Assuming the solution of (11) to take the form (13), we compute

$$\frac{d}{dt}\, [\, e^{g_1A_1} e^{g_2A_2} \cdots e^{g_dA_d} \,] = \dot{g}_1 A_1 \, [\, e^{g_1A_1} e^{g_2A_2} \cdots e^{g_dA_d} \,] \tag{14}$$

$$+ \, e^{g_1A_1} \dot{g}_2 A_2 \, [\, e^{g_2A_2} \cdots e^{g_dA_d} \,] + \cdots + [\, e^{g_1A_1} \cdots e^{g_{d-1}A_{d-1}} \,] \, \dot{g}_d \, A_d \, e^{g_d A_d} \, .$$

To proceed, we need to collect the common factor $[\, e^{g_1A_1} e^{g_2A_2} \cdots e^{g_dA_d} \,]$ on the right side to be able to equate common coefficients of the basis elements $\{ A_j \; j = 1,\ldots,d \}$ on both sides of (11). Since the A_i do not commute in general, we use the Baker-Campbell-Hausdorff formula:

$$e^{A_i t} \, A_j = \sum_{k=0}^{\infty} \frac{t^k}{k!} \, \mathrm{ad}^k_{A_i} \, A_j \, e^{A_i t} \tag{15}$$

where

$$\mathrm{ad}^0_A \; B = B , \tag{16}$$

$$\mathrm{ad}^{k+1}_A \; B = [\, A \,, \mathrm{ad}^k_A \; B \,] = [\, A \,, [\, A \,, [\, A \, \cdots [\, A \,, B \,]] \cdots]] \quad (k+1 \text{ times}).$$

Since the Lie algebra is finite dimensional, (15) can be written

$$e^{A_i t} \, A_j = \sum_{k=1}^{d} c_k^{ij} \, A_k \, e^{A_i t} \tag{17}$$

where the structure constants c_k^{ij} are computed from (15). This formula allows us to move the A_i's past the $e^{A_j t}$'s and collect the common product of exponentials on the right. Equating coefficients of each A_j in (11) yields a set of differential equations for the $g_j(t)$'s called the *Wei-Norman equations* of the form

$$\begin{bmatrix} dg_1(t) \\ dg_2(t) \\ \cdot \\ \cdot \\ \cdot \\ dg_d(t) \end{bmatrix} = \begin{bmatrix} f_1(g) \\ f_2(g) \\ \cdot \\ \cdot \\ \cdot \\ f_d(g) \end{bmatrix} dt + \begin{bmatrix} h_1(g) \\ h_2(g) \\ \cdot \\ \cdot \\ \cdot \\ h_d(g) \end{bmatrix} dy(t) \tag{18}$$

where the f_j and h_k are nonlinear functions of $g = [g_1, \cdots, g_d]$. If the Lie algebra generated by A and B is solvable, then there is a basis for the Lie algebra and an ordering of this basis, for which the Wei-Norman representation is global (for $t \geq 0$) [26]. The significance of this property in the context of nonlinear filtering problems is developed in detail in [3, 27].

2.2. Symbolic algorithm structure

To implement the approximate nonlinear filter using this methodology and MACSYMA, one must program the following six step procedure:

(M1) Compute the differential operators A and B in (11) in terms of the original functions in the filtering problem model (5), modulo the appropriate power of the small parameter ϵ, if one is present in the model.

(M2) Compute basis elements $A_1, A_2, \ldots, A_d$ for the Lie algebra generated by A and B.

(M3) Compute ODE's for the $g_1, \ldots, g_d$ functions in (13) by substituting (13) into (11), differentiating, and inverting the resulting matrix equation.

(F1) Solve the ODE's for the g_i's numerically using an appropriate numerical integration scheme for stochastic ODE's [28].

(M4) Compute the approximate conditional mean $\hat{x}_t^\epsilon$ (the best mean square estimator of x_t^ϵ given Y_t^ϵ) by integrating the conditional density.

(F2) Represent the conditional mean computed in (M4) in FORTRAN code in terms of the actual measurements $\{ y_s^\epsilon, \ s \leq t \}$ as input data to the code.

The steps labeled (M1)-(M4) are done in the MACSYMA part of the code. The step (F1) is best carried out in FORTRAN; and so, the MACSYMA step (M3) which computes the symbolic ODE's has an auxiliary component which writes the FORTRAN code to numerically integrate the ODE's.[5] The result of step (F2) is the output of the system - that is, a FORTRAN code to numerically evaluate the best estimate of the state in terms of numerical measurement data. All the steps (M1) - (M4) and (F1) (F2) are carried out automatically after the system model (the constants a, b, and c and the function $\tilde{h}(x)$ in (4) or the functions f, g, and h in (2)) are specified.

In sections 3.1 and 3.2 we provide the listings from sample sessions with this portion of the system.

2.3. Automatic computation of the likelihood ratio

There are several other approaches to computation of approximate nonlinear filters, including direct evaluation of the conditional density by application of "standard" numerical methods to solve the Zakai equation, i.e., its robust form [28]. Another approach which has some advantages in certain situations is direct evaluation of the "likelihood ratio" associated with the estimation problem. The setup we have implemented using MACSYMA is based on the approximation schemes (for scalar systems) developed in [7] and enhanced in W.E. Hopkins' dissertation [8]. See also [29].

The procedure is as follows: the solution to the Zakai equation (2) associated with the filtering problem (1) may be written in terms of a function space integral

$$u(t,x) = E_x \{ e^{\int_0^t h(x(s))\, dy(s) - \frac{1}{2} \int_0^t h^2(x(s))\, ds} \; p_0(x(t)) \} \tag{19}$$

[5]At this point in the system it is possible to have MACSYMA offer a choice of numerical integration procedures to be (automatically) coded. Our system does not now have this capability.

where E_x is expectation over the paths of (1) (the state equation) starting at $x(0) = 0$. The mathematical technique is evaluation of the function space integral (19) by quadrature approximations.

The first step is to regard (19)[6] as a "stochastic Wiener integral" and apply the formulas in [7]. This leads to an approximation of the function space integral by an n-fold ordinary integral with an error (in mean square) of $O(n^{-2})$. The next step is to reduce the n-fold integral to a recursive sequence of one dimensional integrals which may be evaluated by Gaussian quadratures. Finally, the algebraic structure of this sequential process is used to express the approximation formulas in the most computationally advantageous form. This three step procedure is complicated and tedious to implement by hand; it is, however, ideal for symbolic computation.

To describe the formulas in [7], let $C_0([0,t])$ be the space of $\mathbb{R}$-valued, continuous functions $x(s)$ on $[0,t]$ with $x(0) = 0$ and let $dW(\cdot)$ be Wiener measure on C_0. If $F: C_0 \to \mathbb{R}$ is a smooth functional, the Wiener integral

$$I = \int_{C_0} F(x)\, dW(x) \tag{20}$$

is defined as the sequential limit[7]

$$I = \lim_{\substack{\max\limits_{1 \le j \le n} |t_j - t_{j-1}| \to 0 \\ n \to \infty}} \int_{\mathbb{R}} \cdots \int_{\mathbb{R}} da_1 \cdots da_n \, F(z_{sx}) \cdot \prod_{j=1}^{n} \frac{e^{[-(a_j - a_{j-1})^2/2(t_j - t_{j-1})]}}{[2\pi(t_j - t_{j-1})]^{n-2}} \tag{21}$$

where $0 < t_1 < t_2 < \cdots < t_n = t$ and z_{sx} is a polynominal function on $[0,t]$ passing through x at $s = 0$ and a_j at t_j, $j = 1,2,\ldots,n$.

In [31] Chorin developed some approximation formulas for Wiener integrals using parabolas to interpolate the Wiener paths and expansions of the nonlinear functional F in a Taylor series with the quadrature formula adjusted to optimize the approximation of the first two terms.

---

[6]Actually a transformed version corresponding to the robust formulation of the filtering problem. See (29).

[7]See Cameron [30] for details and a discussion of the subtleties of Wiener and Feymann integrals.

Chorin's formulas are of the form

$$\int_C F(x)\, dW(x) \cong \pi^{-n/2} \int_{\mathbb{R}^n} F_n(u_1,\ldots,u_n)\, e^{-u_1^2 - \cdots - u_n^2}\, du_1\, du_2 \cdots du_n + O(n^{-2}) \tag{22}$$

where F_n is a simple form based on a "rectangular rule" in specific cases.

In [7] Chorin's formulas were extended to cover cases like (19) where a random (Ito) process, $\{ y(s), 0 \le s \le t \}$, is involved. The result required for the filtering problem takes the following form: Let $y(\cdot)$ be an Ito process on an interval $[0,T]$ and suppose $V(\cdot):\mathbb{R} \to \mathbb{R}$ is a smooth function whose fourth derivative satisfies

$$\int_t^{t+1/n} | V^{(4)}(x(s)) |^2\, ds = O(n^{-1}) \tag{23}$$

for all $t \in [0,T]$ and any continuous function $x(\cdot)$. Then for any $t \in [0,T]$

$$\begin{aligned} I &= \int_C \exp [\int_0^t V(x(s))\, dy(s)]\, dW(s) \\ &= (2\pi)^{-n/2} \int_{\mathbb{R}^n} \{ \exp [\sum_{i=1}^n V(x_{i-1} + \frac{u_n t}{\sqrt{2n}})\, \Delta y_{i-1}] \} \\ &\quad \cdot \exp [\frac{1}{2} (-u_1^2 \cdots - u_n^2)]\, du_1 \cdots du_n + e_n \end{aligned} \tag{24}$$

where

$$x_i = t\,(u_1 + \cdots + u_i)/n \;, i = 1,2,\ldots,n$$

$$t_i = \frac{it}{n}, \quad \Delta y_{i-1} = y(t_i) - y(t_{i-1})$$

and the approximation error is

$$(Ee_n^2)^{\frac{1}{2}} = O(n^{-2}).$$

The remarkable feature of this formula, as noted in [31] is that it is no more complicated in structure nor does it require more computing effort than the standard "rectangular rule" which has accuracy $O(n^{-1})$. Moreover, evaluation of the n-fold integral in (24) may be reduced to a sequence of one dimensional integrals which are recursive in the increments Δy_{i-1}.

The simple form of (24), the accuracy of the approximation, and the recursive evaluation procedure are consequences of the fact that the underlying measure is Wiener measure. Since the process $x(t)$ in (1) is not a Wiener process, it is necessary to make a change of coordinates or a change of measure (i.e., a Girsanov transformation) in (1) to take advantage of this structure. This is the starting point of the algorithm we have implemented.

Refering to (1), suppose f and g are smooth (see [8, 29] for details) and

$$g(x) > g_0 > 0 \text{ for some } g_0 \text{ and all } x \in \mathbb{R}$$

$$\int_0^{\infty} \frac{dx}{g(x)} = \int_{-\infty}^{0} \frac{dx}{g(x)} = +\infty .$$

Consider the change of coordinates in (1)

$$z_t = \phi[x_t] = \int_0^{x_t} \frac{dx}{g(x)} . \tag{25}$$

Using Ito's formula

$$dz_t = (\frac{f}{g} - \frac{1}{2} g')(\phi^{-1}[z_t])\, dt + dw_t \tag{26}$$

$$dy_t = h(\phi^{-1}[z_t])\, dt + dv_t .$$

The associated Zakai equation (Stranovich calculus) is

$$du(t,z) = \{ \frac{1}{2} \partial_{zz} u + [\frac{1}{2} g' - \frac{f}{g}](\phi^{-1}(z))\, \partial_z u$$

$$+ [g(\frac{1}{2} g'' - (\frac{f}{g})')- \frac{1}{2} h^2](\phi^{-1}(z))\, u \}\, dt + h(\phi^{-1}(z))\, u\, dy_t \tag{27}$$

We can eliminate the first order term in (27) using the exponential transformation

$$v(t,z) = u(t,z)\, e^{-\psi(z)} \tag{28}$$

$$\psi(z) = \int_0^{z} (\frac{f}{g} - \frac{1}{2} g')(\phi^{-1}(x))\, dx .$$

The equation for $v(t,z)$ is (Stratonovich calculus)

$$dv(t,z) = [\frac{1}{2}\partial_{zz} v(t,z) - V(z)v(t,z)]\,dt + H(z)v(t,z)\,dy_t \tag{29}$$

where

$$V(z) = \frac{1}{2}[h^2 + (\frac{f}{g} - \frac{1}{2}g')^2 + g(\frac{f}{g} - \frac{1}{2}g')'](\phi^{-1}(z)) \tag{30}$$

$$H(z) = h(\phi^{-1}(z))$$

Since the Laplacian in (29) is "isolated," we can use the formulas in [7] to evaluate it. This is the point of the coordinate change and the exponential transformation.[8]

Using the Feynmann-Kac formula (i.e., the Kallianpur-Streibel formula), we can write the solution to (29) as

$$v(t,z) = E_z\{\exp[\int_0^t H(z(s))\,dy(s) - \frac{1}{2}\int_0^t H^2(z(s))\,ds] \tag{31}$$

$$\bullet \exp[-\int_0^t V(z(s))\,ds + \mu_0(z(t))]\}$$

where E_z is expectation over Brownian paths starting at $z(0) = z$, and we have assumed the initial density satisfies $p_0(x) = \exp[\mu_0(x)]$.

Applying the approximation formulas, we can write

$$v(t,z) = I_n(t,z) + O(n^{-2}) \tag{32}$$

where

$$I_n(t,z) = c_n(t)\int_{-\infty}^{\infty} \exp[(\mu_0 - \psi)(\phi^{-1}(w_{n-1})) + g(t,w_{n-1},y_{n-1})] \tag{33}$$

$$\bullet\, I_{n-1}(t,z,w_{n-1})\,dw_{n-1}$$

with $c_n(t) = (2^{1/2n}\sqrt{n})/(t\sqrt{2\pi})$, and

$$g(t,w,y) = H(w)\,y - \frac{1}{2}H^2(w)\frac{t}{n} - V(w)\frac{t}{n}.$$

The integrals $I_{n-1}, \cdots, I_1$ are defined by the recursion

[8] See [8, 29] for discussions of alternative procedures and the vector case.

$$I_1(t,z,w_1) = c_n(t) \int_{-\infty}^{\infty} \exp\left\{ -\frac{1}{2}\frac{n}{t^2}\left[2(w_0 - z)^2 + (w_1 - w_0)^2 \right] + g(t, w_0, y_0) \tag{34}$$

$$\cdot$$
$$\cdot$$
$$\cdot$$

$$I_k(t,z,w_k) = c_n(t) \int_{-\infty}^{\infty} \exp\left\{ -\frac{1}{2}\frac{n}{t^2}(w_k - w_{k-1})^2 + g(t, w_{k-1}, y_{k-1}) \right\} \tag{34k}$$

$$\cdot\, I_{k-1}(t,z,w_{k-1})\, dw_{k-1}\,, \quad k = 2,3,\ldots,n-1\,.$$

The system (34) is recursive in the observations $y_0, y_1, \ldots, y_{n-1}$; I_k depends on $y_0, y_1, \ldots, y_{k-1}$, and is computed from I_{k-1} and y_{k-1}. Also, the integrals $I_1, I_2, \ldots, I_{n-1}$ are independent of the initial data; consequently, (34) has the form of a Green's representation

$$I_n(t,z) = \int_{-\infty}^{\infty} q_0(w)\, G(t,z;0,w)\, dw\,. \tag{35}$$

That is, $c_n(t)\exp\{ g(t, w_{n-1}, y_{n-1}) \}\, I_{n-1}(t,z,w_{n-1})$ approximates the fundamental solution of the partial differential equation (29) for the un-normalized conditional density.

The numerical evaluation of (34) permits considerable flexibility. The numerical implementation here, which is based on [8] emphasizes the computational advantage of the n-step recursion over direct evaluation of the n-fold integration (24) (adapted to (31)). Suppose we require evaluation of $v(t,z)$ in (32) at points $z = a_1, a_2, \ldots, a_m$ for some $m \geq 1$ and some prespecified observation times $t = t_1, t_2, \ldots, t_N = T$. Then the quadrature rule for one-dimensional integrals of the form

$$I(z,w) = \int_{-\infty}^{\infty} e^{-\phi(v)}\, F(z,w,v)\, dv \tag{36}$$

with ϕ a given weight function may be used; that is,

$$I(z,w) \simeq \sum_{i=1}^{m} A_i\, F(z,w,a_i)\,. \tag{37}$$

Here A_i are weights associated with the rule, and a_i are chosen in accordance with the prescription of the quadrature rule. By using the same rule for all the I_k no interpolation is needed to evaluate I_{k+1} in terms of I_k. The Gaussian functions in (34) suggest that Gaussian Quadrature [32, 33] is the method of choice for this problem.

Modifying the notation, we can streamline the recursion in (34) and make its underlying structure more evident. This is useful in designing the automatic code generation facility. Let $\alpha_i \in (0,1)$, $i = 1,2,\ldots,n-1$ and define

$$J_1 = (t/\sqrt{n})\,\mathrm{diag}\,[\sqrt{2}/3, 1, \cdots, 1, \sqrt{2}] \in \mathbb{R}^{n \times n}$$

$$J_2 = diag\,[(1-\alpha_i)^{-\frac{1}{2}}] \in \mathbb{R}^{n \times n}$$

$$J = J_1 \cdot J_2 \triangleq \mathrm{diag}\,[J_i]$$

$$\tilde{c}_n(z,t) = \left[2\frac{\sqrt{2}}{3} \frac{\prod_{i=1}^{n-1} (1-\alpha_i)^{\frac{1}{2}} e^{-\frac{nz^2}{t^2}}}{2^{\frac{n}{2}}} \right]^{\frac{1}{n}}$$

$$\tilde{a}_{ii} = \frac{\alpha_i}{(1-\alpha_i)}$$

$$\tilde{a}_{i,i+1} = \begin{cases} -(1/\sqrt{t})/\sqrt{(1-\alpha_0)(1-\alpha_1)} & i = 0 \\ -(\frac{1}{2})/\sqrt{(1-\alpha_i)(1-\alpha_{i+1})} & 1 \le i \le n-3 \\ -(1/\sqrt{2}/\sqrt{(1-\alpha_{n-2})(1-\alpha_{n-1})} & i = n-2 \end{cases} \tag{38}$$

$$r^1(a_i, t) = \exp\,[H(J_i A_i)]$$

$$r^2(a_i, t) = \exp\,[-\frac{1}{2} H^2(J_i a_i)\, t/n - V(J_i a_i) t/n]$$

$$q_0(a_0, z, t) = \exp\,[(2\sqrt{2n/3}/t\sqrt{(1-\alpha_0)}\, z\, a_0]$$

$$S_k(a_k, a_{k+1}) = \exp\,[\tilde{a}_{k,k+1} a_k a_{k+1} - \tilde{a}_{k,k} a_k^2]\; 0 \le k \le n-2$$

$$S_{n-1}(a_{n-1}) = \exp\,[(\mu_0 - \psi)(\phi^{-1}(J_{n-1} a_{n-1}))]$$

For fixed α_i, the m-point Gaussian Quadrature formula for each I_k may be expressed as

$$I_k \simeq \sum_{i=1}^{m} A_i\, f(a_i) \tag{39}$$

with weights $A_i > 0$ and a_i zeros of the mth Hermite polynominal generated by the recursion

$$H_0(x) = 1, \quad H_1(x) = 2x \tag{40}$$

$$H_{m+1}(x) = 2x\, H_m(x) - 2m\, H_{m-1}(x) .$$

Let $H_{m(x)} = k_m x^m + k_{m-1} x^{m-1} + \cdots + k_1 x + k_0$, then

$$A_i = \frac{k_{m+1}}{k_m} \cdot \frac{1}{H_{m+1}(a_i) \cdot H'_m(a_i)} \tag{41}$$

with a_i the roots of $H_m(x)$. Using this (39) becomes

$$I_{k+1}(a_j, a_i) = \tilde{c}_n(a_i, t) \, [\, SS_k(a_j, a_i) + e_k \,] \tag{42}$$

where e_k is the error term and

$$SS_k(a_j, a_i) = \sum_{l=1}^{m} A_l\, \gamma^1(a_l, t)^{y_k}\, \gamma^2(a_l, t)\, S_k(a_l, a_j)\, I_k(a_l, a_i) \tag{43}$$

The functions SS_k are given by the recursive formula

$$SS_k(a_j, a_i) = \sum_{l=1}^{m} A_l\, \gamma^1(a_l, t)^{y_k}\, \gamma^2(a_l, t)\, S_k(a_l, a_j)\, \tilde{c}_n(a_i, t)\, SS_{k-1}(a_l, a_i) \tag{44}$$

$$SS_0(a_j, a_i) = \sum_{l=1}^{m} A_l\, \gamma^1(a_l, t)^{y_k}\, \gamma^2(a_l, t)\, S_k(a_l, a_j)\, q_k(a_l, a_i, t)$$

A complete derivation of this algorithm for computing the conditional density may be found in [8] Chapter 7.

The complexity of the algorithm is necessary to exploit the structure of system dynamics in the numerical implementation. This advantage must be balanced against the involved coding procedures necessary to implement the algorithm. By using a symbolic program module to write the numerical FORTRAN code, we can relieve the user of this burden.

The MACSYMA module[9] incorporating this algorithm starts from the expressions for f, g, and h, the initial density $p_0(x)$ and the integers n and m (which determine the precision of the numerical approximation) and produces a FORTRAN code which computes the integrals $I_n(t, a_1), I_n(t, a_2), \cdots, I_n(t, a_m)$ in (33) at the points a_i, $i = 1,2,\ldots,n$ with the numerical values of the observations $y_0, y_1, \ldots, y_n$ as data. In effect, the MACSYMA code sets up the

[9]Written by I. Yan.

formula (44) appropriately initialized and then produces a FORTRAN code which evaluates the sequence of integrals (42), and from these the values of $I_n(t, a_j)$.

In section 3.3 we provide listings from sample sessions with this portion of the system.

2.4. Testing well-posedness of the model

The system also includes a test for *well-posedness* which consists of checking relative growth conditions on the coefficients *f*, *g*, and *h* for existence and uniqueness of a (classical) solution to the *robust form* of the Zakai equation. The criteria used are those developed in the paper [6]. As is the case in the other components of the system, complicated analysis is required to setup the existence and uniqueness test. This analysis is done "automatically" by the program. Since the existence and uniqueness theory for nonlinear filtering problems is not complete, the program cannot give a definitive analysis of a large class of problems. Rather, it uses the classical existence and uniqueness theorems for parabolic partial differential equations as adapted to a class of filtering models with smooth coefficient functions. It is not unlikely that more sophisticated techniques could be accomodated by simple modifications of the program.

The setup is as follows:[10] If $u(t, x)$ is the solution to the Zakai equation (2), then under the exponential transformation

$$V(t,x) = e^{[-h(x)y(t)]}\, u(t,x) \tag{45}$$

we find that $V(t, x)$ satisfies the "ordinary" parabolic PDE

$$\begin{aligned} V_t(t,x) &= A(x)\, V_{xx}(t,x) + B(x)\, V_x(t,x) + C(t,x)\, V(t,x) \\ V(0,x) &= p_0(x), \quad 0 \le t \le T \end{aligned} \tag{46}$$

where

$$A(x) = \frac{1}{2} g^2(x)$$

$$B(t,x) = 2g(x)\, g_x(x) - f(x) + 2A(x)\, h_x(x)\, y(t)$$

[10]Only scalar problems are treated in the current version of the program (written by I. Yan).

$$C(t,x) = g_x^2(x) + g(x)\,g_{xx}(x) - f_x(x) - \frac{1}{2}h^2(x) \tag{47}$$
$$+ [2g(x)\,g_x(x) - f(x) + 2A(x)\,h_x(x)\,y(t)]\,h_x(x)\,y(t)$$
$$+ A(x)\,[h_{xx}(x)\,y(t) + h_x^2(x)\,y^2(t)]$$

For each "given" path $\{ y(s), 0 \leq s \leq T \}$ (46) may be regarded as a classical PDE. Under various conditions on the relative growth of f, g, and h and their derivatives the theorems in [6] show that (46) has a unique fundamental solution, and if the initial density $p_0(x)$ falls off rapidly enough as $||x|| \rightarrow \infty$, then (46) has a unique solution which also decays rapidly and the asymptotic decay rate can be estimated precisely. The relative growth conditions preclude the occurence of "explosions" in the $\{ x(\cdot) \}$ process. They take a particularly simple form when the coefficients f, g, and h are polynominals; and our symbolic computational algorithms have special subroutines to exploit this case when it occurs.

The program proceeds in three steps all done in symbolic terms: First, it computes the pathwise-robust form of the Zakai equation using the exponential transformation (45). Second, it checks whether or not the coefficient functions f, g and h are polynomials. If so, then the simplified polynomial criteria in [6], which essentially involve comparing exponents of the highest order terms in certain combinations of the coefficients, are used to test well-posedness. If the coefficients are not polynominals, then the growth of the terms involving combinations of them are tested using the general criteria in [6]. If the relative growth tests succeed (there are several), then the program specifies a class of functions to which the solution function belongs. If one of the tests fails, the program indicates this to the user.

In section 3.4 we provide a sample of the output of this system. This module does not generate FORTRAN code.

3. Sample sessions with the filtering module

In the examples which follow we shall use the system to compute analytical expressions for the conditional densities in two specific filtering problems based on their underlying algebraic structure. The first is the *weak quadratic sensor* treated in the paper [5]. This is one of the few examples in which the estimation algebra and the Wei-Norman series representation of the conditional density have been worked out by hand. It is necessary to have such examples when programming in MACSYMA to validate the program. We shall also have the system carry out the same calculations for the a version of the so-called "Benes problem," which is one of the few nonlinear filtering problems which is known to have a finite dimensional recursive filter [34]. The estimation algebra for this problem is known to have the same Lie algebraic structure as the Kalman filter algebra, and this is one form of equivalence between the two problems. To illustrate the way the system handles vector problems, we shall combine the Benes problem in a diagonal system with a simple Kalman filtering problem.

3.1. The weak-quadratic sensor problem

In this example the program will be asked to analyze a simple filtering problem

$$dx_t = dw_t \tag{48}$$

$$dy_t = x_t + \epsilon\, x_t^{\,2} + dv_t$$

The system asks the user to declare the system as scalar or vector and then to enter the functions f , g , and h in a standard format. It then sets up the Zakai equation, identifies the two operators in the equation which will be used to generate the estimation algebra, and proceeds to find the representation of the conditional density to the appropriate power of ϵ, in the form of a Wei-Norman series. The system returns a basis for the estimation algebra (modulo the designated power of ϵ) and various intermediate expressions of interest, including the matrix of structure constants for the estimation algebra and a version of the Baker-Campbell-Hausdorff formula. Its "final result" is a set of stochastic ordinary differential equations for the functions appearing in the exponentials of the Wei-Norman series. The final output of the program is a FORTRAN code integrating these equations and computing numerical expressions for the conditional mean. The

FORTRAN code has a very "tedious" structure; it is not the kind of code that one would like to develop and enter by hand and then debug and revise each time the problem definition is modified. The complexity of the the expressions serves to emphasize the value of automatic generation of code.

```
c
c  This is a routine that solves for the conditional estimate of
c     x at t, given the observations y up to time t.  It is written
c     by Vaxima
c
c
c  Program Variables
c
c      xinput   real(4000)
c               contains the actual x values which are to be estimated for
c               times at instants of 0.001 sec.
c
c      observ   real(4000)
c               contains the sample observations to be filtered again at
c               0.001 sec time intervals.
c
c      estim    real (4000)
c               contains the calculated estimate given the observations up to
c               that time instant.
c
      real xinput(4002), var(4002), observ(4002), estim(4002), g(10), ng
     1   (10), x0xfx, x1xfx, x2xfx, x3xfx, x4xfx
      integer i, j, k
      do  300   ii = 1 , 300 , 1
      j=1
      k=0
      size=0.001
      xinput(0)=1
      observ(0)=1
      estim(0)=1
      do  100   i = 1 , 10 , 1
         g(i)=0
 100   continue
 110   continue
      if(g(1) .gt. 2.5) goto 120
         call nexty(xinput,observ,1,size,j,k)
         deltay=observ(j)-observ(k)
c
         sinhg1=sinh(1.0*g(1))
         sinhg2=sinh(2.0*g(1))
         coshg1=cosh(1.0*g(1))
         coshg2=cosh(2.0*g(1))
         ng(1)=size+g(1)
c
c
         ng(2)=coshg2*deltay+coshg1*deltay-1.0*deltay+g(2)
c
c
         ng(3)=-2.0*deltay*sinhg2+2.0*deltay*sinhg1+g(3)
c
c
         ng(4)=coshg1*deltay+g(4)
c
c
         ng(5)=g(5)-1.0*deltay*sinhg1
```

```
c
c
      ng(6)=coshg2*deltay-2.0*coshg1*deltay+deltay+g(6)
c
c
      cst1=-coshg1*sinhg2*size+coshg2*sinhg1*size+2.0*coshg1*sinhg1*
     1   size-sinhg1*size+coshg1*deltay**2*sinhg2
      cst2=-coshg2*deltay**2*sinhg1-2.0*coshg1*deltay**2*sinhg1+delt
     1   ay**2*sinhg1-2.0*g(2)*deltay*sinhg1-g(3)*coshg1*deltay
      ng(7)=cst2+cst1+g(7)
c
c
      cst3=-sinhg1*sinhg2*size+sinhg1**2*size-coshg1*coshg2*size+2.0
     1   *coshg1**2*size-coshg1*size
      cst4=deltay**2*sinhg1*sinhg2-deltay**2*sinhg1**2-g(3)*deltay*s
     1   inhg1+coshg1*coshg2*deltay**2-2.0*coshg1**2*deltay**2
      cst5=coshg1*deltay**2+2.0*g(4)*coshg2*deltay-4.0*g(4)*coshg1*d
     1   eltay+2.0*g(4)*deltay+g(8)
      ng(8)=cst5+cst4+cst3
c
c
      ng(9)=0.5*coshg1*sinhg1*size-0.5*coshg1*deltay**2*sinhg1-1.0*g
     1   (4)*deltay*sinhg1+g(9)
c
c
      cst6=-g(4)*sinhg1*sinhg2*size+g(5)*coshg1*sinhg2*size+g(4)*sin
     1   hg1**2*size+g(2)*sinhg1**2*size-g(5)*coshg2*sinhg1*size
      cst7=-2.0*g(5)*coshg1*sinhg1*size+g(3)*coshg1*sinhg1*size+g(5)
     1   *sinhg1*size-g(4)*coshg1*coshg2*size+2.0*g(4)*coshg1**2*siz
     2   e
      cst8=-g(4)*coshg1*size+g(4)*deltay**2*sinhg1*sinhg2-g(5)*coshg
     1   1*deltay**2*sinhg2-deltay*sinhg2-g(4)*deltay**2*sinhg1**2
      cst9=-g(2)*deltay**2*sinhg1**2+g(5)*coshg2*deltay**2*sinhg1+2.
     1   0*g(5)*coshg1*deltay**2*sinhg1-g(3)*coshg1*deltay**2*sinhg1
     2   -g(5)*deltay**2*sinhg1
      cst10=2.0*g(2)*g(5)*deltay*sinhg1-g(3)*g(4)*deltay*sinhg1+delt
     1   ay*sinhg1+g(4)*coshg1*coshg2*deltay**2-2.0*g(4)*coshg1**2*d
     2   eltay**2
      cst11=g(4)*coshg1*deltay**2+g(4)**2*coshg2*deltay+g(3)*g(5)*co
     1   shg1*deltay-2.0*g(4)**2*coshg1*deltay+g(4)**2*deltay
      ng(10)=cst9+cst8+cst7+cst6+cst11+cst10+g(10)
c
c
    do 105  i = 1 , 10 , 1
        g(i)=ng(i)
105  continue
c
      x0xfx=x0fx(g)
      x1xfx=x1fx(g)
      x2xfx=x2fx(g)
      x3xfx=x3fx(g)
      x4xfx=x4fx(g)
      sinhg1=sinh(1.0*g(1))
      sinhg2=sinh(2.0*g(1))
      sinhg3=sinh(3.0*g(1))
```

```
      coshg1=cosh(1.0*g(1))
      coshg3=cosh(3.0*g(1))
      coshg2=cosh(2.0*g(1))
      cst12=-0.25*coshg1**2*sinhg3*x3xfx/sinhg1**2-0.083333333333333
     1   33*sinhg3*x3xfx-0.75*sinhg1*x3xfx+0.25*coshg1*coshg3*x3xfx/
     2   sinhg1+1.5*coshg1**2*x3xfx/sinhg1
      cst13=-coshg1*x3xfx/sinhg1+0.08333333333333333*coshg1**3*coshg
     1   3*x3xfx/sinhg1**3-0.75*coshg1**4*x3xfx/sinhg1**3+0.66666666
     2   66666667*coshg1**3*x3xfx/sinhg1**3-0.5*g(5)*coshg1*sinhg3*x
     3   2xfx/sinhg1**2
      cst14=0.5*coshg1*sinhg3*x2xfx/sinhg1**2-g(6)*coshg1*sinhg2*x2x
     1   fx/sinhg1-g(2)*coshg1*sinhg2*x2xfx/sinhg1+0.5*g(3)*coshg1**
     2   2*sinhg2*x2xfx/sinhg1**2+0.5*g(3)*sinhg2*x2xfx
      cst15=0.25*g(5)*coshg3*x2xfx/sinhg1-0.25*coshg3*x2xfx/sinhg1-g
     1   (3)*coshg1*coshg2*x2xfx/sinhg1+2.25*g(5)*coshg1*x2xfx/sinhg
     2   1-2.25*coshg1*x2xfx/sinhg1
      cst16=-g(5)*x2xfx/sinhg1+x2xfx/sinhg1+0.5*g(6)*coshg1**2*coshg
     1   2*x2xfx/sinhg1**2+0.5*g(2)*coshg1**2*coshg2*x2xfx/sinhg1**2
     2   +0.5*g(6)*coshg1**2*x2xfx/sinhg1**2
      cst17=-0.5*g(2)*coshg1**2*x2xfx/sinhg1**2+0.25*g(5)*coshg1**2*
     1   coshg3*x2xfx/sinhg1**3-0.25*coshg1**2*coshg3*x2xfx/sinhg1**
     2   3-2.25*g(5)*coshg1**3*x2xfx/sinhg1**3+2.25*coshg1**3*x2xfx/
     3   sinhg1**3
      cst18=2.0*g(5)*coshg1**2*x2xfx/sinhg1**3-2.0*coshg1**2*x2xfx/s
     1   inhg1**3+0.5*g(6)*coshg2*x2xfx+0.5*g(2)*coshg2*x2xfx-0.5*g(
     2   6)*x2xfx
      cst19=0.5*g(2)*x2xfx+0.5*coshg1*sinhg3*x1xfx/sinhg1-0.25*g(5)*
     1   *2*sinhg3*x1xfx/sinhg1**2+0.5*g(5)*sinhg3*x1xfx/sinhg1**2-0
     2   .25*sinhg3*x1xfx/sinhg1**2
      cst20=-g(5)*g(6)*sinhg2*x1xfx/sinhg1+g(6)*sinhg2*x1xfx/sinhg1-
     1   g(2)*g(5)*sinhg2*x1xfx/sinhg1+g(2)*sinhg2*x1xfx/sinhg1+g(3)
     2   *g(5)*coshg1*sinhg2*x1xfx/sinhg1**2
      cst21=-g(3)*coshg1*sinhg2*x1xfx/sinhg1**2+g(8)*sinhg1*x1xfx-2.
     1   0*g(4)*g(6)*sinhg1*x1xfx-g(3)*g(5)*coshg2*x1xfx/sinhg1+g(3)
     2   *coshg2*x1xfx/sinhg1
      cst22=-g(8)*coshg1**2*x1xfx/sinhg1+2.0*g(4)*g(6)*coshg1**2*x1x
     1   fx/sinhg1+0.75*g(5)**2*x1xfx/sinhg1-1.5*g(5)*x1xfx/sinhg1+0
     2   .75*x1xfx/sinhg1
      cst23=-0.25*coshg1**2*coshg3*x1xfx/sinhg1**2+g(5)*g(6)*coshg1*
     1   coshg2*x1xfx/sinhg1**2-g(6)*coshg1*coshg2*x1xfx/sinhg1**2+g
     2   (2)*g(5)*coshg1*coshg2*x1xfx/sinhg1**2-g(2)*coshg1*coshg2*x
     3   1xfx/sinhg1**2
      cst24=2.25*coshg1**3*x1xfx/sinhg1**2-2.0*coshg1**2*x1xfx/sinhg
     1   1**2+g(5)*g(6)*coshg1*x1xfx/sinhg1**2-g(6)*coshg1*x1xfx/sin
     2   hg1**2-g(2)*g(5)*coshg1*x1xfx/sinhg1**2
      cst25=g(2)*coshg1*x1xfx/sinhg1**2+0.25*g(5)**2*coshg1*coshg3*x
     1   1xfx/sinhg1**3-0.5*g(5)*coshg1*coshg3*x1xfx/sinhg1**3+0.25*
     2   coshg1*coshg3*x1xfx/sinhg1**3-2.25*g(5)**2*coshg1**2*x1xfx/
     3   sinhg1**3
      cst26=4.5*g(5)*coshg1**2*x1xfx/sinhg1**3-2.25*coshg1**2*x1xfx/
     1   sinhg1**3+2.0*g(5)**2*coshg1*x1xfx/sinhg1**3-4.0*g(5)*coshg
     2   1*x1xfx/sinhg1**3+2.0*coshg1*x1xfx/sinhg1**3
      cst27=-0.25*coshg3*x1xfx-2.25*coshg1*x1xfx+x1xfx+0.25*g(5)*sin
     1   hg3*x0xfx/sinhg1-0.25*sinhg3*x0xfx/sinhg1
      cst28=-0.5*g(3)*coshg1*sinhg2*x0xfx/sinhg1+0.5*g(3)*g(5)**2*si
```

```
     1      nhg2*x0xfx/sinhg1**2-g(3)*g(5)*sinhg2*x0xfx/sinhg1**2+0.5*g
     2      (3)*sinhg2*x0xfx/sinhg1**2+0.5*g(6)*sinhg2*x0xfx
        cst29=0.5*g(2)*sinhg2*x0xfx-0.5*g(6)*coshg1*coshg2*x0xfx/sinhg
     1      1-0.5*g(2)*coshg1*coshg2*x0xfx/sinhg1-g(5)*g(8)*coshg1*x0xf
     2      x/sinhg1+g(8)*coshg1*x0xfx/sinhg1
        cst30=2.0*g(4)*g(5)*g(6)*coshg1*x0xfx/sinhg1-2.0*g(4)*g(6)*cos
     1      hg1*x0xfx/sinhg1-0.5*g(6)*coshg1*x0xfx/sinhg1+0.5*g(2)*cosh
     2      g1*x0xfx/sinhg1-0.25*g(5)*coshg1*coshg3*x0xfx/sinhg1**2
        cst31=0.25*coshg1*coshg3*x0xfx/sinhg1**2+0.5*g(5)**2*g(6)*cosh
     1      g2*x0xfx/sinhg1**2-g(5)*g(6)*coshg2*x0xfx/sinhg1**2+0.5*g(6
     2      )*coshg2*x0xfx/sinhg1**2+0.5*g(2)*g(5)**2*coshg2*x0xfx/sinh
     3      g1**2
        cst32=-g(2)*g(5)*coshg2*x0xfx/sinhg1**2+0.5*g(2)*coshg2*x0xfx/
     1      sinhg1**2+2.25*g(5)*coshg1**2*x0xfx/sinhg1**2-2.25*coshg1**
     2      2*x0xfx/sinhg1**2-2.0*g(5)*coshg1*x0xfx/sinhg1**2
        cst33=2.0*coshg1*x0xfx/sinhg1**2+0.5*g(5)**2*g(6)*x0xfx/sinhg1
     1      **2-g(5)*g(6)*x0xfx/sinhg1**2+0.5*g(6)*x0xfx/sinhg1**2-0.5*
     2      g(2)*g(5)**2*x0xfx/sinhg1**2
        cst34=g(2)*g(5)*x0xfx/sinhg1**2-0.5*g(2)*x0xfx/sinhg1**2+0.083
     1      33333333333333*g(5)**3*coshg3*x0xfx/sinhg1**3-0.25*g(5)**2*
     2      coshg3*x0xfx/sinhg1**3+0.25*g(5)*coshg3*x0xfx/sinhg1**3
        cst35=-0.08333333333333333*coshg3*x0xfx/sinhg1**3-0.75*g(5)**3
     1      *coshg1*x0xfx/sinhg1**3+2.25*g(5)**2*coshg1*x0xfx/sinhg1**3
     2      -2.25*g(5)*coshg1*x0xfx/sinhg1**3+0.75*coshg1*x0xfx/sinhg1*
     3      *3
        cst36=0.6666666666666667*g(5)**3*x0xfx/sinhg1**3-2.0*g(5)**2*x
     1      0xfx/sinhg1**3+2.0*g(5)*x0xfx/sinhg1**3-0.6666666666666667*
     2      x0xfx/sinhg1**3+0.5*g(3)*coshg2*x0xfx
        cst37=g(10)*x0xfx-g(4)*g(8)*x0xfx+g(7)*x0xfx+g(4)**2*g(6)*x0xf
     1      x-0.75*g(5)*x0xfx
        u1=-0.5*g(3)*x0xfx+0.75*x0xfx+cst37+cst36+cst35+cst34+cst33+cs
     1      t32+cst31+cst30+cst29+cst28+cst27+cst26+cst25+cst24+cst23+c
     2      st22+cst21+cst20+cst19+cst18+cst17+cst16+cst15+cst14+cst13+
     3      cst12
c
        cst38=-0.25*coshg1**2*sinhg3*x4xfx/sinhg1**2-0.083333333333333
     1      33*sinhg3*x4xfx-0.75*sinhg1*x4xfx+0.25*coshg1*coshg3*x4xfx/
     2      sinhg1+1.5*coshg1**2*x4xfx/sinhg1
        cst39=-coshg1*x4xfx/sinhg1+0.08333333333333333*coshg1**3*coshg
     1      3*x4xfx/sinhg1**3-0.75*coshg1**4*x4xfx/sinhg1**3+0.66666666
     2      66666667*coshg1**3*x4xfx/sinhg1**3-0.5*g(5)*coshg1*sinhg3*x
     3      3xfx/sinhg1**2
        cst40=0.5*coshg1*sinhg3*x3xfx/sinhg1**2-g(6)*coshg1*sinhg2*x3x
     1      fx/sinhg1-g(2)*coshg1*sinhg2*x3xfx/sinhg1+0.5*g(3)*coshg1**
     2      2*sinhg2*x3xfx/sinhg1**2+0.5*g(3)*sinhg2*x3xfx
        cst41=0.25*g(5)*coshg3*x3xfx/sinhg1-0.25*coshg3*x3xfx/sinhg1-g
     1      (3)*coshg1*coshg2*x3xfx/sinhg1+2.25*g(5)*coshg1*x3xfx/sinhg
     2      1-2.25*coshg1*x3xfx/sinhg1
        cst42=-g(5)*x3xfx/sinhg1+x3xfx/sinhg1+0.5*g(6)*coshg1**2*coshg
     1      2*x3xfx/sinhg1**2+0.5*g(2)*coshg1**2*coshg2*x3xfx/sinhg1**2
     2      +0.5*g(6)*coshg1**2*x3xfx/sinhg1**2
        cst43=-0.5*g(2)*coshg1**2*x3xfx/sinhg1**2+0.25*g(5)*coshg1**2*
     1      coshg3*x3xfx/sinhg1**3-0.25*coshg1**2*coshg3*x3xfx/sinhg1**
     2      3-2.25*g(5)*coshg1**3*x3xfx/sinhg1**3+2.25*coshg1**3*x3xfx/
     3      sinhg1**3
```

```
      cst44=2.0*g(5)*coshg1**2*x3xfx/sinhg1**3-2.0*coshg1**2*x3xfx/s
     1   inhg1**3+0.5*g(6)*coshg2*x3xfx+0.5*g(2)*coshg2*x3xfx-0.5*g(
     2   6)*x3xfx
      cst45=0.5*g(2)*x3xfx+0.5*coshg1*sinhg3*x2xfx/sinhg1-0.25*g(5)*
     1   *2*sinhg3*x2xfx/sinhg1**2+0.5*g(5)*sinhg3*x2xfx/sinhg1**2-0
     2   .25*sinhg3*x2xfx/sinhg1**2
      cst46=-g(5)*g(6)*sinhg2*x2xfx/sinhg1+g(6)*sinhg2*x2xfx/sinhg1-
     1   g(2)*g(5)*sinhg2*x2xfx/sinhg1+g(2)*sinhg2*x2xfx/sinhg1+g(3)
     2   *g(5)*coshg1*sinhg2*x2xfx/sinhg1**2
      cst47=-g(3)*coshg1*sinhg2*x2xfx/sinhg1**2+g(8)*sinhg1*x2xfx-2.
     1   0*g(4)*g(6)*sinhg1*x2xfx-g(3)*g(5)*coshg2*x2xfx/sinhg1+g(3)
     2   *coshg2*x2xfx/sinhg1
      cst48=-g(8)*coshg1**2*x2xfx/sinhg1+2.0*g(4)*g(6)*coshg1**2*x2x
     1   fx/sinhg1+0.75*g(5)**2*x2xfx/sinhg1-1.5*g(5)*x2xfx/sinhg1+0
     2   .75*x2xfx/sinhg1
      cst49=-0.25*coshg1**2*coshg3*x2xfx/sinhg1**2+g(5)*g(6)*coshg1*
     1   coshg2*x2xfx/sinhg1**2-g(6)*coshg1*coshg2*x2xfx/sinhg1**2+g
     2   (2)*g(5)*coshg1*coshg2*x2xfx/sinhg1**2-g(2)*coshg1*coshg2*x
     3   2xfx/sinhg1**2
      cst50=2.25*coshg1**3*x2xfx/sinhg1**2-2.0*coshg1**2*x2xfx/sinhg
     1   1**2+g(5)*g(6)*coshg1*x2xfx/sinhg1**2-g(6)*coshg1*x2xfx/sin
     2   hg1**2-g(2)*g(5)*coshg1*x2xfx/sinhg1**2
      cst51=g(2)*coshg1*x2xfx/sinhg1**2+0.25*g(5)**2*coshg1*coshg3*x
     1   2xfx/sinhg1**3-0.5*g(5)*coshg1*coshg3*x2xfx/sinhg1**3+0.25*
     2   coshg1*coshg3*x2xfx/sinhg1**3-2.25*g(5)**2*coshg1**2*x2xfx/
     3   sinhg1**3
      cst52=4.5*g(5)*coshg1**2*x2xfx/sinhg1**3-2.25*coshg1**2*x2xfx/
     1   sinhg1**3+2.0*g(5)**2*coshg1*x2xfx/sinhg1**3-4.0*g(5)*coshg
     2   1*x2xfx/sinhg1**3+2.0*coshg1*x2xfx/sinhg1**3
      cst53=-0.25*coshg3*x2xfx-2.25*coshg1*x2xfx+x2xfx+0.25*g(5)*sin
     1   hg3*x1xfx/sinhg1-0.25*sinhg3*x1xfx/sinhg1
      cst54=-0.5*g(3)*coshg1*sinhg2*x1xfx/sinhg1+0.5*g(3)*g(5)**2*si
     1   nhg2*x1xfx/sinhg1**2-g(3)*g(5)*sinhg2*x1xfx/sinhg1**2+0.5*g
     2   (3)*sinhg2*x1xfx/sinhg1**2+0.5*g(6)*sinhg2*x1xfx
      cst55=0.5*g(2)*sinhg2*x1xfx-0.5*g(6)*coshg1*coshg2*x1xfx/sinhg
     1   1-0.5*g(2)*coshg1*coshg2*x1xfx/sinhg1-g(5)*g(8)*coshg1*x1xf
     2   x/sinhg1+g(8)*coshg1*x1xfx/sinhg1
      cst56=2.0*g(4)*g(5)*g(6)*coshg1*x1xfx/sinhg1-2.0*g(4)*g(6)*cos
     1   hg1*x1xfx/sinhg1-0.5*g(6)*coshg1*x1xfx/sinhg1+0.5*g(2)*cosh
     2   g1*x1xfx/sinhg1-0.25*g(5)*coshg1*coshg3*x1xfx/sinhg1**2
      cst57=0.25*coshg1*coshg3*x1xfx/sinhg1**2+0.5*g(5)**2*g(6)*cosh
     1   g2*x1xfx/sinhg1**2-g(5)*g(6)*coshg2*x1xfx/sinhg1**2+0.5*g(6
     2   )*coshg2*x1xfx/sinhg1**2+0.5*g(2)*g(5)**2*coshg2*x1xfx/sinh
     3   g1**2
      cst58=-g(2)*g(5)*coshg2*x1xfx/sinhg1**2+0.5*g(2)*coshg2*x1xfx/
     1   sinhg1**2+2.25*g(5)*coshg1**2*x1xfx/sinhg1**2-2.25*coshg1**
     2   2*x1xfx/sinhg1**2-2.0*g(5)*coshg1*x1xfx/sinhg1**2
      cst59=2.0*coshg1*x1xfx/sinhg1**2+0.5*g(5)**2*g(6)*x1xfx/sinhg1
     1   **2-g(5)*g(6)*x1xfx/sinhg1**2+0.5*g(6)*x1xfx/sinhg1**2-0.5*
     2   g(2)*g(5)**2*x1xfx/sinhg1**2
      cst60=g(2)*g(5)*x1xfx/sinhg1**2-0.5*g(2)*x1xfx/sinhg1**2+0.083
     1   33333333333333*g(5)**3*coshg3*x1xfx/sinhg1**3-0.25*g(5)**2*
     2   coshg3*x1xfx/sinhg1**3+0.25*g(5)*coshg3*x1xfx/sinhg1**3
      cst61=-0.08333333333333333*coshg3*x1xfx/sinhg1**3-0.75*g(5)**3
     1   *coshg1*x1xfx/sinhg1**3+2.25*g(5)**2*coshg1*x1xfx/sinhg1**3
```

```
     2        -2.25*g(5)*coshg1*x1xfx/sinhg1**3+0.75*coshg1*x1xfx/sinhg1*
     3        *3
          cst62=0.6666666666666667*g(5)**3*x1xfx/sinhg1**3-2.0*g(5)**2*x
     1        1xfx/sinhg1**3+2.0*g(5)*x1xfx/sinhg1**3-0.6666666666666667*
     2        x1xfx/sinhg1**3+0.5*g(3)*coshg2*x1xfx
          cst63=g(10)*x1xfx-g(4)*g(8)*x1xfx+g(7)*x1xfx+g(4)**2*g(6)*x1xf
     1        x-0.75*g(5)*x1xfx
          u1x=-0.5*g(3)*x1xfx+0.75*x1xfx+cst63+cst62+cst61+cst60+cst59+c
     1        st58+cst57+cst56+cst55+cst54+cst53+cst52+cst51+cst50+cst49+
     2        cst48+cst47+cst46+cst45+cst44+cst43+cst42+cst41+cst40+cst39
     3        +cst38
c
          estim(j)=(u1x-u1*x1xfx/x0xfx)/x0xfx+x1xfx/x0xfx
          var(j)=(xinput(j)-estim(j))**2+var(j)
          k=j
          j=j+1
       go to  110
120    continue
300    continue
       do  400   i = 1 , 4000 , 1
       var(i)=var(i)/300.0
       y=i*size
       print*,y,var(i)
400    continue
       stop
       end

       real function x0fx(g)
       real g(1)
       x0fx=1/sqrt(cosh(g(1)))
       return
       end

       real function x1fx(g)
       real g(1)
       x1fx=(1-g(5))*cosh(g(1))**((-3.0)/2.0)
       return
       end

       real function x2fx(g)
       real g(1)
       x2fx=cosh(g(1))**((-5.0)/2.0)*(cosh(g(1))*sinh(g(1))+g(5)**2-2*g(5
     1   )+1)
       return
       end

       real function x3fx(g)
       real g(1)
       x3fx=(1-g(5))*cosh(g(1))**((-7.0)/2.0)*(3*cosh(g(1))*sinh(g(1))+g(
     1   5)**2-2*g(5)+1)
       return
       end

       real function x4fx(g)
       real g(1)
```

```
 x4fx=cosh(g(1))**((-9.0)/2.0)*(3*cosh(g(1))**2*sinh(g(1))**2+6*g(5
1   )**2*cosh(g(1))*sinh(g(1))-12*g(5)*cosh(g(1))*sinh(g(1))+6*cosh
2   (g(1))*sinh(g(1))+g(5)**4-4*g(5)**3+6*g(5)**2-4*g(5)+1)
 return
 end

 subroutine nexty(xinput,observ,e,size,j,k)
 real xinput(1), observ(1), e, size
 integer j, k
 xinput(j)=grand(0)*sqrt(size)+xinput(k)
 observ(j)=(e*xinput(j)**2+xinput(j))*size+grand(0)*sqrt(size)+obse
1   rv(k)
 return
 end
```

3.2. A vector system including the Benes problem

To illustrate the facility of the system to deal with vector problems we shall treat the following example

$$\begin{aligned} dx_{1t} &= x_{1t} + x_{2t} + dw_{1t} \\ dx_{2t} &= \tanh x_{2t} + dw_{2t} \\ dy_{1t} &= x_{1t} + dv_{1t} \\ dy_{2t} &= x_{2t} + dv_{2t} \end{aligned} \tag{49}$$

The system goes through the same steps as before, computing the estimation algebra which is a product of two versions of the estimation Lie algebra for the Kalman filtering problem [21, 34]. The transcript which follows contains no surprises (fortunately); but it does illustrate the way the system deals with vector quantities. FORTRAN code was not generated for this example.

3.3. Computing the likelihood ratio

In this example we shall have the system generate a FORTRAN code to compute the likelihood ratio for the following system

$$
\begin{aligned}
dx_t &= -x_t\, dt + dw_t \\
dy_t &= x_t^2\, dt + dv_t \\
p_0(x) &= c\, e^{-x^2}.
\end{aligned}
\tag{50}
$$

To execute the program it is necessary to specify the functions f, g, and h, the initial density $p_0(x)$, n, the order of the approximation, and m, the degree of the Hermite polynominals used in the Gaussian quadrature formula. A sample of the FORTRAN code written by the system and its numerical output for the case $n = 4$, $m = 5$ follows. Note that the FORTRAN makes use of double precision arithmetic, and that the likelihood ratio is "un-normalized;" hence, small numerical values are produced in the execution. While the FORTRAN code is relatively simple for this example, it is once again the kind of "tedious" code one would not like to write and revise by hand.

```
      program gaussianint
      integer mm,nn
      double precision y(30),weight(30),rr1(30),aa(30)
      double precision cait,sqot
      double precision cs(30,30),rr2(30),ss(30,30),qq0(30)
      double precision csnn(30),ssnn(30),iinn(30)
      common csnn,ssnn,iinn
      common mm,nn,y,weight,rr1,aa,cait,cs,rr2,ss,qq0
      sqot(xx)=sqrt(real(xx))
      nn=4
      mm=5
      read(5,5) t
      write(6,7) t
7     format('  t=',d40.25)
      do 10 i=1,nn
      read(5,5) y(i)
5     format(f10.3)
10    continue
      weight(1)=1.0/1600000.0
      weight(2)=1.0/1600000.0
      weight(3)=1.0/4800000.0
      weight(4)=1.0/4800000.0
      weight(5)=1.0/1200000.0
      aa(1)=-sqot(5)/sqot(2)
```

```
      aa(2)=sqot(5)/sqot(2)
      aa(3)=-sqot(15)/sqot(2)
      aa(4)=sqot(15)/sqot(2)
      aa(5)=0
      do 13 i=1,mm
      write(6,12) i,aa(i)
12    format('  aa(',i2,')=',d40.25)
13    continue
      rr1(1)=exp(5.0/2.0)
      rr1(2)=exp(5.0/2.0)
      rr1(3)=exp(15.0/2.0)
      rr1(4)=exp(15.0/2.0)
      rr1(5)=1
      ssnn(1)=exp((-747417557.0)*t**2/4.253116844126304E+16)
      ssnn(2)=exp((-747417557.0)*t**2/4.253116844126304E+16)
      ssnn(3)=exp((-94769023.0)*t**2/4.253116844126304E+16)
      ssnn(4)=exp((-94769023.0)*t**2/4.253116844126304E+16)
      ssnn(5)=1
      do 200 i=1,mm
      do 20 j=1,mm
      call numerical(i,j,t,aa(i),aa(j),cs(i,j))
      write(6,15) i,j,cs(i,j)
15    format(' cs(',i2,',',i2,')=',d40.25)
20    continue
200   continue
      do 500 i=1,mm
      csnn(i)=0
      do 400 j=1,mm
      csnn(i)=cait*rr1(j)**y(nn)*rr2(j)*ssnn(j)*weight(j)*cs(j,i)+csn
     1  n(i)
400   continue
      iinn(i)=cait*csnn(i)
      write(6,450) i,iinn(i)
450   format('  iinn(y,aa(',i2,'),t)=',d40.25)
500   continue
      stop
      end

      subroutine numerical(i,j,t,ai,aj,ccss)
      integer mm,nn,l
      integer i,j
      double precision t,ai,aj,ccss,tcs(30)
      double precision csnn(30),ssnn(30),iinn(30)
      double precision y(30),weight(30),rr1(30),aa(30)
      double precision cait,sqot
      double precision cs(30,30),rr2(30),ss(30,30),qq0(30)
      common csnn,ssnn,iinn
      common mm,nn,y,weight,rr1,aa,cait,cs,rr2,ss,qq0
      sqot(xx)=sqrt(real(xx))
      cait=314794333.0*2**(1.0/8.0)*3**((-1.0)/8.0)*exp(-ai**2/t**2)/
     1  434722062.0
      rr2(1)=exp((-7.0)*t/4.0)
      rr2(2)=exp((-7.0)*t/4.0)
      rr2(3)=exp((-119.0)*t/8.0)
      rr2(4)=exp((-119.0)*t/8.0)
```

```
      rr2(5)=exp(t/8.0)
      ss(1,1)=exp(250706837.0*sqot(5)*aj/(423322282.0*sqot(2)))
      ss(1,2)=exp((-250706837.0)*sqot(5)*aj/(423322282.0*sqot(2)))
      ss(1,3)=exp(250706837.0*sqot(15)*aj/(423322282.0*sqot(2)))
      ss(1,4)=exp((-250706837.0)*sqot(15)*aj/(423322282.0*sqot(2)))
      ss(1,5)=1
      ss(2,1)=exp(497549753.0*sqot(5)*aj/(541902816.0*sqot(2)))
      ss(2,2)=exp((-497549753.0)*sqot(5)*aj/(541902816.0*sqot(2)))
      ss(2,3)=exp(497549753.0*sqot(15)*aj/(541902816.0*sqot(2)))
      ss(2,4)=exp((-497549753.0)*sqot(15)*aj/(541902816.0*sqot(2)))
      ss(2,5)=1
      ss(3,1)=exp(345403687.0*sqot(5)*aj/(933856711.0*sqot(2)))
      ss(3,2)=exp((-345403687.0)*sqot(5)*aj/(933856711.0*sqot(2)))
      ss(3,3)=exp(345403687.0*sqot(15)*aj/(933856711.0*sqot(2)))
      ss(3,4)=exp((-345403687.0)*sqot(15)*aj/(933856711.0*sqot(2)))
      ss(3,5)=1
      qq0(1)=exp((-526060093.0)*sqot(5)*ai/(124766106.0*sqot(3)*t))
      tcs(1)=0
      do 11 l=1,mm
      tcs(1)=qq0(1)*ss(1,l)*rr1(l)**y(1)*rr2(l)*weight(l)+tcs(1)
11     continue
      do 101 k=2,nn-1
      tcs(k)=0
      do 51 l=1,mm
      tcs(k)=cait*tcs(k-1)*ss(k,l)*rr1(l)**y(k)*rr2(l)*weight(l)+tcs(
     1  k)
51     continue
101    continue
      ccss=tcs(nn-1)
      return
      end
```

---

n = 4 m = 5 f(x) = - x g(x) = 1 h(x) = x^2 p0(x) = exp(-x^2)

```
 aa( 1)=     -0.1581138711771437976860710d+01
 aa( 2)=      0.1581138711771437976860710d+01
 aa( 3)=     -0.2738612743201017774019590d+01
 aa( 4)=      0.2738612743201017774019590d+01
 aa( 5)=      0.                       d+00

t=0.3
 iinn(y,aa( 1),t)=      0.4685290991761690171379939d-09
 iinn(y,aa( 2),t)=      0.4685290991761690504446847d-09
 iinn(y,aa( 3),t)=      0.1462871214290576982186565d-06
 iinn(y,aa( 4),t)=      0.1462871214290577071004407d-06
 iinn(y,aa( 5),t)=      0.1620165719704102758047526d-11

t=0.5
```

```
  iinn(y,aa( 1),t)=      0.1197086675310372023695038d-09
  iinn(y,aa( 2),t)=      0.1197086675310372023695038d-09
  iinn(y,aa( 3),t)=      0.2241371836876454248788093d-07
  iinn(y,aa( 4),t)=      0.2241371836876454248788093d-07
  iinn(y,aa( 5),t)=      0.7804265806432602148312583d-12

t=1
  iinn(y,aa( 1),t)=      0.6105581910761162967027360d-12
  iinn(y,aa( 2),t)=      0.6105581910761162967027360d-12
  iinn(y,aa( 3),t)=      0.2788197530515623301372585d-10
  iinn(y,aa( 4),t)=      0.2788197530515623190350283d-10
  iinn(y,aa( 5),t)=      0.2678150147248882728678865d-13

t=3
  iinn(y,aa( 1),t)=      0.2596809689821663214015501d-19
  iinn(y,aa( 2),t)=      0.2596809689821663214015501d-19
  iinn(y,aa( 3),t)=      0.2469985217357907270052664d-18
  iinn(y,aa( 4),t)=      0.2469985217357907270052664d-18
  iinn(y,aa( 5),t)=      0.6083668813757230453553859d-20

t=5
  iinn(y,aa( 1),t)=      0.1474701869704486778545061d-22
  iinn(y,aa( 2),t)=      0.1474701869704486778545061d-22
  iinn(y,aa( 3),t)=      0.2432379509452652632806036d-22
  iinn(y,aa( 4),t)=      0.2432379509452652632806036d-22
  iinn(y,aa( 5),t)=      0.1261550153998446471836914d-22
```

(remainder omitted)

3.4. The existence and uniqueness test

In the example below we have the system test the well-posedness of the robust from of the Zakai equation for a version of the weak quadratic sensor problem. Unlike the portion of the code which derives an approximate filtering structure by expanding in terms of the small parameter, this portion of the system treats the problem as posed entirely in symbolic terms. It uses no approximations. The example treated is

$$dx_t = -x_t + dw_t \tag{51}$$

$$dy_t = [1 + \epsilon x^2] \, dt + dv_t \, .$$

The system first requests the functions f, g, and h from the user. It then converts the Zakai equation to robust form using the exponential transformation. It also tests the functions to see if they are polynominals, or not. In this case they are, and so the system uses simplified forms of the existence and uniqueness criteria in [6]. The parameters q, r, s reported next in the session are the exponents of the highest powers of x occuring in the functions f, g, and h; the variables $-ff$, gg, hh are the coefficients of those terms. The next lines summarize the various tests on the coefficients and the exponents which apply to check well-posedness in this case. For example, **test1** corresponds to the condition

$$ff > 0 \text{ and } r \leq 1 \text{ and } q > r \text{ and } s \geq 1 \text{ and } q \geq 1 \tag{52}$$

The MACSYMA code reproduced in section 4.3 lists the criteria for the case of polynominal coefficients based on the results in [6] p. 209, equations (3.1) - (3.6). Notice that the system makes use of the tail estimates in [6] to specify the class of functions in which the solution lies.

4. Samples of the MACSYMA code development

The MACSYMA code implementing the nonlinear filtering methodology described in sections 2.1 and 2.2 was developed[11] in modules consisting of MACSYMA functions, or *function blocks*[12] implementing the various steps in the methodology. Modular structure permits the incorporation of a command language interface, or a highly structured interface depending on user experience. We shall explain the structure of some key function blocks in this portion of the system to illustrate the nature of MACSYMA programming.

4.1. Lie basis generator

One of the basic function blocks is **gen_basis** which generates a version of the estimation algebra associated with the problem. The code for this function is shown in Figure 1. It accepts as input *prebasis* which is a list of the differential operators appearing in the Zakai equation. It applies a function to these operators called **lie_bracket** (not shown) which does the Lie bracket operation (for matrix valued partial differential operators). The output of this function is screened for new basis elements which are added to the basis list. The basis generator also counts the elements in this list, and displays the list ten elements at a time, simultaneously asking the user if he wishes to continue the computation.[13]

Within **gen_basis** the function **find_coeff** is used to isolate the coefficients of the components in each Lie bracket that match previously generated basis elements. It is necessary to "strip" these coefficients to test for linear independence of the terms. Another function, **de_comp** is used to decompose a given basis list to ensure that it has linearly independent elements. The other commands, "factor," "endcons," "length," etc., appearing in the block are basic

[11]By A. LaVigna, D. MacEnany, and I. Yan with the assistance of P. Kumar and K. Paul.

[12]A function *block* in MACSYMA is analogous to a subroutine in FORTRAN or a procedure in PASCAL. It allows variables to be defined locally and permits access to variables which are global to the system. See [35] and [36].

[13]This constitutes a simplistic "test" for problems with infinite dimensional estimation algebras which avoids infinite loops. If the coefficient functions f, g, and h in the original problem (1) are polynominal functions of x, then the estimation algebra will be a subalgebra of the Weyl algebra of differential operators with polynominal coefficients. In this case the techniques used in [20] that is, testing basis elements of the estimation algebra as generated to see if they contain the generators of the Weyl algebra, may be used as a check for infinite dimensionality of the estimation algebra. Marcus [3] calls this the *homomorphism principle*. It would be relatively straightforward to enhance the existing MACSYMA code to include this procedure.

```
gen_basis(prebasis,order,flag):= block (
[brack,lengthpre,coeff,ttt,test],
if (flag) then test:10 else test:30,
lengthpre:length(prebasis),
for j:1 thru lengthpre do
(for i:j+1 while (length(prebasis)+1>i) do
(if (integerp(length(prebasis)/test)) then
(ttt:read("The length of the basis is now",
length(prebasis),"and the basis is",prebasis,
"Do you wish to continue? (yes; , no;)"),
if (ttt=no) then
(display(prebasis),
error("        "))
else
(test:test+10,
disp("Computation continuing") )),
brack:lie_bracket(prebasis[j],prebasis[i],order),
if ( brack # 0 ) then
(for k:1 thru length(prebasis) do
(coeff:find_coeff(brack,prebasis[k]),
brack:expand(brack-coeff*prebasis[k]) ),
if (brack # 0) then
(brack:factor(brack),
brack:brack/numfactor(brack),
prebasis:endcons(expand(brack),prebasis) )))),
prebasis:de_comp(prebasis),
prebasis );
```

Figure 1. MACSYMA code for Lie basis generation.

MACSYMA functions. All "custom functions" developed in this project are indicated by underscore functions in the code.

4.2. Transition matrix

The second block we shall describe is **find_transition_matrix** (see Figure 2). This block plays a key role in the computation of the Baker-Campbell-Hausdorff formula (15). Since MACSYMA cannot sum arbitrary infinite series (directly), this formula is evaluated by computing the matrix of structure constants of the Lie algebra (indexed by a particular operator, say A_1) and then exponentiating that matrix. This explains the lines referring to the transition matrix and the exponentiation of that matrix in the sample sessions displayed in the previous section. This procedure is used twice (in some cases): first to compute the BCH formula; and second (when appropriate) to compute the Green's function for the unperturbed system, which is, in effect $\exp[A_0 t]$ with A_0 the operator in (6) with ϵ set equal to zero. This explains the second stage of exponentiating the transition matrix which occurs in the sample sessions.

In the listing *dimb* is the dimension of the basis, *order* is the order of the estimation, *tbrack* is the sum of operators resulting from a Lie bracket of the first operator in the "basis" with another operator, and *tmx* is the transition matrix (of structure constants) associated with the first operator in the basis. The matrix exponential of the transition matrix is computed by another block which evaluates the inverse Laplace transform (a standard MACSYMA function) of the resolvent of the matrix.

The other components of the system are MACSYMA blocks for **lie_bracket, lie_ad** which computes the ad_L operator, **wei_norman** which computes the Wei-Norman expansion, and other blocks which rearrange the elements and compute the differential equations for the Wei-Norman

```
find_transition_matrix(basis,order):= block(
[tmx,dimb,tbrack,entry],
dimb:length(basis),
tmx:zeromatrix(dimb-1,dimb-1),
for i:2 thru dimb do
(tbrack:lie_bracket(basis[1],basis[i],order),
if (tbrack # 0 ) then
(for j:2 thru dimb do
(entry:find_coeff(tbrack,basis[j]),
tbrack:expand(tbrack-entry*basis[j]),
if (entry#0) then setelmx(entry,i-1,j-1,tmx))),
if tbrack#0 then error("error in find_transition_matrix")),
tmx);
```

Figure 2. MACSYMA code for finding the matrix of structure constants.

functions. Still other components implement the numerical scheme[14] for integrating the Wei-Norman equations and computing the conditional statistics of the estimate. Finally, we use a module developed at INRIA to write the FORTRAN code into a file.

4.3. Well-posedness for the polynominal case

After the coefficient functions for the filtering problem are entered into the system and the robust form of the Zakai equation has been generated, the system tests the coefficient functions to determine whether or not they are polynominals. If so, the system tests for well-posedness using a sequence of tests and options from [6] p. 209, equations (3.1) - (3.6). A portion of the code which does the test and computes the class of functions in which a solution exists is given in Figure 3. This is part of one of several function blocks involved in this portion of the code. The remainder of the **polycriterion()** function block tests the truth of other combinations of **test1, test2a, test2aa,** etc. The structure is similar to that presented in the Figure.

5. Commentary and conclusions

A basic theme of this work is that many of the powerful techniques of control theory can be put at the disposal of a practicing engineer by encoding them in a facility like MACSYMA and making the system interfaces convenient and flexible. The system discussed here has some elements of these features, but it is obviously a prototype. The module for stochastic control via the Bellman equation discussed in [10, 9] achieves these objectives much more fully; and it defines the current "standard" for the other modules under development.

The system for nonlinear filtering suggests other uses for an "expert system" based on MACSYMA. Since the intermediate computations display a great deal of technical information about the structure of a given filtering problem, it is possible to use the system for basic research on nonlinear filtering. For example, one could imbed the system in a logical loop which would

[14]We use the Mil'stein method [37] as described in [28] to integrate the nonlinear stochastic ODE's for the Wei-Norman functions.

```
polycriterion():=block (

    q:hipow(f,x),
    q:(q+1)/2,
    tryg:g, r:0,
    for i:1 step 1 unless numberp(tryg)
        do ( r:r+1, tryg:tryg/(1+x^2)^(1/2) ),
    s:hipow(h,x),
    display(q,r,s),

    ff:-coeff(f,x,2*q-1),
    gg:coeff(g,(1+x^2),r/2),
    hh:coeff(h,x,s),
    display(ff,gg,hh),

    test1:  ff>0 and r<=1 and q>r and s>=1 and q>=1,
    test2a:  r>=0 and r<=1 and q>r and s>=1 and q>=1,
    test2aa: r=1 and s#0 and q>=2,
    test2ab: r#1 and true,
    test2b:  r>1 and q>r+(1/2)*s and s>=1 and q>=2,
    display(test1,test2a,test2aa,test2ab,test2b),

    if test1 and test2a then
     ( phi1: (ff/gg)*x^(2*(q-r))/(2*(q-r)),
      if (r>=1) or ( r<1 and r+s>2*q-1)    then phi2: hh*x^s  else
      if r<1 and r+s>2*q-1    then phi2:(hh/gg)*x^(s-r+1)/(s-r+1) else
      display(" phi2 is undefined. "),

      if r+s>2*q-1 then eata1:0 else
      if r+s<2*q-1 then eata1:1 else
      if r+s=2*q-1 then eata1:(ff/gg)/(hh^2+ff^2/gg^2)^(1/2) else
      display(" eatd1 is undefined. "),

      if r=1 and r+s<2*q-1 then eata2:0 else
      if r=1 and r+s>2*q-1 then eata2:s*gg else
      if r=1 and r+s=2*q-1 then
            eata2:(s*gg*hh)/(hh^2+ff^2/gg^2)^(1/2) else
      if r<1 and r+s<=2*q-1 then eata2:0 else
      if r<1 and r+s>2*q-1 then eata2:1  else
      display(" eata2 is undefined. "),
      display(phi1,phi2,eata1,eata2) ) ,

  if test1 and test2b
  then display("THE ROBUST DMZ EQUATION HAS BOUNDED UNIQUE SOLUTION!")
```

Figure 3. A portion of the function block **polycriterion()** for well-posedness.

have it make a systematic search among models for problems with some specific structure; e.g., coefficient functions which are polynominals of degree n or less, searching for specific models which have a finite dimensional estimation algebra. The search could be structured by imposing certain relationships on the functions derived perhaps from conditions like those in [34, 38]. We have made some preliminary investigations of this type, but have not formulated a procedure which carries out the analysis in a constructive way.

It is also possible to use the system as a basis for the comparison of numerical algorithms for the study of specific filtering problems. For example, it is a comparatively straightforward procedure to generate MACSYMA blocks to implement other methods for constructing approximate solutions to given classes of filtering problems, including the methods of Sussmann [39] and the direct integration methods developed by Pardoux and Talay [28]. Our results in this direction will be reported on another occasion.

References

[1] M. Hazewinkel and J.C. Willems, eds., *Stochastic Systems: The Mathematics of Filtering and Identification and Applications*, Reidel, Dordrecht, The Netherlands (1981).

[2] M.H.A. Davis and S.I. Marcus, "An introduction to nonlinear filtering," pp. 565-572 in *Stochastic Systems: The Mathematics of Filtering and Identification and Applications*, ed. M. Hazewinkel and J.C. Willems, Reidel, Dordrecht, The Netherlands (1981).

[3] S.I. Marcus, "Algebraic and geometric methods in nonlinear filtering," *SIAM J. Control and Optimization* vol. 22, pp. 817-844 (1984).

[4] C.-H. Liu, "Applications of algebraic and approximation methods in nonlinear filtering," Ph.D. Thesis, Electrical Engineering Department, University of Texas (1981).

[5] G.L. Blankenship, C.-H. Liu, and S.I. Marcus , "Asymptotic expansions and Lie algebras for some nonlinear filtering problems," *IEEE Trans. Automatic Control* vol. AC-28, pp. 787-797 (1983).

[6] J.S. Baras, G.L. Blankenship, and W.E. Hopkins, Jr., "Existence, Uniqueness, and Asymptotic Behavior of Solutions to a Class of Zakai Equations with Unbounded Coefficients," *IEEE Transactions on Automatic Control* vol. AC-28, pp. 203-214 (1983).

[7] G.L. Blankenship and J.S. Baras, "Accurate evaluation of stochastic Wiener integrals with applications to scattering in random media and to nonlinear filtering," *SIAM J. Appl. Math.* vol. 41, pp. 518-552 (1981).

[8] W.E. Hopkins Jr., "Nonlinear Filtering of Nondegenerate Diffusions with Unbounded Coefficients," Ph.D. Dissertation, Electrical Engineering Department, University of Maryland, College Park (November 1982).

[9] J.P. Quadrat, C. Gomez, and A. Sulem, "Computer Algebra as a Tool for Solving Optimal

Control Problems," INRIA Report (1984 to appear).

[10] C. Gomez, J.P. Quadrat, A. Sulem, G.L. Blankenship, P. Kumar, A. LaVigna, D.C. MacEnany, K. Paul, and I. Yan, "An Expert System for Control and Signal Processing with Automatic FORTRAN Code Generation," *Proc. IEEE Conf. on Decision and Control*, pp. 716-723 (December 1984).

[11] F. Hayes-Roth, D.A. Waterman, and D.B. Lenat, eds., *Building Expert Systems*, Addison-Wesley, Reading, Mass. (1983).

[12] C. Gomez, J.P. Quadrat, A. Sulem, G.L. Blankenship, P. Kumar, A. LaVigna, D.C. MacEnany, K. Paul, and I. Yan, *An Expert System for Control and Signal Processing with Automatic FORTRAN Code Generation*. to appear.

[13] J.P. Quadrat, C. Gomez, and A. Sulem, "Towards an expert system in stochastic control: optimization in the class of local feedbacks," INRIA Report (1985 to appear).

[14] J.P. Quadrat, C. Gomez, and A. Sulem, "Towards an expert system in stochastic control: the Hamilton-Jacobi part," in *Lecture Notes in Control and Information Sciences*, ed. A. Bensoussan and J.L. Lions, Springer-Verlag, Nice (June 1984).

[15] E.S. Armstrong, *ORACLS. A Design System for Linear Multivariable Control*, Marcel Dekker, Inc., New York, N.Y. (1980).

[16] D. G. Mayne, W. T. Nye, E. Polak, P. Siegel, and T. Wuu, "DELIGHT-MIMO: An interactive, optimization-based multivariable control System Design Package," *Control System Magazine* vol. 2, no. 4, (1982).

[17] M.K.H. Fan, W.T. Nye, and A.L. Tits, "DELIGHT.MaryLin User's Guide," Electrical Engineering Department, University of Maryland, College Park, Maryland (in preparation).

[18] A. J. Laub, "Numerical linear algebra aspects of control design computations," *IEEE Trans. Automatic Control* vol. AC-30, pp. 97-108 (1985).

[19] M. Hazewinkel, "On deformations, approximations and nonlinear filtering," *Systems and Control Letters* vol. 1, pp. 32-36 (1981).

[20] M. Hazewinkel and S.I. Marcus, "On Lie algebras and finite dimensional filtering," *Stochastics* vol. 7, pp. 29-62 (1982).

[21] R.W. Brockett, "Remarks on finite dimensional nonlinear estimation," *Analyse des Systemes, Asterisque* vol. 75-76, pp. 47-55 (1980).

[22] S.K. Mitter, "On the analogy between mathematical problems of non-linear filtering and quantum physics," *Richerche di Automatica* vol. 10, pp. 163-216 (1980).

[23] C.-H. Liu and S.I. Marcus, "The Lie algebraic structure of a class of finite dimensional nonlinear filters," pp. 277-297 in *Algebraic and Geometric Methods in Linear Systems Theory (Lectures in Applied Math.)*, ed. C.I. Byrnes and C.F. Martin, American Mathematics Society, Providence (1980).

[24] D. Ocone, "Nonlinear filtering problems with finite dimensional estimation algebras," *Proc. Joint Automatic Control Conf.*, (1980).

[25] M. Chaleyat-Maurel and D. Michel, "Un theoreme de non existence de filtre de dimension finite," *C.R. Acad. Sc. Paris* vol. 296, pp. 933-936 (1983).

[26] J. Wei and E. Norman, "On the global representations of the solutions of linear differential equations as a product of exponentials," *Proc. Amer. Math. Soc.* vol. 15, pp. 327-334 (1964).

[27] R.W. Brockett, "Nonlinear systems and nonlinear estimation theory," in *Stochastic Systems: The Mathematics of Filtering and Identification and Applications*, ed. M. Hazewinkel and J.C. Willems, Reidel, Dordrecht, The Netherlands (1981).

[28] E. Pardoux and D. Talay, "Discretization and simulation of stochastic differential equations," Publication de Mathematiques Appliquees Marseille - Toulon (1983).

[29] W.E. Hopkins, Jr., G.L. Blankenship, and J.S. Baras, "Accurate evaluation of conditional

densities arising in the nonlinear filtering of diffusion processes," pp. 54-68 in *Proc. IFIP Workshop on Recent Advances in Filtering and Optimization*, ed. W. Fleming and L. Gorostiza, Springer-Verlag, New York (1982).

[30] R.H. Cameron, "A family of integrals serving to connect the Wiener and Feymann integrals," *J. Math. Phys.* vol. 39, pp. 126-140 (1960).

[31] A. Chorin, "Accurate evaluation of Wiener integrals," *Math. Comp.* vol. 27, pp. 1-15 (1973).

[32] A.H. Stroud and D. Secrest, *Gaussian Quadrature Formulas*, Prentice-Hall, Englewood Cliffs (1966).

[33] A.H. Stroud, *Approximate Calculation of Multiple Integrals*, Prentice-Hall, Englewood-Cliffs (1971).

[34] V. Benes, "Exact finite dimensional filters for certain diffusions with nonlinear drift," *Stochastics* vol. 5, pp. 65-92. (1981).

[35] Mathlab Group MIT, *MACSYMA Reference Manual.* January 1983.

[36] R.H. Rand, *Computer Algebra in Applied Mathematics: An Introduction to MACSYMA*, Pitman, Marshfield, MA (1984).

[37] G.N. Mil'stein, "A method of second order accuracy integration of stochastic differential equations," *Theory of Prob. and Appl.* vol. 23, p. 396 (1976).

[38] W.S. Wong, "New classes of finite-dimensional nonlinear filters," *Systems and Control Letters* vol. 3, pp. 155-164 (1983).

[39] H. Sussmann, "Approximate finite dimensional filters for some nonlinear problems," *Stochastics* vol. 7, pp. 183-203 (1982).

GENERALIZED SOLUTION OF SOME STOCHASTIC P.D.E.'S IN TURBULENT DIFFUSION*

P.L. Chow
Department of Mathematics
Wayne State University
Detroit, Michigan 48202 U.S.A.

ABSTRACT. The paper is concerned with the generalized solution of some parabolic equations with a random drift, as a model in turbulent diffusion. According to Hida's theory, the solution is regarded as a, possibly generalized, Brownian functional. The solution is constructed in the form of an infinite series of generalized Brownian functionals of ascending degrees. The method of solution is based on a Fourier transform or an iteration procedure, depending on whether the coefficients are spatially homogeneous or not. A singular perturbation and regularity of the solution for the spatially homogeneous case are also studied. Furthermore the solution is discussed by means of Hopf's functional differential equation for this problem.

0. INTRODUCTION

In a turbulent flow, let x_t denote the position of a marked fluid particle with the random velocity $v_t(\omega) = v(x_t,t,\omega)$. Then we have

$$x_t = x + \int_0^t v_s(\omega)\,ds .$$

In a simple model for turbulent diffusion, Taylor [1] considered the problem in one-dimension and assumed the velocity is stationary with the mean $Ev_t = 0$ and the correlation function

$$R_\tau = E\{v_t v_{t+\tau}\} .$$

Therefore, in terms of R_τ , we have

$$\sigma_t^2 = \sigma^2\{x_t\} = 2\int_0^t \int_0^s R_\tau\,d\tau\,ds .$$

Clearly, for small t , $\sigma_t \sim \sigma_0 t$, with $\sigma_0 = \sqrt{R_0}$.

*This work was supported in part by the NSF grant DMS-01998.

S. Albeverio et al. (eds.), Stochastic Processes in Physics and Engineering, 63–74.

If $\rho_o = \int_o^\infty R_\tau d\tau < \infty$, then, for large t, $\frac{d}{dt}\sigma_t^2 \sim 2\rho_o$,

or

$$\sigma_t \sim \sqrt{2\rho_o}\, t^{\frac{1}{2}} .$$

The above results provided the theoretical basis for Taylor to explain the experimental observations by Richardson on the diffusion of smoke emitted from a fixed point in a uniform wind. He concluded that, at small distances from the point of emission, the smoke surface (containing the standard deviations of the smoke particles) is a cone, while it becomes a paraboloid at larger distances. We notice that, in effect, the process x_t was approximated by a diffusion process, the Brownian motion for large t.

Now let us consider a cloud of smoke particles or other passive substance transported by the turbulent flow. Denote by $u(x,t,\omega)$ the concentration of the substance. For an incompressible fluid, the function u satisfies [2]

$$\text{(1)} \qquad \begin{cases} \dfrac{\partial u}{\partial t} + v(x,t,\omega)\cdot\nabla u = \dfrac{1}{2}\nu\nabla^2 u, \\ u(x,o,\omega) = \delta(x) \end{cases}$$

where ν is the molecular diffusivity and $\delta(x)$ is the Dirac delta function. That is, the initial concentration is a point source at the origin. On physical grounds, the parameter ν is small.

The paper is concerned with the solution of the equation (1) by a long-time diffusion approximation. In this case, the velocity field is given by a white noise

$$v(x,t,\omega) = \dot{\eta}_t(x,\omega),$$

which is a formal time-derivative of η_t, with η_t defined by the Itô equation

$$\text{(2)} \qquad \begin{cases} d\eta_t = a(x,t)dt + \sigma(x,t)dB_t \\ \eta_o = x_o . \end{cases}$$

Therefore, instead of (1), we are interested in the solution of the problem:

$$\text{(3)} \qquad \begin{cases} \dfrac{\partial u}{\partial t} + \dot{\eta}_t(x,\omega)\circ\nabla u = \dfrac{1}{2}\nu\nabla^2 u, \\ u(x,o,\omega) = \delta(x), \end{cases}$$

where the symbol "$\circ$" in the drift term stands for the symmetric (or Stratonovich) product. While the diffusion approximation simplifies the computational problems, it also creates some serious mathematical questions as to be seen. The solution of (3) will be constructed as a Brownian functional. As such, the regularity of the solution and its dependence on ν will be studied. In addition the Hopf's differential equation for the characteristic functional of the solution will be derived and analyzed.

For notational simplicity, we will only treat the problem in one-space dimension. The preliminaries on (ordinary or generalized) Brownian functionals can be found in Hida's book [3] or his article in this volume. We will follow his notations as far as possible. Since a full paper containing the results of this article will be published elsewhere [4] the detailed proofs are not to be given here.

1. EXAMPLE OF GENERALIZED SOLUTION

As an example, let us consider the simple one-dimensional problem:

$$(1.1)\qquad \begin{cases} \dfrac{\partial u}{\partial t} + \dot{B}_t \circ \dfrac{\partial u}{\partial x} = \dfrac{1}{2}\,\varepsilon\, \dfrac{\partial^2 u}{\partial x^2}\,, \quad t > o\,,\ x \in \mathbb{R}\,, \\ u(x,o) = \delta(x)\,, \end{cases}$$

where we set $\nu = \varepsilon$, a small parameter. By the stochastic calculus, it is easy to show that the solution of (1.1) is given by

$$(1.2)\qquad u(x,t) = q^{\varepsilon}_{x,t}(\dot{B}) = p(x-B_t, \varepsilon t)\,,$$

where

$$(1.3)\qquad p(x,t) = \frac{1}{\sqrt{2\pi t}}\, e^{-\frac{x^2}{2t}}$$

Thus, it is easy to see that $q^{\varepsilon}_{x,t} \in (L^2)$ for $t\,,\ \varepsilon > o$, where $(L^2) = L^2(S',\mu)$ is the space of L^2-functionals with respect to the white-noise distribution μ. In this case the solution is said to be ordinary or regular. But, as $\varepsilon \downarrow o$, $p(x,\varepsilon t) \to \delta(x)$ in distribution so that

$$(1.4)\qquad q^{\varepsilon}_{x,t}(\dot{B}) \to \delta(x-B_t)$$

which is a generalized Brownian functional (G.B.F.) known as Donsker's delta function defined by Kuo [5]. Therefore, at least formally, $u(x,t) = \delta(x-B_t)$ is the solution of the reduced equation of (1.1):

$$(1.5)\qquad \begin{cases} \dfrac{\partial u}{\partial t} + \dot{B}_t \circ \dfrac{\partial u}{\partial x} = o\,, \\ u(x,o) = \delta(x)\,. \end{cases}$$

Since the solution is a G.B.F., it will be termed as a generalized solution. The remarkable change in the solution behavior as $\varepsilon \downarrow o$ is due to the singular perturbation.

By changing the Stratonovich to Itô product [6], the equation (1.1) can be rewritten as

$$(1.6) \quad \begin{cases} \frac{\partial u}{\partial t} + \dot{B}_t \cdot \frac{\partial u}{\partial x} = \frac{1}{2} \kappa^\varepsilon \frac{\partial^2 u}{\partial x^2}, & t > 0, \ x \in \mathbb{R} \\ u(x,o) = \delta(x) . \end{cases}$$

where $\kappa^\varepsilon = (1+\varepsilon\nu)$. Thus, for a parabolic Itô equation, such as (1.6), the folution is, in general, a generalized solution unless κ^ε is sufficiently large, ($\kappa^\varepsilon > 1$ in this case). In what follows, as a matter of convenience, we shall discuss the random parabolic equations in Itô's form.

2. SOLUTION OF RANDOM PARABOLIC EQUATIONS

Let us consider the parabolic Itô equation of the form

$$(2.1) \quad \begin{cases} \frac{\partial u}{\partial t} + \dot{\eta}_t \cdot \frac{\partial u}{\partial x} = \frac{1}{2} \kappa^\varepsilon \frac{\partial^2 u}{\partial x^2}, & t > o, \ x \in \mathbb{R}, \\ u(x,o) = \delta(x) , \end{cases}$$

where $\eta_t(x)$ satisfies the stochastic equation (2), and

$$(2.2) \quad \kappa^\varepsilon(x,t) = \kappa(x,t) + \varepsilon\nu(x,t) , \text{ with } \kappa,\nu > o .$$

Since the analytical techniques differ for different types of coefficients, the results will be presented as two separate cases, depending on whether the coefficients are spatially homogeneous or not.

2.1. Spatially Homogeneous Equations

Suppose the coefficients depend on t only so that η_t and κ^ε do not depend on x . Then, by the Fourier transform $\hat{u}(\lambda,t) = \int u(x,t) e^{i\lambda x} dx$, the transformed equation of (2.1) can be easily solved to give

$$(2.3) \quad \hat{u}(\lambda,t) = \exp\cdot\{ i\lambda a_t + i\lambda <\dot{B},\sigma>_t \sqrt{\tau_t} - \frac{1}{2} \lambda^2 \theta_t^\varepsilon \},$$

where

$$(2.1) \quad \begin{cases} a_t = x_o + \int_o^t a(s)ds , \\ <\dot{B},\sigma>_t = \frac{1}{\sqrt{\tau_t}} \int_o^t \sigma(s)dB_s , \text{ with } \tau_t = \int_o^t \sigma^2(s)ds \\ \sigma_t^\varepsilon = \kappa_t^\varepsilon - \tau_t , \text{ and} \\ \kappa_t^\varepsilon = \int_o^t \kappa^\varepsilon(s)ds = \int_o^t [\kappa(s) + \varepsilon\nu(S)]ds \text{ with } \kappa_t = \kappa_t^o . \end{cases}$$

Let $H_n(x)$ denote the Hermite polynomial of degree n which can be generated by the formula (p. 311, [3])

$$e^{2tx-x^2} = \sum_{n=0}^{\infty} \frac{t^n}{n!} H_n(x) .$$

By applying this formula to (2.3) and the inverse Fourier transform, the solution $u = q^{\varepsilon}_{x,t}$ has the series representation

$$(2.5) \qquad q^{\varepsilon}_{x,t}(\dot{B}) = p(x-a_t,\kappa^{\varepsilon}_t) \sum_{n=0}^{\infty} \frac{1}{n!2^n} (r^{\varepsilon}_t)^{n/2} H_n\left(\frac{x-a_t}{\sqrt{2\kappa^{\varepsilon}_t}}\right) H_n\left(\frac{<B,\sigma>_t}{\sqrt{2}}\right) ,$$

where $r^{\varepsilon}_t = \tau_t/\kappa^{\varepsilon}_t$.

Recall the Wiener-Itô decomposition $(L^2) = \sum_{n=0}^{\infty} \oplus \mathcal{H}_n$.

We can show the series converges in (L^2) if $r^{\varepsilon}_t < 1$, by noting $\| H_n\left(\frac{<\cdot,\sigma>_t}{\sqrt{2}}\right) \|^2_{\mathcal{H}_n} = n!2^n$ and the estimate (p.173,[5]), for $n \geq 2$

$$(2.6) \qquad \sup_{\substack{t>0 \\ x\in\mathbb{R}}} |(\sqrt{\pi}\, n!\, 2^n)^{-1/2} e^{-\frac{x^2}{4t}} H_n\left(\frac{x}{\sqrt{2t}}\right)| = O(n^{-1/12}) .$$

This is so because the n-th term

$$\| p(x-a_t,\kappa^{\varepsilon}_t) \frac{(r^{\varepsilon}_t)^{n/2}}{n!2^n} H_n\left(\frac{x-a_t}{\sqrt{2\kappa^{\varepsilon}_t}}\right) H_n\left(\frac{<\dot{B},\sigma>_t}{\sqrt{2}}\right) \|^2_{\mathcal{H}_n}$$

$$\leq C(\kappa^{\varepsilon}_t)^{-1} p(x,\kappa^{\varepsilon}_t)(r^{\varepsilon}_t)^n \qquad \text{for } n > n_0$$

where C and n_0 are some positive numbers. In this case, the series represents an ordinary solution. However, if $r^{\varepsilon}_t \geq 1$, such solution may not exist. In particular consider the special case $\kappa(t) = \nu^2(t)$ so that

$$(2.7) \qquad \kappa^{\varepsilon}_t = \tau_t + \varepsilon\nu_t , \quad \text{with } \nu_t = \int_0^t \nu(s)ds .$$

Then, clearly, the series solution (2.5) converges for $\varepsilon > 0$, but diverges in (L^2) as $\varepsilon \to 0$, $(\kappa^{\varepsilon}_t \to \tau_t)$. Thus, as in the previous example, we expect $q^{\varepsilon}_{x,t}$ converges to a generalized solution

$$(2.8) \qquad q_{x,t}(\dot{B}) = p(x-a_t,\tau_t) \sum_{n=0}^{\infty} \frac{1}{n!2^n} H_n\left(\frac{x-a_t}{\sqrt{2\tau_t}}\right) H_n\left(\frac{<\dot{B},\sigma>_t}{\sqrt{2}}\right)$$

of the reduced equation of (2.1). To verify $q_{x,t} \in (L^2)^-$, the space of G.B.F.'s, we make use of the face $(L^2)^- = \sum_{n=o}^{\infty} \oplus H_n^{(-n)}$ and the estimate (Lemma 1,[5])

$$(2.9) \qquad \| H_n\left(\frac{\langle\cdot,\sigma\rangle_t}{\sqrt{2}}\right) \|^2_{H_n^{(-n)}} \leqslant 2(2t)^n\, n!\sigma_n \quad \text{for} \quad n \geqslant 2 ,$$

where $\sigma_n = 2\pi^{n/2}/\Gamma\left(\frac{n}{2}\right)$, and $\Gamma(x) = \int_o^\infty t^{x-1} e^{-t}\, dt$ denotes the Gamma function. In view of (2.9) and (2.6), it can be shown that the n-th term $q_{x,t}^{(n)}$ in the RHS of (2.8) has the bound

$$\| q_{x,t}^{(n)} \|^2_{H_n^{(-n)}} \leqslant C \frac{\sigma_n t^n}{n!2^n \tau_t} p(x-a_t,\tau_t) \quad \text{for} \quad n > n_o ,$$

which, together with the Stirling's formula $\Gamma(n) \sim \sqrt{2\pi}\, e^{-n} n^{n-\frac{1}{2}}$, implies the series in (2.8) converges in $(L^2)^-$. Hence the series solution (2.8) is a generalized solution of the problem.

$$(2.10) \qquad \begin{cases} \dfrac{\partial u}{\partial t} + \dot{\eta}(t)\cdot \dfrac{\partial u}{\partial x} = \dfrac{1}{2}\sigma^2(t)\dfrac{\partial^2 u}{\partial x^2} \\ u(x,o) = \delta(x) \end{cases}$$

Let us summarize the results so far as a theorem.

Theorem 1. Let the coefficients of the random parabolic equation (2.1) be continuous functions of t alone. If $(\tau_t/\kappa_t^\varepsilon) < 1$, then it has a unique solution $q_{x,t}^\varepsilon$ which has the series representation (2.5) as a Brownian functional with $\varepsilon, t > o$. Under the assumption (2.7), the functional $q_{x,t}^\varepsilon$ converges as $\varepsilon \downarrow o$ to the generalized solution $q_{x,t}$ given by (2.8) for the reduced problem (2.10).

We note that, setting $a = o$ in (2.8), we get

$$(2.11) \qquad \delta(x-\xi_t) = p(x,\tau_t) \sum_{n=o}^{\infty} \frac{1}{n!2^n} H_n\left(\frac{x}{\sqrt{2\tau_t}}\right) H_n\left(\frac{\xi_t}{\sqrt{2}}\right) ,$$

where $\xi_t = \int_o^t \sigma(s)\, dB_s$. The above series generalizes Kuo's representation for Donsker's delta function. It is also possible to show the regularity result:

For every φ $(L^2)^t$, the function $g_{\varphi,t}(x) = \int_{S*} \varphi q^{\varepsilon}_{x,t} d\mu$ is a C^{∞}- function of x.

We should point out that the study on regularity question has been inspired by an interesting paper by Kallianpur and Kuo [7].

2.2. Spatially Nonhomogeneous Equation

To simplify the notations, we consider the following special case

$$\begin{cases} \frac{\partial u}{\partial t} + \dot{\xi}_t(x) \cdot \frac{\partial u}{\partial x} = \frac{1}{2} \nu \frac{\partial^2 u}{\partial x^2} , \\ u(x,o) = \delta(x) , \end{cases} \tag{2.12}$$

where ν is a positive constant and, as before

$$\xi_t(x) = \int_o^t \sigma(x,s) dB_s . \tag{2.13}$$

Since $g(x,t) = p(x,\nu t)$ is the Green's function for the heat equation, the system (2.12) can be rewritten as

$$u(x,t) = g(x,t) - \int_o^t \int g(x-y,t-s)\sigma(y,s) \frac{\partial u(y,s)}{\partial y} dy dB_s . \tag{2.14}$$

Since we are interested in a generalized solution, the above stochastic integral should be interpreted in the Kubo sense [8]. The solution of the equation (2.14) may be constructed by an iteration procedure. To this end, let

$$u_o(x,t) = g(x,t) ,$$

and, for $n \geqslant 1$, define

$$u_{n+1}(x,t) = g(x,t) - \int_o^t \int g(x-y,t-s)\sigma(y,s) \frac{u_n(y,s)}{\partial y} dy dB_s .$$

Therefore we have

$$u_n(x,t) = g(x,t) + \sum_{\kappa=1}^{n} (-1)^{\kappa} I_{\kappa}(x,t) , \tag{2.15}$$

where

$$\begin{aligned} I_{\kappa}(x,t) &= \left(\int_o^t \int\right) \cdots \left(\int_o^{t_{\kappa-1}} \int\right) [g(x-x_1,t-t_1)\sigma(x_1,t_1)] \\ &\times [g_x(x_1-x_2,t_1-t_2)\sigma(x_2,t_2)] \cdots [g_x(x_{\kappa-1}-x_{\kappa},t_{\kappa-1}-t_{\kappa})\sigma(x_{\kappa},t_{\kappa})] \\ &\times g_x(x_{\kappa},t_{\kappa}) (dx, dB_{t_1}) \cdots (dx_{\kappa} dB_{t_{\kappa}}) . \end{aligned} \tag{2.16}$$

with $g_x(y,t) = \frac{\partial}{\partial x} g(x,t)\big|_{x=y}$.

Since the integrand is singular in t's, the above iterated stochastic integral cannot be defined in an L^2-sense. However it is possible to define it as a G.B.F.. To see why, let us rewrite (2.16) as

$$(2.17)\qquad I_\kappa(x,t) = \int_o^t \cdots \int^{t_{\kappa-1}} F^{(\kappa)}_{x,t}(t_1,\ldots,t_\kappa)\,dB_{t_1}\cdots dB_{t_\kappa} ,$$

where

$$(2.18)\qquad \begin{aligned} F^{(\kappa)}_{x,t}(t_1,\ldots,t_\kappa) &= \int_{\mathbb{R}^\kappa} [g(x-x_1,t-t_1)\sigma(x_1,t_1)]\cdots[g_x(x_{\kappa-1}-x_\kappa,t_{\kappa-1}-t_\kappa) \\ &\quad\times \sigma(x_\kappa,t_\kappa)]g_x(x_\kappa,t_\kappa)\,dx_1\cdots dx_\kappa\; I_t(t_1,\ldots,t_\kappa) . \end{aligned}$$

Here $I_t(t_1,\ldots,t_\kappa)$ denotes the indicator function of the set $\{o \leqslant t_\kappa \leqslant \cdots \leqslant t_1 \leqslant t\} \subset \mathbb{R}^\kappa$. It is not difficult to show that, if σ is bounded and continuous,

$$(2.19)\qquad |F^{(\kappa)}_{x,t}(t_1,\ldots,t_\kappa)| \leqslant A^n p(x,\nu t)\{(t_1-t_2)\cdots(t_{\kappa-1}-t_\kappa)t_\kappa\}^{-\frac{1}{2}} ,$$

for any $\nu_1 < \nu$ and some $A > o$.

Let $H^m_\kappa = H^m(\mathbb{R}^\kappa)$ be the Sobolev space of order m, with the norm $\|F\|_{H^m_\kappa} = \left\{\int_{\mathbb{R}^\kappa}(1+|\lambda|^2)^m|\hat{F}(\lambda)|^2 d\lambda\right\}^{\frac{1}{2}}$, where $\hat{F}$ denotes the Faurier transform of F on $\mathbb{R}^\kappa$. Then, by definition, we can check that

$$(2.20)\qquad \|F^{(\kappa)}_{x,t}\|^2_{H^{-\alpha_\kappa}_\kappa} \leqslant 2\sigma_\kappa \|F^{(\kappa)}_{x,t}\|^2_{L^1(\mathbb{R}^\kappa)} \quad \text{for } \alpha_\kappa = \frac{\kappa+1}{2} .$$

In view of (2.17), we also have

$$(2.21)\qquad \|I_\kappa(x,t)\|_{H^{(-\kappa)}_\kappa} \leqslant \|F^{(\kappa)}_{x,t}\|_{H^{-\alpha_\kappa}_\kappa} .$$

By invoking (2.19) and a Dirichlet integral (p. 258, [9]), the following inequality holds

$$(2.22)\qquad \|F^{(\kappa)}_{x,t}\|_{L^1(\mathbb{R}^\kappa)} \leqslant 2A(\pi t)^{\kappa/2} \Big/ \kappa\,\Gamma\!\left(\frac{\kappa}{2}\right) .$$

Therefore $I^{(\kappa)}_{x,t} \in H^{(-\kappa)}_\kappa$ is well defined for all κ, as asserted.

To show the sequence of iterates u_n defined by (2.15) converges in $(L^2)^-$, we note that

$$\| u_n(x,t) \|^2_{(L^2)^-} \leq g^2(x,t) + \sum_{\kappa=1}^{n} \| I(x,t) \|^2_{H_\kappa}(-\kappa) .$$

But, with the aid of (2.20) - (2.22), it is straightforward to show, as $n \to \infty$, the above infinite series converges for $t > o$ and uniformly in x over $\mathbb{R}$. Therefore we have the following theorem.

Theorem 2. If $\sigma \in L^\infty(\mathbb{R} \times \mathbb{R}^+)$, the equation (2.12) has a generalized solution $u = q_{x,t}(\dot{B})$ which can be represented as

$$q_{x,t}(\dot{B}) = p(x,\nu t) + \sum_{\kappa=1}^{\infty} (-1)^\kappa I_\kappa(x,t) ,$$

where I_κ's are iterated stochastic integrals defined by (2.16). Furthermore, for each $t > o$, the above series converges in $(L^2)^-$ uniformly in x over $\mathbb{R}$.

We remark that the above construction of solution to (2.12) can be extended to a more general equation, where ξ_t is replaced by η_t given by (2) and the RHS by a Sturm-Liouville type of operator, (for details, see [4]).

3. METHOD OF HOPF'S EQUATION

A constructive method of determining the solution measure of a stochastic differential equation is to solve the associated Hopf's equation for the characteristic functional of the solution process [10]. To be specific let $u = q_t(x,\dot{B})$ denote the solution process for the parabolic Itô equation (2.12). Define the characteristic functional

$$(3.1) \quad \varphi_t(\lambda) = E\{e^{i<q_t,\lambda>}\} = \int_{S^*} e^{i<q_t,\lambda>} d\mu, \quad \lambda \in S$$

From (2.12), we have

$$(3.2) \quad d<q_t,\lambda> = \frac{1}{2}\nu<\partial^2 q_t,\lambda> dt - <\sigma\partial q_t,\lambda> dB_t ,$$

where $\partial = \frac{\partial}{\partial x}$, $\partial^2 = \frac{\partial^2}{\partial x^2}$. Then apply Itô's calculus to $e^{i<q_t,\lambda>}$ and take the expected value to get

$$(3.3) \quad \frac{\partial\varphi_t(\lambda)}{\partial t} = E\left\{\frac{1}{2}\nu<\partial^2 q_t,\lambda> - \frac{1}{2}<\sigma\partial \quad \lambda>^2) e^{i<q_t,\lambda>}\right\} .$$

Since the variational derivative $\frac{\delta\varphi_t(\lambda)}{\delta\lambda(x)} = iE\left\{q_t(x)e^{i<q_t,\lambda>}\right\}$ and so on, the RHS of (3.3) can be expressed in terms of φ_t so that it satisfies the Hopf's equation

$$(3.4)\quad \begin{cases} \dfrac{\partial\varphi_t(\lambda)}{\partial t} = \dfrac{\nu}{2}\displaystyle\int \lambda(x)\,\frac{\partial^2}{\partial x^2}\,\frac{\partial\varphi_t(\lambda)}{\partial\lambda(x)}\,dx \\ \qquad\qquad + \dfrac{1}{2}\displaystyle\iint R(x,y,t)\lambda(x)\lambda(y)\,\frac{\partial^2}{\partial x\partial y}\,\frac{\delta^2\varphi_t(\lambda)}{\delta\lambda(x)\delta\lambda(y)}\,dxdy\,, \\ \varphi_o(\lambda) = e^{i\lambda(o)} \end{cases}$$

where $R(x,y,t) = \sigma(x,t)\sigma(y,t)$.

Note that the variational-derivative terms are homogeneous in λ . The functional equation (3.4) admits a power series solution

$$(3.5)\quad \varphi_t(\lambda) = 1 + \sum_{n=1}^{\infty} \frac{i^n}{n!}\int_{\mathbb{R}^n}\Gamma_t^{(n)}(x_1,\ldots,x_n)\lambda(x_1)\ldots\lambda(x_n)\,dx_1\ldots dx_n\,,$$

where $\Gamma_t^{(n)}(x_1,\ldots,x_n) = E\{q_t(x_1)\ldots q_t(x_n)\}$ is the n-th moment of the solution. By substituting (3.5) into (3.4), it is easy to see that $\Gamma_t^{(n)}$ satisfies the following sequence of diffusion equations:

$$(3.6)\quad \begin{cases} \dfrac{\partial\Gamma_t^{(1)}}{\partial t} = \dfrac{1}{2}\nu\,\dfrac{\partial^2\Gamma_t^{(1)}}{\partial x^2}\,, \\ \Gamma_o^{(1)}(x) = \delta(x)\,, \end{cases}$$

and

$$(3.7)\quad \begin{cases} \dfrac{\partial\Gamma_t^{(n)}}{\partial t} = \dfrac{\nu}{2}\displaystyle\sum_{\kappa=1}^{\infty}\frac{\partial^2}{\partial x_\kappa^2}\,\Gamma_t^{(n)} + \frac{1}{2}\sum_{j,\kappa=1}^{n} R(x_j,x_\kappa,t)\,\frac{\partial^2}{\partial x_j\partial x_\kappa}\,\Gamma_t^{(n)}\,, \\ \Gamma_o^{(n)}(x_1,\ldots,x_n) = \delta(x_1)\ldots\delta(x_n)\,, \quad \text{for } n \geqslant 2\,. \end{cases}$$

By appealing to the theory of fundamental solution to a parabolic equation (pp. 23-24, [11]), the equations (3.6) and (3.7) have unique smooth solutions for $t > o$, provided that the function σ , hence R be bounded and uniformly continuous. In fact, for any $\nu_1 < \nu$, there exists a constant $A > o$ such that

$$|\Gamma_t^{(n)}(x_1,\ldots,x_n)| < A^n(2\pi\nu_1 t)^{-n/2}\exp.\{-\frac{1}{2\nu_1 t}(x_1^2+\ldots+x_n^2)\}.$$

Hence, from (3.5)

$$|\varphi_t(\lambda)| \leqslant 1 + \sum_{n=1}^{\infty} \frac{A^n \|\lambda\|_\infty^n}{n!} = \exp\{A\|\lambda\|_\infty\} .$$

Therefore we have shown that

Theorem 3. Suppose the covariance $R(x,y,t) = \sigma(x,t)\sigma(y,t)$ be bounded and uniformly Hölder continuous over $\mathbb{R}^2 \times \mathbb{R}^+$. Then all the moments of the solution to the parabolic equation (2.12) exist and satisfy (3.6) and (3.7). Further the associated characteristic functional $\varphi_t(\lambda)$ can be represented as a functional power series (3.5), which converges uniformly for $t > o$ and λ in any bounded subset of $C(\mathbb{R})$.

4. REFERENCES

[1] Taylor, G.I., Diffusion by continuous movements, Proc. London Math. Soc., 20 (1921), 196-212.

[2] Chow, P.L., Stochastic partial differential equations in turbulence related problems, in Prob. Analy. and Related Topics, Vol. 1, ed. by A.T. Bharucha-Reid, Acad. Press, New York (1978), 1-43.

[3] Hida, T., Brownian motion, Springer-Verlag, New York (1980).

[4] Chow, P.L., Generalized solution of some parabolic equations with a random drift, (1986), submitted.

[5] Kuo, H.H., Donsker's delta function as a generalized Brownian function and its application, Proc. Conf. on Theory and Appl. of Random Fields, Lecture Notes in Control and Inf. Sci., Vol. 49, Springer-Verlag, New York (1983).

[6] Ikeda, N. and S. Watanabe, Stochastic Differential Equations and Diffusion Processes, North-Holland, Amsterdam (1981).

[7] Kallianpur, G. and H.H. Kuo, Regularity property of Donsker's delta function, Appl. Math. and Optim., 12 (1984), 89-95.

[8] Kubo, I., Itô's lemma for generalized Brownian functionals, Proc. Conf. on Theory and Appl. of Random Fields, Lecture Notes in Control and Inf. Sci., Vol. 49, Springer-Verlag, New York (1983), 156-166.

[9] Whittaker, E.T. and G.N. Watson, A Course of Modern Analysis, Cambridge Univ. Press, London-N.Y. (1967).

[10] Chow, P.L., Function-space differential equations associated with a stochastic partial differential equation, Indiana Univ. Math. J., 25 (1976), 609-627.

[11] Friedman, A., Partial Differential Equations of Parabolic Type, Prentice-Hall, Edgewood Cliffs, N.J., (1964).

NONSTANDARD TECHNIQUES IN STOCHASTIC OPTIMAL CONTROL THEORY

Nigel J. Cutland
Department of Pure Mathematics
University of Hull
HULL HU6 7RX
England

ABSTRACT

The use of nonstandard techniques in stochastic optimal control theory is illustrated by discussing two controlled systems, the first with observations restricted to be a cumulative digital read-out, the second with complete observations and singular noise. The main ingredient in each case is the Loeb construction of standard measures from nonstandard measures, and the application of this to stochastic analysis as developed by Anderson and Keisler.

1. INTRODUCTION

A general stochastic controlled system may be described by a stochastic evolution equation of the form

$$dx_t = f(t,x,u_t)dt + g(t,x,u_t)db_t$$

where $x_t \in \mathbf{R}^n$ is the state and b is an m-dimensional Brownian motion. In this equation x denotes the past $(x_s)_{s\leq t}$ and u is a control taking values in a compact metric space M. The control u_t at time t may depend on some or all of the information about the past of x.

S. Albeverio et al. (eds.), Stochastic Processes in Physics and Engineering, 75–90.

Associated with each control there will be a cost $J(u)$.

The first problem of optimal control is whether there is a control whose cost is J_0, the infimum of all possible costs. Nonstandard techniques suggest the following approach: let (u_n) be a sequence of controls with $J(u_n) \to J_0$. Take an infinite natural number N, and let $U = u_N$; this is a nonstandard control and under reasonable assumptions we would expect that $J(U) \approx J_0$. Now try to show that U can be converted to a standard control u, with $J(u) \approx J(U)$. Then $J(u) \approx J_0$, so $J(u) = J_0$ because both quantities are standard. (In this discussion $\approx$ means 'infinitely close to', a fundamental notion of nonstandard analysis).

In this paper we describe in outline some of the controlled systems where the above approach works; references are given for the source papers, where full details can be found. We begin with a brief sketch of the basic ideas of nonstandard analysis that are needed.

2. NONSTANDARD ANALYSIS

Nonstandard analysis begins by extending the reals $\mathbf{R}$ to a set of hyperreals $^*\mathbf{R} \supset \mathbf{R}$, which contains both infinite and infinitesimal elements. The following picture indicates how to think of this extension.

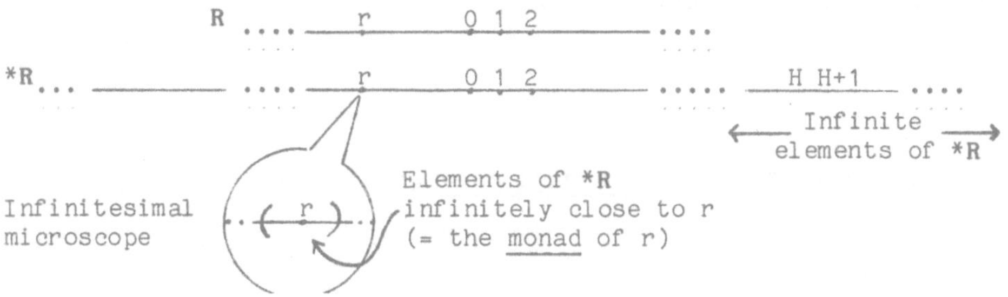

One way to construct ***R** is to set ***R** = $\mathbf{R}^{\mathbf{N}}/D$ where D is a non-principal ultrafilter on **N**. It is best for practical purposes to assume ***R** as given axiomatically as in [5] or [8]. The essential feature of ***R** is that it has 'the same properties as **R**' (in a sense that can be made precise). In particular, ***R** is an ordered field. With ***R** given (or constructed) we make the following definitions.

Definition For $x, y \in$ ***R**

(a) x is infinite if $|x| > n$, all $n \in \mathbf{N}$

(b) x is infinitesimal if $|x| < \frac{1}{n}$, all $n \in \mathbf{N}$

(c) $x \approx y$ (x is infinitely close to y) if $x - y$ is infinitesimal.

The next lemma is very important:

Standard Part Lemma If $x \in {}^*\mathbf{R}$ is finite there is a unique $r \in \mathbf{R}$ with $x \approx r$. We say r is the standard part of x, written ${}^\circ x$ or $st(x)$.

In order to discuss continuity, we need the following notion, which is the nonstandard counterpart of standard continuity.

Definition A function $F : {}^*[0,1] \to {}^*\mathbf{R}$ is S-continuous if for all $t_1, t_2 \in {}^*[0,1]$ ($= \{x \in {}^*\mathbf{R}: 0 \leq x \leq 1\}$)

$$t_1 \approx t_2 \quad \Rightarrow \quad F(t_1) \approx F(t_2).$$

The standard part lemma has the following counterpart for S-continuous functions.

Lemma If F is S-continuous and nonstandard and $F(0)$ is finite there is a unique $f \in C[0,1]$ such that

$$f({}^\circ t) = {}^\circ(F(t)) \qquad \text{all } t \in {}^*[0,1].$$

This can be envisaged as follows:

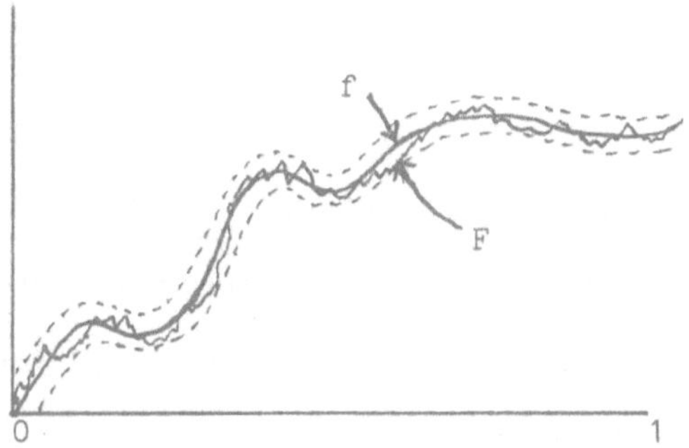

The dotted lines indicate the tube around f containing all the points that are an infinitesimal distance from f.

To develop nonstandard analysis for structures other than the reals, a 'nonstandard universe' must be constructed, in which every standard mathematical object A has an extension *A. Again this can be assumed as given axiomatically. Objects in the nonstandard universe are called either nonstandard or internal. For details consult [5] or [8] or [10]. With such a nonstandard universe standing above the world of standard mathematics, we have a rich interplay between the two worlds. For our purposes, two aspects of this should be mentioned. First, given any metric space M the idea of a standard part map st: $*M \to M$ can be defined in the obvious way. (It is not defined on all of *M unless M is compact.) Second, it is possible to construct certain rich measure spaces known as Loeb spaces, which play an important role in our approach to stochastic control theory. They are constructed as follows.

Suppose that $\mathbf{\Omega}_0 = (\Omega, \mathcal{A}, \nu)$ is an internal (or nonstandard) finitely additive probability space; i.e. $\mathcal{A}$ is an algebra of subsets of Ω and $\nu: \mathcal{A} \to {}^*[0,1]$ is finitely additive. The Loeb measure ν_L on $\sigma(\mathcal{A})$ is constructed by setting $\nu_L(A) = {}^\circ(\nu(A))$ for $A \in \mathcal{A}$, and then using the Caratheodory extension theorem (which applies trivially because $\mathcal{A}$ contains no non-trivial countable union of sets from $\mathcal{A}$). The probability space $\mathbf{\Omega} = (\Omega, \sigma(\mathcal{A}), \nu_L)$ is a Loeb space; it is a standard probability space.

In the next two sections we describe two stochastic control systems where Loeb space techniques give the existence of optimal controls.

3. PARTIALLY OBSERVED SYSTEMS WITH DIGITAL READ-OUT

In the paper [6] we considered stochastic control systems of the form

$$(3.1) \quad \begin{cases} dx_t = f(t,x,y, u(t,y))dt + g(t,x,y,u(t,y))db_t \\ dy_t = \bar{f}(t,x,y, u(t,y))dt + \bar{g}(t,y)d\bar{b}_t \\ x_0 = 0, \; y_0 = 0 \end{cases} \quad (0 \le t \le 1)$$

where $x_t \in \mathbf{R}^d$, $y_t \in \mathbf{R}^m$, and b, $\bar{b}$ are independent Brownian motions of dimension d, m respectively. We think of x_t as the state at time t, and y_t as the (noisy) observation at time t. The assumptions made on f, $\overline{f}$, g, $\overline{g}$ are of the usual kind required to ensure that the equations in (3.1) have a continuous solution that is weakly unique for each control u. Such a solution can be obtained using the Girsanov measure change technique to add a drift to the solution to the equations without drift:

$$(3.2)\quad \begin{cases} d\xi_t = g(t,x,\eta,u(t,\eta))db_t \\ d\eta_t = \overline{g}(t,\eta)d\overline{b}_t. \end{cases}$$

Controls A control u is assumed to take values in a compact metric space M. The space $\mathcal{U}$ of admissible controls is defined to be the class of measurable functions $u:[0,1] \times \mathcal{C}^m \to M$ that are adapted to a fixed information filtration $I = (I_t)_{t\leq 1}$. Such a filtration is generated by a cumulative digital read-out of the observations as follows. This is a function $r: [0,1] \times \mathcal{C}^m \to \mathbb{N}$ (where $\mathcal{C}^m = C^m[0,1]$) such that

(a) r_t is measurable, each t;

(b) if $y = z$ on $[0,t]$ then $r_t(y) = r_t(z)$ (i.e. r is adapted);

(c) if $r_t(y) = r_t(z)$ then $r_s(y) = r_s(z)$ all $s\leq t$ (cummulation).

Such a read-out generates an information filtration by setting $I_t = \sigma(r_t)$. Conversely, given a filtration $I = (I_t)_{t\leq 1}$ with the properties

(a) $I_s \subseteq I_t \subseteq F_t$ $(s \leq t)$

(where (F_t) is the natural filtration on $\mathcal{C}^m$),

(b) I_1 is generated by a countable number of atoms,

then there is a cumulative digital read-out r that generates I. Thus $\mathcal{U}$ may be defined alternatively as the class of functions u of the form $u_t(y) = u_t(r_t(y))$. (The filtration I is assumed to be the same for all controls).

The class $\mathcal{U}$ is extended to the class $\mathcal{V}$ of relaxed controls - i.e. controls taking values in $\mathcal{M}(M)$, the set of probability measures on M. A measure ν is interpreted in the drift by defining

$$f(t,x,y,\nu) = \int_M f(t,x,y,a)d\nu(a)$$

and in the diffusion g by setting

$$g(t,x,y,\nu) = \left(\int_M g^2(t,x,y,a)d\nu(a)\right)^{1/2}$$

(so g is assumed to be positive definite symmetric).

Cost We assume for each control $v \in \mathcal{V}$ a cost $J(v)$ of the form

$$J(v) = E\left(\int_0^1 h(t,x^v,y^v, v(t,y^v))dt + \bar{h}(x^v,y^v)\right)$$

where (x^v,y^v) is the solution to (3.1) for control v. Denote by J_0 the optimal cost $\inf J(\mathcal{V})$.

The main results in [6] for the system described above are as follows:

. Theorem

(a) There is a topology that makes $\mathcal{V}$ compact, and for which ordinary step controls are dense. (A step control is one that constant on each of a finite number of intervals of time).

(b) The measure μ^v induced on $\mathcal{C}^n$ (where $n = d+m$) by the ution to (3.1) for control v is continuous as a function of v th respect to the weak* topology on $\mathcal{M}(\mathcal{C}^n)$).

(c) The cost function $J(v)$ is continuous in v.

(d) There is an optimal control $v \in \mathcal{V}$ with $J(v) = J_0$.

(e) With a suitable convexity condition on the coefficients, there is an optimal control $u \in \mathcal{U}$.

We will sketch the main ingredients in the proof of these ults; at the heart is the idea mentioned in the introduction - ely taking a nonstandard control V and converting it to a ndard control v with $J(V) = J(v)$.

Notice first that each control $v \in V$ determines (and is ermined by) a sequence of deterministic controls $(v_i)_{i \in \mathbf{N}}$ given by

$$v_i(t) = v(t,y) \qquad \text{for } y \in D_i$$

re $\{D_i\}_{i \in \mathbf{N}}$ are the atoms of I_1. The topology on V is defined to the weak* topology with subbase of open sets of the form

$$\{v: |v_i(g)|<\varepsilon\} \quad (i \in \mathbf{N},\ g \in \mathcal{K},\ \varepsilon>0)$$

re $\mathcal{K}$ is the set of measurable bounded functions g of the form $[0,1] \times M \to \mathbf{R}$ with $g(t,\cdot)$ continuous, and

$v_i(g) = \int_0^1 g(t, v_i(t))dt.$

The compactness of $\mathcal{V}$ is established as follows. Let $V \in {}^*\mathcal{V}$ be a nonstandard control; the nonstandard criterion for compactness of $\mathcal{V}$ is that there is $v \in \mathcal{V}$ with $v \approx V$. To construct v from V, first define the sequence of deterministic controls $(V_i)_{i\in\mathbf{N}}$ as above (using $^*D_i$ of course). For each $\tau \in {}^*[0,1]$, $V_i(\tau)$ is a nonstandard probability on $^*M$; this defines a standard probability μ on M by

$$\mu(B) = (V_i(\tau))_L \ (st^{-1}(B))$$

where st: $^*M \to M$ is the standard part mapping. It was shown in [3] that μ is in fact the standard part $°(V_i(\tau))$ of the measure $V_i(\tau)$ in the weak* topology on $\mathcal{M}(M)$. Now a sequence of deterministic controls $(v_i)_{i\in\mathbf{N}}$ is obtained roughly by defining for each $t \in [0,1]$

$$v_i(t) = \text{average of } °(V_i(\tau)) \text{ for } \tau \approx t.$$

(In detail, each v_i is obtained by disintegration of the measure q defined on $[0,1] \times M$ by

$$q(A) = Q_L(st^{-1}(A))$$

where $$Q(C \times D) = \int_C V_i(\tau)(D)d\tau.\ )$$

It can be shown that the v_i are consistent so that they can be "glued together" to define an admissible relaxed control v; it is now routine to show that $v = °V$ in the topology on $\mathcal{V}$.

For part (b) of Theorem 3.3, the nonstandard criterion for continuity requires that for a control V as above, with $v = °V$, we have $\mu^V \approx \mu^v$, where μ^V is the internal measure on $^*\mathcal{C}^n$ given by the solution to the internal version of the equations (3.1):

$$(3.4)\quad \begin{cases} dX_\tau = {}^*f(\tau,X,Y,V(\tau,Y))d\tau + {}^*g(t,X,Y,V(\tau,Y))dB\tau \\ dY_\tau = {}^*\overline{f}(\tau,X,Y,V(\tau,Y))d\tau + {}^*\overline{g}(t,Y)d\overline{B}\tau. \end{cases}$$

(Here B, $\overline{B}$ are internal Brownian motions.) The solution (X^V, Y^V) to (3.4) is an internal *continuous process on some probability space Ω_0. By Keisler's important S-continuity theorem, we see that for almost all ω in Ω (the Loeb space construced from Ω_0) the paths of (X^V, Y^V) are S-continuous, and so we can define a continuous process $(x^V, y^V) = ({}^\circ X^V, {}^\circ Y^V)$ on Ω. It is clear that the measure on $\mathcal{C}^n$ induced by (x^V, y^V) is ${}^\circ\mu^V$. The rest of the proof consists in showing that (x^V, y^V) actually gives a solution to (3.2)

for control v, for suitable Brownian motions b and $\overline{b}$ on Ω. This being so, we have ${}^\circ\mu^V = \mu^v$ as required. The details of this part of the proof are a little technical so we refer the reader at this point to the article [6].

Part (c) of Theorem 3.3 follows rather easily by observing that $J(V) \approx$ the cost for the process (x^V, y^V)

$$= J(v)$$

since $(x^V, y^V) = (x^v, y^v)$. The final part of the theorem is established by routine standard measurable selection techniques.

4. COMPLETELY OBSERVED SYSTEMS WITH SINGULAR NOISE

In contrast to the restricted observation pattern considered in the previous section, here we discuss systems with complete observations in which the noise may be singular. These take the form

$$(4.1)\quad \begin{cases} dx_t = f(t,x,u_t)dt + g(t,x)db_t, & (0 \le t \le 1) \\ x_0 = 0; \end{cases}$$

where $x_t \in \mathbf{R}^d$ and b is an m-dimensional Brownian motion. For simplicity f, g are assumed to be bounded, but g may be singular. Both coefficients are continuous in each argument except possibly t.

The control u_t takes values in a compact metric space M as before; the maximum possible information is used in the control, expressed by assuming that $u_t = u_t(x,b)$, with u_t adapted. The wider class of relaxed controls is defined similarly.

For such systems there is a problem about existence and uniqueness of solutions. In the paper [7] a way round this is suggested, and the existence of optimal controls is established by Loeb-space methods somewhat similar to those described in the previous section.

It is assumed in [7] that the cost for a process x is given by E(h(x)) for some bounded continuous function h. Then we define a cost function J(u,x,b) for any control u (of the form above) and continuous process x,b on a space Ω by

$$J(v,x,b) = \begin{cases} E(h(x)) & \text{if } x \text{ is a solution to (4.1) on } \Omega \\ & \text{with control } v \text{ and driving Brownian motion } b. \\ \infty & \text{otherwise.} \end{cases}$$

Putting $J_0 = \inf J(v,x,b)$, where v is a relaxed control and x, b are processes, it is easy to see that J_0 is finite (simply take v = constant and show that (4.1) has a solution). The main results of

[7] are then:

4.2 <u>Theorem</u> (a) The set of costs $J(v,x,b)$ is compact.

(b) There is an optimal triple (v,x,b) such that $J(v,x,b) = J_0$, with v a relaxed control.

(c) With a suitable convexity condition on f there is an optimal triple (u,x,b) with u an ordinary control.

For (a), take an internal triple (V,X,B) such that B is an internal Brownian motion, V is an internal control and X satisfies the internal equation

(4.3) $dX_\tau = {}^*f(\tau,X,V_\tau(X,B))d\tau + {}^*g(\tau,X)dB_\tau.$

Then X,B live on an internal space Ω_0; on the Loeb space Ω obtained from Ω_0, both B and X are almost surely S-continuous (as in the previous section) and $b = {}^\circ B$ is a Brownian motion. Moreover $J(V,X,B) \approx E(h(x))$ where $x = {}^\circ X$. Thus it is sufficient to find a control v such that x is a solution to (4.1) for v and b; for then we have

$$J(v,x,b) = E(h(x)) \approx J(V,X,B)$$

which shows that the set of costs is compact, using the nonstandard criterion for compactness.

Notice that the nonstandard approach to Itô integration developed by Anderson [2] ensures that from (4.3) we have

$$(4.4)\quad x_t - \int_0^t g(s,x)db_s = {}^\circ\!\!\int_0^t {}^*f(\tau,X,V_\tau(X,B))d\tau \quad \text{a.s.},$$

so the problem reduces to that of finding a control v such that

$$(4.5)\quad \int_0^t f(s,x,v_s(x,b))ds = {}^\circ\!\!\int_0^t {}^*f(\tau,X,V_\tau(X,B))d\tau \quad \text{a.s.}$$

Now observe that from (4.4) the right-hand side is adapted to (x,b), so there is adapted $\theta(t,x,b)$ with

$$(4.6)\quad {}^\circ\!\!\int_0^t {}^*f(\tau,X,V_\tau(X,B))d\tau = \int_0^t \theta(s,x,b)ds \qquad \text{a.a.}$$

So (by Loeb theory)

$$(4.7)\quad \int_{st^{-1}(A)} {}^{\circ *}f(\tau,X,V_\tau(X,B))(d\tau \times d\mu^{X,B})_L = \int_A \theta(s,x,b)ds \times d\mu^{x,b}$$

for Borel $A \subseteq [0,1] \times \mathcal{C}^{d+m}$; where $\mu^{x,b}$ = measure on $\mathcal{C}^{d+m}$ induced by (x,m) and similarly for $\mu^{X,B}$.

The last stage of the proof is to constrct a control v such that for a sufficiently large class of Borel sets D

$$(4.8)\quad \int_{st^{-1}(D)} {}^{\circ *}f(\tau,X,V_\tau(X,B))(d\tau \times d\mu^{X,B})_L$$

$$= \int_D f(s,x,v_s(x,b))ds \times d\mu^{x,b}.$$

With this done, it will be seen (from 4.7 and 4.8) that

(4.9) $\theta(s,x,b) = f(s,x,v_s(x,b))$ a.s. $(ds \times d\mu^{x,b})$.

Combining this with (4.6) we establish equation (4.5) as required.

In preparation for the construction of v, denote by $\mathcal{D}$ the <u>adapted σ-field</u> on $[0,1] \times \mathcal{C}^{d+m}$; i.e. $D \in \mathcal{D}$ if and only if D is Borel and $D_t \in F_t$ for each t (see [4]). Then a function $\alpha(t,x,b)$ is adapted, measurable if and only if it is $\mathcal{D}$-measurable; so the function θ above <u>is</u> $\mathcal{D}$-measurable.

Now to construct v, define Q, a nonstandard measure on $*[0,1] \times *\mathcal{C}^{d+m} \times *M$ by

$$Q(C \times E) = \int_C V_\tau(X,B)(E)d\tau \times d\mu^{X,B}.$$

Let

$$q(A) = Q_L(st^{-1}(A))$$

for Borel $A \subseteq [0,1] \times \mathcal{C}^{d+m}$. Finally disintegrate q to obtain $\mathcal{D}$-measurable $v_t(x,b)$ with

$$q(D \times F) = \int_D v_t(x,b)(F)\; dt \times d\mu^{x,b}$$

for $D \in \mathcal{D}$. It is now routine to establish (4.8) for $D \in \mathcal{D}$; then the equation (4.9) is obtained by noting that both sides are $\mathcal{D}$-measurable and using 4.7 and 4.8. This completes the proof of Theorem 4.2(a). Parts (b) and (c) follow routinely.

REFERENCES

[1] Albeverio S., Fenstad J.-E., Hoegh-Krohn R. & Lindstrøm, T.L., Nonstandard Methods in Stochastic Analysis and Mathematical Physics, Academic Press, New York, to appear.

[2] Anderson, R.M., A non-standard representation for Brownian motion and Itô integration, Israel J. Math. **25** (1976), 15-46.

[3] Anderson, R.M. and Rashid, S., A nonstandard characterisation of weak convergence, Proc. Amer. Math. Soc. **69** (1978), 327-332.

[4] Benes V.E., Existence of optimal stochastic control laws, SIAM J. Control **9** (1971), 446-472.

[5] Cutland N.J., Nonstandard measure theory and its applications, Bull. London Math. Soc. **15** (1983), 529-589.

[6] Cutland N.J., Infinitesimal methods in control theory: deterministic and stochastic, Acta Appl. Math. **5** (1986), 105-135.

[7] Cutland N.J., Optimal control for stochastic systems with singular noise, Systems & Control Letters **7** (1986), 55-59.

[8] Keisler H.J., Foundations of Infinitesimal Calculus, Prindle, Webber & Schmidt, 1976.

[9] Keisler H.J., An infinitesimal approach to stochastic analysis, Memoirs of the AMS **197**, AMS, Providence 1984.

[10] Hurd A.E. & Loeb P., An Introduction to Nonstandard Real Analysis, Academic Press, 1985.

RECURSIVE ESTIMATION OF ORDER AND PARAMETERS OF A PILOT PLANT TURBO GENERATOR

T. Fresewinkel
Ruhr-University Bochum
Lehrstuhl für Elektrische Steuerung und Regelung
P.O. Box 102 148
D-4630 Bochum 1
Germany

ABSTRACT. A discrete transfer function with unknown order and parameters of a dynamical, stochastic system is given. For a simultaneous on-line estimation of the model order and the parameters a special parametrization of the transfer function in terms of poles and residuals is chosen. With the knowledge of these parameters it is possible to compute dominance values to determine on-line the order as well as the time delay of the system. The used recursive identification algorithms are of nonlinear regression type and they are applied to two different identification error formulations: the generalized error and the output error method.

As an example the methods are applied to estimate parameters, time delay and order of a pilot plant turbo generator set.

1. INTRODUCTION

For the identification of a good system model as well as for the design of an efficient controller it is necessary to have an a-priori estimate of the system structure. In the case of a single input/single output (SISO) system, the structure is completely determined by its order and time delay. Nearly all methods for estimating the order of a system are off-line methods. These off-line methods can be separated into three different groups (Unbehauen, 1974, 1985).

a) *a-priori methods* depend on a certain matrix of input-output data of the system. The system order is determined by the rank of the data matrix.

b) *error test methods* evaluate the prediction error of the parameter identification. To get an estimate of the system order or the time delay, several experiments with different order and time delay must be applied to the system.

c) *polynomial tests* are based on the estimated transfer function. One

S. Albeverio et al. (eds.), Stochastic Processes in Physics and Engineering, 91–105.

possibility to determine the system order is to cancel poles and zeros of the transfer function. Another method requires a decomposition of the transfer function into poles and residuals to determine which poles are responsible for the system dynamics (Unbehauen, 1985; Litz, 1979).

This last method contains a possibility for on-line determination of the system order and time delay. The idea is to parameterize the system model in terms of the poles and residuals, i.e. the partial fractions (PFs) of the transfer function. Clearly, these parameters are combined in a nonlinear way. But the advantage is that it is possible to decide with the knowledge of the quantity of the residuals and poles, which PF is necessary to model the dynamic behaviour of the system (Unbehauen, 1985; Litz, 1979). This idea was first introduced by Keviczky et al. (1979, 1983). They proposed a method for estimating poles and residuals by an iterative off-line Maximum-Likelihood (ML) estimator. After the identification it is possible to decide which PF is necessary to describe the dynamic behaviour. This method was extended to on-line application by Bokor (1985). The identified model was based on the minimization of the output error, i.e. the difference between the measured system output and the output of an estimated parallel model.

In this paper another set of models should be identified with the same parameterization, but based on the minimization of the equation prediction error, i.e. the difference between input and output signals filtered by the estimated numerator and denominator polynomials, respectively. The algorithms are introduced in chapter 2. The criterions to determine the system structure given by Bokor (1985) are improved in chapter 3 by using the dominance concept of Litz (1979) and Unbehauen (1985). In chapter 4 the results are verified and compared with the recursive ML method using a pilot plant turbo generator.

2. THE IDENTIFICATION ALGORITHMS

The single-input/single-output system which whould be identified is parameterized in terms of first and second order PFs containing the real and complex poles of the dynamic system

$$Y(z) = \left[K + \sum_{i=1}^{m} \frac{B_i(z)}{A_i(z)} \right] z^{-d} U(z) + R(z). \tag{1}$$

K is a constant gain and $B_i(z)$ and $A_i(z)$ are polynomials of order n_i. The polynomial order n_i may take the values "1" or "2" for real or complex poles, respectively. The exact system order is given by

$$n = \sum_{i=1}^{m} n_i. \tag{2}$$

The systems time delay is d. r(k) is some noise. The first identification algorithm to estimate all unknown parameters is based on the output error formulation according to fig. 1 (Ljung, 1984). The error to be minimized is the difference between the measured system output and the computed model output.

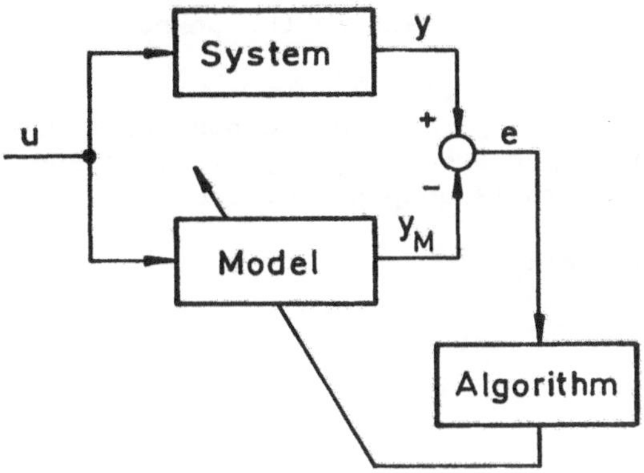

Figure 1. Output error method

Let us assume here that the system noise r(k) is white noise ε(k). Then it follows from eq.(1) that

$$\varepsilon(k) = r(k) = y(k) - y_M(k) \tag{3a}$$

with

$$y_M(k) = \mathfrak{Z}^{-1}\left\{\left[K + \sum_{i=1}^{m} \frac{B_i(z)}{A_i(z)}\right] z^{-d}\, U(z)\right\}. \tag{3b}$$

The symbol "$\mathfrak{Z}^{-1}$" denotes the inverse z-transformed. In general case it is possible to introduce additionaly a noise filter to relate r(k) to ε(k). The unknown polynomial coefficients can be arranged in a parameter vector

$$\underline{p} = [K,\ a_{11},\ldots,a_{1n_1},\ b_{11},\ldots,b_{1n_1},\ a_{21},\ldots,b_{mn_m}]^T. \tag{4}$$

To get the identification error equation the exact parameters, the number of PFs and the time delay are replaced by their estimates of the last sampling step k-1. This will be indicated by writing $\hat{\underline{p}}(k-1)$, $\hat{K}$, $\hat{m}$ and $\hat{d}$, respectively. The error to be minimized then becomes

$$e(k) = \mathcal{Z}^{-1}\left\{ Y(z) - \left[K + \sum_{i=1}^{\hat{m}} \frac{B_i(z)}{A_i(z)} \right] z^{-\hat{d}}\, U(z) \right\} \Bigg|_{\underline{p}=\hat{\underline{p}}(k-1)}, \tag{5}$$

i.e. eq.(3a) computed with the estimated parameters.

The identification error is nonlinear in the unknown coefficients of the polynomials.

So, for the estimation of the parameters, a nonlinear regression type algorithm can be applied, published e.g. by Albert (1969) or Sorenson (1980). The parameter update is computed according to

$$\hat{\underline{p}}(k) = \hat{\underline{p}}(k-1) + \underline{q}(k)\ e(k) \tag{6a}$$

$$\underline{q}(k) = \underline{P}(k)\ \underline{m}(k) \tag{6b}$$

$$\underline{m}(k) = \frac{-\partial e(k)}{\partial \hat{\underline{p}}(k-1)} \tag{6c}$$

$$\underline{P}(k) = (\lambda(k)\ \underline{P}^{-1}(k-1) + \underline{m}(k)\ \underline{m}^T(k))^{-1}. \tag{6d}$$

$\lambda(k)$ is a weighting factor to accelerate the convergence of the algorithm or to handle slowly time varying systems. For our special parameterization the elements of the vector of derivatives in equation (6c) are defined by

$$\frac{\partial e(k)}{\partial \hat{K}(k-1)} = u(k-d) \tag{7a}$$

$$\frac{\partial e(k)}{\partial \hat{a}_{ij}(k-1)} = -\mathcal{Z}^{-1}\left\{ \frac{B_i(z)}{A_i^2(z)}\, z^{-d-j}\ U(z) \right\} \Bigg|_{\underline{p}=\hat{\underline{p}}(k-1)} \tag{7b}$$

$$\frac{\partial e(k)}{\partial \hat{b}_{ij}(k-1)} = \mathcal{Z}^{-1}\left\{ \frac{1}{A_i(z)}\, z^{-d-j}\ U(z) \right\} \Bigg|_{p=\hat{p}(k-1)}. \tag{7c}$$

This algorithm has been introduced first by Bokor (1985).

A second method to estimate the unknown parameters is based on the generalized error formulation according to fig. 2 (Fresewinkel, 1986). If eq.(1) is multiplied by the polynomial

$$A(z) = \prod_{i=1}^{m} A_i(z) \tag{8}$$

we get

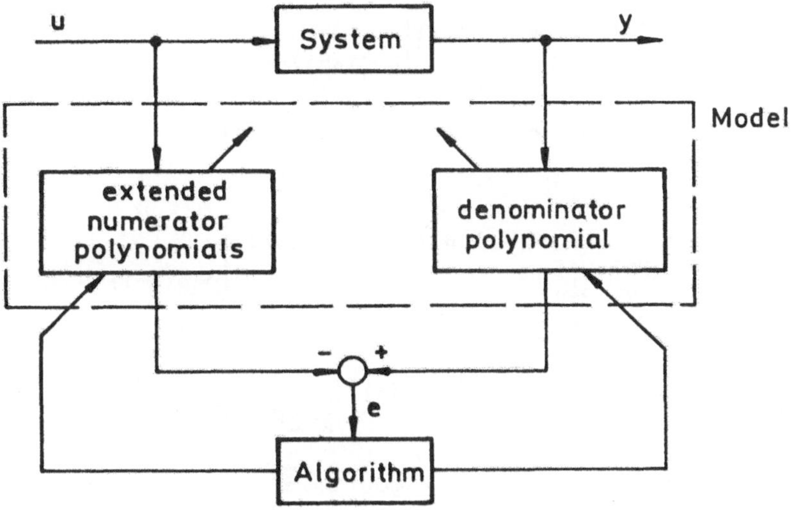

Figure 2. Generalized error method

$$A(z)\ Y(z) = [A(z)\ K + \sum_{i=1}^{m} A_{ri}(z)\ B_i(z)]\ z^{-d}\ U(z) + A(z)\ R(z) \tag{9}$$

with the reduced polynomials

$$A_{ri}(z) = \prod_{\substack{j=1 \\ j \neq i}}^{m} A_j(z).$$

Introducing the variable

$$\varepsilon(z) = A(z)\ R(z) \tag{10}$$

eq.(9) results in

$$\varepsilon(z) = A(z)\ Y(z) - [A(z)\ K + \sum_{i=1}^{m} A_{ri}(z)\ B_i(z)]\ z^{-d}\ U(z). \tag{11}$$

Replacing the unknown parameters by their estimates gives the generalized identification error

$$e(k) = \mathcal{Z}^{-1}\ \{A(z)\ Y(z) - [A(z)\ K + \sum_{i=1}^{\hat{m}} A_{ri}(z)\ B_i(z)]\ z^{-\hat{d}}\ U(z)\}\Big|_{\underline{p}=\underline{\hat{p}}(k-1)}.$$

To eq.(11) the same nonlinear regression type algorithm of eqs.(6a-d) can be applied. For this parameterization the elements of the vector of derivatives is given by

$$\frac{\partial e(k)}{\partial \hat{K}(k-1)} = \mathcal{Z}^{-1}\{A(z)\,z^{-\hat{d}}\,U(z)\}\Big|_{\underline{p}=\hat{\underline{p}}(k-1)} \tag{12a}$$

$$\frac{\partial e(k)}{\partial \hat{a}_{ij}(k-1)} = -\mathcal{Z}^{-1}\{z^{-j}[A_{ri}(z)(Y(z)-K\,z^{-\hat{d}}\,U(z)) - \sum_{\substack{\mu=1\\ \mu\neq i}}^{\hat{m}} \prod_{\substack{\nu=1\\ \nu\neq\mu\\ \nu\neq i}}^{\hat{m}} A_\nu(z)\,B_\mu(z)\,z^{-\hat{d}}\,U(z)]\}\Big|_{\underline{p}=\hat{\underline{p}}(k-1)} \tag{12b}$$

$$\frac{\partial e(k)}{\partial \hat{b}_{ij}(k-1)} = \mathcal{Z}^{-1}\{A_{ri}(z)\,z^{-\hat{d}-j}\,U(z)\}\Big|_{\underline{p}=\hat{\underline{p}}(k-1)}. \tag{12c}$$

So with the general identification algorithm of eqs.(6a-d) and the special derivatives of eqs.(7a-c) and (11a-c) we have two different algorithms for estimating the unknown parameters.

3. DETERMINATION OF ORDER AND TIME DELAY

The first problem examined in this chapter is how to decide during the identification experiment which PF is necessary to describe the dynamic behaviour of the system and which PF can be cancelled. The second problem is the determination of the time delay.

3.1. Order Determination

The rules given by Bokor (1985) for cancelling the i-th partial fraction are that the system order can be reduced by n_i if for any two estimated polynomials of the same order the inequalities

$$|\hat{A}_i(z) - \hat{A}_j(z)| < \rho(z),\ i \neq j \tag{13a}$$

or

$$|\hat{B}_i(z)| < \delta(z) \tag{13b}$$

are fulfilled. The absolut value of a polynomial is here defined as

$$|C(z)| := |c_o| + |c_1|q^{-1} + \ldots + |c_n|q^{-n}. \tag{13c}$$

$\rho(z)$ and $\delta(z)$ are polynomials with small positive coefficients:

$$\rho(z) = \sum_{i=0}^{n} |\rho_i| z^{-i} \tag{14a}$$

$$\delta(z) = \sum_{i=0}^{n} |\delta_i| z^{-i}. \tag{14b}$$

It can be shown that both conditions may be not sufficient in certain cases. Let us suppose that we identified two PFs with

$$\hat{A}_i(z) = \hat{A}_j(z) + \rho(z).$$

The sum of the two PFs is

$$\hat{G}_{ij}(z) = \frac{\hat{B}_i(z)}{\hat{A}_i(z)} + \frac{\hat{B}_j(z)}{\hat{A}_j(z)}$$

$$= \frac{(\hat{B}_i(z) + \hat{B}_j(z))}{(\hat{A}_i(z) + \varepsilon(z))} (1 + \rho(z)\ C_{ij}(z)) \tag{15}$$

with

$$C_{ij}(z) = \frac{\hat{B}_i(z)}{\hat{A}_i(z)(\hat{B}_i(z) + \hat{B}_j(z))}.$$

The term $\rho(z)\ C_{ij}(z)$ can be neglected if it is small enough. This is the case only if the poles of the polynomials $\hat{A}_i(z)$ and $\hat{A}_j(z)$ are nearly equal and single. If they are multiple or very close together, the coefficients of the polynomials $\hat{B}_i(z)$ and $\hat{B}_j(z)$ will have nearly the same magnitude but different sign, so that the sum of both will be very small and $C_{ij}(z)$ will become large.

So necessary conditions for reducing the system order by n_i are that

$$|\hat{A}_i(z) - \hat{A}_j(z)| < \rho(z)$$

and

$$|\hat{B}_i(z) + \hat{B}_j(z)| \gg \rho(z). \tag{16}$$

The condition (13b) is not sufficient, too. Let us suppose again that we identified two PFs with

$$|\hat{B}_j(z)| \approx |\hat{B}_i(z)| < \delta(z).$$

The gain of each PF is

$$g_\ell = \frac{\hat{B}_\ell(1)}{\hat{A}_\ell(1)}, \quad \ell = i, j.$$

If $\hat{A}_i(z)$ has poles in the neighbourhood of 1 and $\hat{A}_j(z)$ has poles in the neighbourhood of 0, then the gain of PF i is much larger than the gain of PF j

$$g_i \gg g_j,$$

so that only the j-th PF can be cancelled.

These considerations show that it is necessary to define some normalized value for the dominance of a PF. This normalized value, which will be used here to decide whether a PF is dominant or not was introduced for continuous systems by Litz (1979) and extended for discrete systems by Unbehauen (1985). If the identified system is stable, its gain is

$$g = \lim_{k\to\infty} \frac{y(k)}{u(k)} = \hat{K} + \sum_{i=1}^{\hat{m}} \frac{\hat{B}_i(1)}{\hat{A}_i(1)} . \tag{17}$$

The dominance of the i-th PF is defined by (Unbehauen 1985)

$$d_i = \frac{\hat{B}_i(1)}{\hat{A}_i(1)g} . \tag{18a}$$

The same can be defined for the constant value

$$d_k = \frac{\hat{K}}{g} . \tag{18b}$$

For second order PFs with a large overshoot it may be not sufficient to consider only the gain of these partial fractions. In this case it is necessary to follow the idea of Litz (1979) to determine an upper bound for the dominance of a second order PF. This upper bound in the discrete case is

$$d_{i,max} = \frac{1}{g} c_i \geq d_i$$

where c_i is given by

$$c_i = \frac{|r_i|}{|1-p_i|} .$$

r_i and p_i are the complex residuals and poles, respectively of the i-th second order PF

$$\frac{\hat{B}_i(z)}{\hat{A}_i(z)} = \frac{r_i z^{-1}}{1-p_i z^{-1}} + \frac{\bar{r}_i z^{-1}}{1-\bar{p}_i z^{-1}} .$$

3.2. Time Delay Determination

To determine the time delay of a system a similar dominance criterion can be defined. As a discrete time delay may be considered as an increasing of the system order, the estimated system can be represented by

$$\hat{G}(z) = \left(\frac{\hat{K}}{\hat{A}(z)} + \sum_{i=1}^{\hat{n}} \frac{\hat{b}'_i}{\hat{A}(z)} z^{-i}\right) z^{-\hat{d}} \tag{19}$$

with the coefficients $\hat{K} = \hat{b}_i = 0$ for $i=1,\dots,d-\hat{d}$ and a pole of multiplicity $d-\hat{d}$ at the origin. $d-\hat{d}$ is the difference between the time delay of the real system and the model. Each term of the sum in eq.(19) is responsible for a part of the step response of this system. Normalized by the gain g of the system each term has a dominance of

$$d_{di} = \frac{1}{g} \frac{\hat{b}'_i}{\hat{A}(1)} \qquad i = 1,\dots,n \tag{20a}$$

$$d_{dk} = \frac{1}{g} \frac{\hat{K}}{\hat{A}(1)} . \tag{20b}$$

To determine the time delay of the system, parametrized according to eq.(1), we have to compute the coefficient

$$\hat{b}'_1 = \sum_{i=1}^{\hat{m}} (\hat{b}_{i1} + \hat{a}_{i1} \hat{K}) . \tag{21}$$

The time delay $\hat{d}$ has to be increased by 1 if

$$|d_{d1}| < \rho$$

and

$$|d_{dk}| < \rho .$$

The estimated time delay $\hat{d}$ is correct if the inequalities

$$|d_{d1}| > \rho$$

$$|d_{dk}| < \rho$$

are fulfilled.

Summarizing this chapter the following improved criterions can be formulated to determine the system structure on line:

(1) Decrease the system order by n_i if

a) $|\hat{A}_i(z) - \hat{A}_j(z)| < \rho(z)$

and

$|\hat{B}_i(z) + \hat{B}_j(z)| \gg \rho(z)$

b) $|d_i| < \rho$

(2) Increase the system time delay by 1 if

$|d_{d1}| < \rho$

$|d_{dk}| < \rho$

The time delay is correct if

$|d_{d1}| > \rho$

$|d_{dk}| < \rho$

(3) A complex pole is replaced by two real poles if (Bokor 1985)

$$a_{i2} - \left(\frac{a_{i1}}{2}\right)^2 < \rho.$$

In all cases ρ is a small positive number and $\rho(z)$ is a polynomial with small positive coefficients.

4. EXAMPLE

As an example a pilot plant turbo generator according to fig. 3 is identified. The system has two inputs. u_1 is the current of the exciting field of the synchron generator, u_2 is the position of the air pressure valve. The outputs of the system are the speed y_1 and the magnitude of the voltage of the generator y_2. During the experiment u_2 was held constant. Only the transfer function G_{11} was identified. For the remaining transfer functions the results are equivalent.

4.1. Experiment 1

In the first experiment G_{11} was identified by the algorithm given in the eqs.(6a-d) and (12a-c). The starting values of the algorithm are given in table 1.

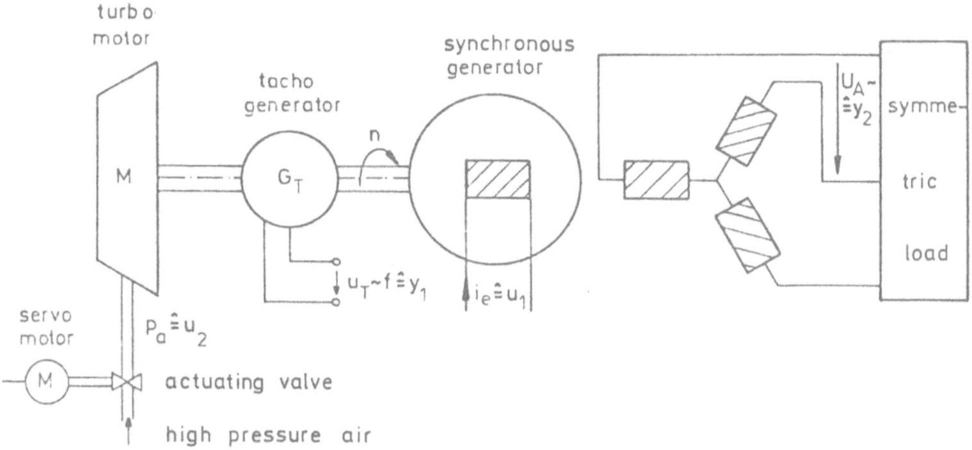

Figure 3. Pilot plant turbo generator

TABLE I. Starting values of experiment 1

parameter vector	$\underline{\hat{p}}(0) = \underline{0}$
covariance matrix	$\underline{P}(0) = 100\ \underline{I}$
sampling time	$T = 0.1$ s
weighting factor	$\lambda = 0.98$
model order	$\hat{n}(0) = 4$
first order PFs	4
second order PFs	0
time delay	$\hat{d}(0) = 0$

The results of the experiment are given in figure 4. At sampling step 250 the time delay was increased by 1, as both d_{dk} and d_{d1} are very small. After the time delay was increased it can be seen that d_{dk} remains small while d_{d1} becomes larger. At step 300 and step 400 the first and the fourth PF were cancelled because their dominance values are small. Figure 4 shows that it is not possible with the criterions given by Bokor (1985) to decide whether PF1 or PF2 is dominant as their numerator parameters are both small and nearly equal. With the improved criterions given in the last chapter this decision is unique.

4.2. Experiment 2

The same experiment was repeated with the recursive ML method proposed by Bokor (1985) which is given by the eqs.(6a-d) and (7a-c). The parameter curves are shown in figure 5. At sampling step 300 the third PF

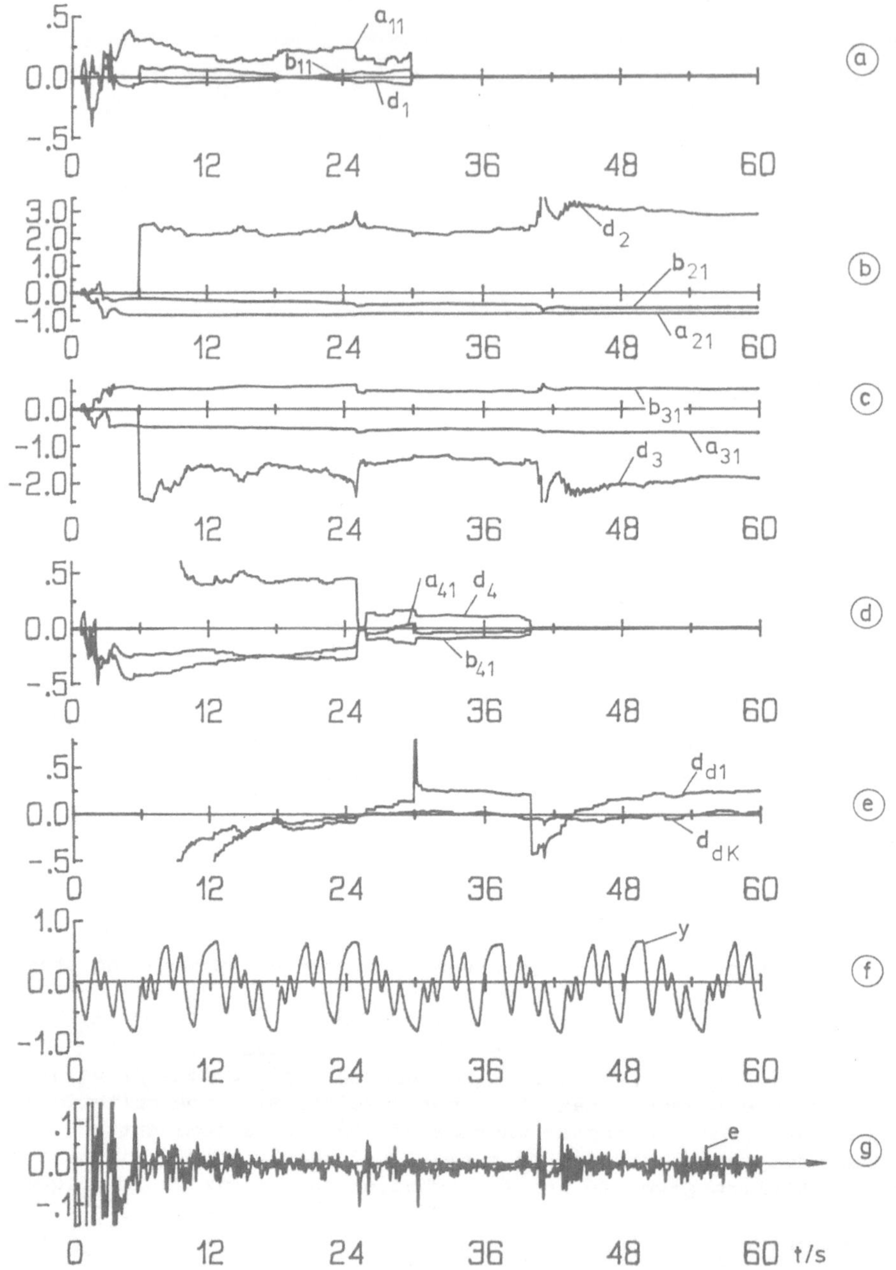

Figure 4. Parameters (a-d), dominance values of time delay (4), measured output (f) and error (g) of experiment 1

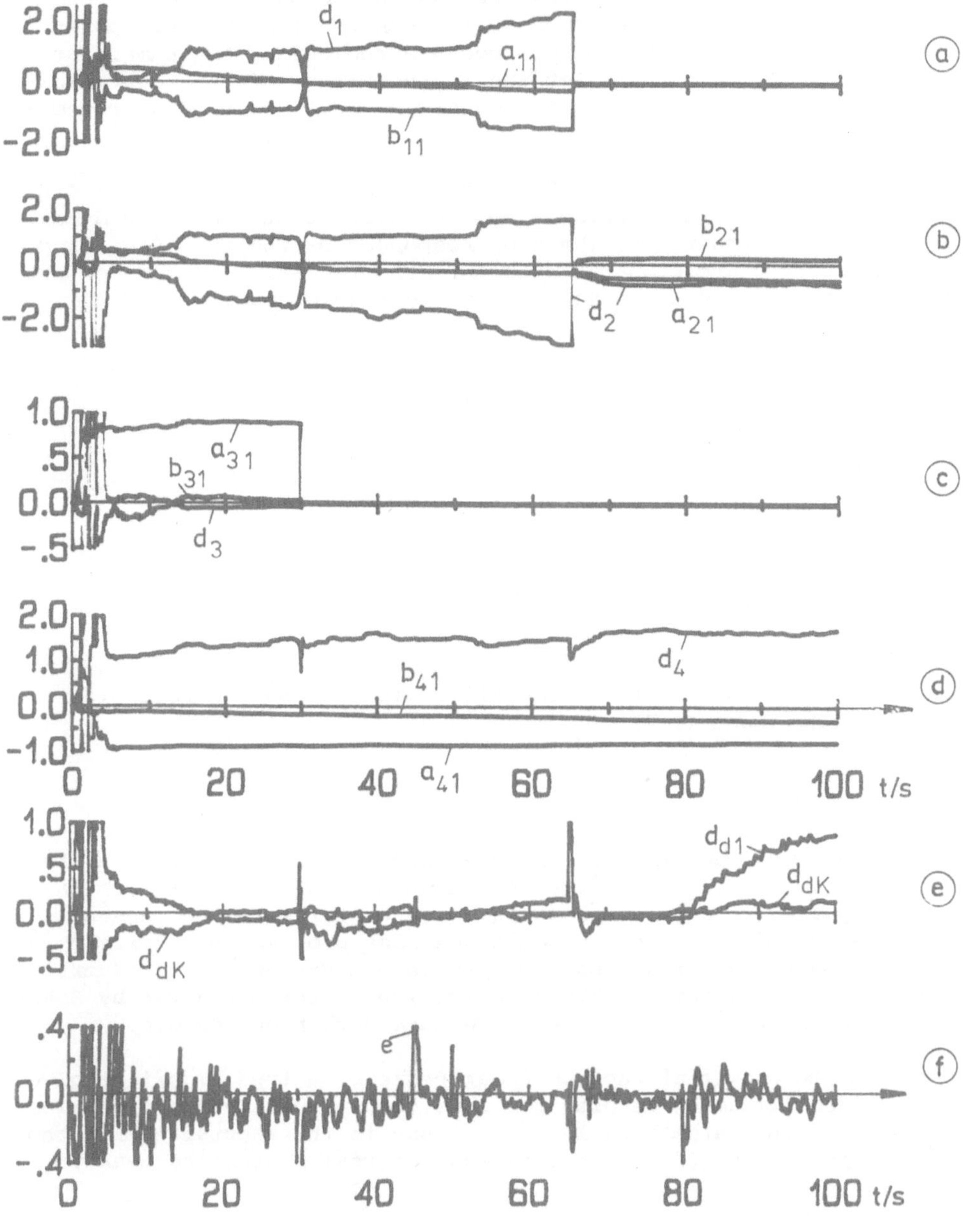

Figure 5. Parameters (a-d), dominance values of time delay (e) and error (f) of experiment 2

is cancelled. At step 650 the first PF and the second PF are grouped together as they have the same poles. It is not unique at this point whether it is allowed to group them because the sum of the numerator parameters $\hat{b}_{21} + \hat{b}_{11}$ is small. But the further curve of the identification error shows that the reduction causes no deterioration of the model. At step 800 the time delay of the model was increased by 1.

The results of experiment 1 and 2 are a second order model with a discrete time delay of one sampling step, which is due to the slow dynamics of the turbo generator. The step responses of the two identified models in figure 6 are compared with the measured step response of the system. As can be seen all three curves are nearly equal.

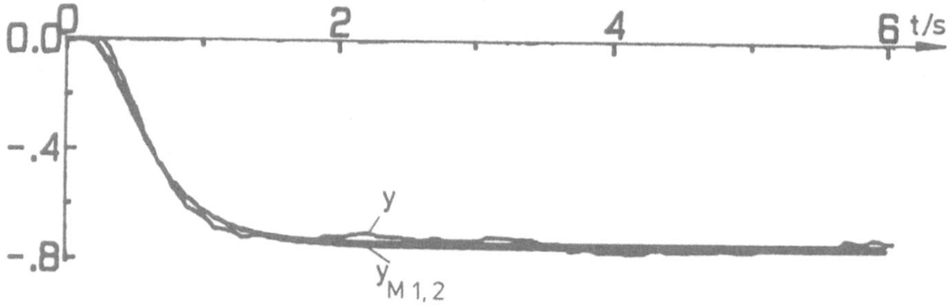

Figure 6. Step responses of the identified models and the real plant (experiment 1, 2)

5. CONCLUSION

In this paper nonlinear regression type identification algorithms based on the generalized error formulation and on the output error formulation were applied to a pilot plant turbo generator set. The chosen parametrization of the model, based on an idea proposed by Keviczky (1979), was such that it was possible to determine the system order, time delay and parameters simultaniously. The criterions given by Bokor (1985) to determine the best model structure have been improved.

The experiments show that the nonlinear regression type algorithm gives very good identification results for this special parametrization. The convergence of the parameters is faster than in the recursive ML method and the step response of both models and the real system are nearly equal.

REFERENCES

Albert, A.E., L.A. Gardner (1969). Stochastic approximation and nonlinear regression. MIT Press, Cambridge.

Bokor, J., L. Keviczky (1983). Structure and parameter estimation of MIMO systems using elementary subsystems representation. Int. J. of Control, 39, No. 5, pp. 965-986.

Bokor, J., L. Keviczky (1985). Recursive structure, parameter and delay time estimation using ESS representations. IFAC Symposium on Identification and System Parameter Estimation, York, pp. 867-872.

Fresewinkel, T. (1986). Simultanious on line identification of system order and parameters. 2nd IFAC Symposium on Stochastic Control, Vilnius.

Keviczky, L., J. Bokor, C. Bânyâsz (1979). A new identification method with special parametrization for model structure determination. IFAC Symposium on Identification and System Parameter Estimation, Darmstadt, pp. 561-567.

Litz, L. (1979). Reduktion der Ordnung linearer Zustandsraummodelle mittels modaler Verfahren. Hochschulverlag Stuttgart.

Ljung, L., T. Söderström (1983). Theory and Practice of Recursive Identification. MIT Press, Cambridge.

Sorenson, H.W. (1980). Parameter estimation. Control and Systems Theory, Vol. 9, Marcel Dekker, New York.

Unbehauen, H., B. Göhring (1974). Test for determining model order in parameter estimation. Automatica, 10, pp. 233-244.

Unbehauen, H. (1985). Regelungstechnik III. Vieweg-Verlag.

ATTAINABILITY AND REVERSIBILITY OF A GOLDEN AGE FOR THE LABOUR-SURPLUS ECONOMY: A STOCHASTIC VARIATIONAL APPROACH

Guillermo L. Gómez M.*
Research Center Bielefeld-Bochum-Stochastics
University of Bielefeld
D-4800 Bielefeld 1
Federal Republic of Germany

KEYWORDS: Optimal economic planning, development and growth economics, stochastic dynamic economics, stochastic optimal control and stochastic analysis.

ABSTRACT. The main objective of the present paper is to work out, by means of a stochastic maximum principle due to A. Bensoussan, development paths along which surplus-labour can be absorbed into production, and to establish patterns of social valuation and an institutional framework which enable the economy to attain and maintain full employment.

In the first two Sections we present the problem and introduce, relying on stochastic differential equations, economic concepts and ideas needed in Section 3 to make an intuitive and precise formulation of the stochastic control problem.

In Section 4 we give a fairly complete derivation of the maximum principle and make detailed economic interpretation of it, in order to keep the links with our original problem alive.

In the last Section, we deal with questions related to the attainability and reversibility of the optimal labour allocation policy obtained.

CONTENTS

* Current address: Institute of Mathematics, University of Erlangen-Nürnberg, Bismarckstr. 1 1/2, D-8520 Erlangen, FRG

S. Albeverio et al. (eds.), Stochastic Processes in Physics and Engineering, 107–148.

1. INTRODUCTION

The main objective of the present study is to establish development paths, patterns of social valuation and an institutional framework suitable for the attainment of a golden age for a labour-surplus economy of the dual type.

A *golden age path* or *steady growth path* is a sort of fictitious dynamic equilibrium characterized by a state of affairs at which capital, labour, output, consumption and investment all grow at a constant rate and the ratio between these variables remains constant.

Since we see here golden ages just as a frame of reference, we are therefore primarily concerned with disequilibria arising under the pressure of an increasing supply of idle labour and with the construction by means of stochastic control techniques of paths along which the economic system can abandon such disequilibria.

Our economic system consists of three sectors obtained with the help of aggregation according to the criterion of *mode of production*.

This criterion relies upon the two following basic concepts: a) *productive forces* which embody whatever contributes to the capacity to produce, and b) *(social) relations of production* characterized by the form of organization, the structure of property, authority, control and hierarchy of the process of production as well as by the (characteristic) attitude of fulfilment growing out of work and the kind of motivation associated with production. We refer the interested reader to Gómez (1984a, 1986a) for a detailed presentation of the model.

According to the control methodology we have chosen, we take as given the *initial* and the *terminal state* of the economic system, where the latter establishes the *macro-goal* towards which the system is expected to move.

Then, we look for: a) *development path* along which the system can attain the macro-goal, b) *behavioural* and *motivational patterns* that set the system to such a path and keep it to it *(irreversibility)*, and c) an *institutional framework* suitable to elicit the appropriate *socio-economical* and *political environment*. The *unknowns* obtained in a)-c) are essential ingredients of goal-adequate controls aimed at increasing *productive capacity*, i.e. skills, know-how and means of production, and at enhancing the level of *efficient performance*. Together with the patterns of efficiency, the fulfilling of productive capacity goals aims at the *full utilization of productive resources*. For the macro-goal is not furthered, whenever labour remains involuntarily idle, if means of production are still available and the desire for output is still unsatisfied.

Now, let us point out that the *market*, like rationing and planning, is one of the major methods of *coordination of resource use* within, between and beyond the modes and covers those processes of *exchange* and *distribution* which arise whenever ultimate *consumption* and *accumulation* does not occur within the producing unit. However, there are situations in which the market, under the prevailing sense of social justice, fails to forward a system of incentives and rewards able to sustain the productive effort at a level adequate to generate the socio-economic progress which is needed to absorb the existing idle labour into production.

To help the system to overcome this impasse, one needs, we believe, governmental intervention.

The present investigation concentrates on the requirements for a goal-conducive motion of the systems. Namely, those which may lead to an increase in the productive capacity and to an adjustment or adaptation of socio-economic reality to that framework taylored by goal-adequate development policies.

Roughly speaking, our strategy consists of two phases:

1.) One increases the productive capacity of the three sectors of the system by means of impulsive controls, which amount to capital injections which in turn enable a new choice of technique at a higher level of employment and bring about a structural change.

2.) One lets the system evolve autonomously until a certain amount of idle work -- which depends upon the size of the capital injection -- has been shifted to more highly productive activities. However, since the capital-labour ratio shall remain fixed until the next impulsion time, any increase in the level of employment shall be conditioned by the formation of (new) capital.

The time elapsed between two consecutive impulsions determines a time period, the duration of which depends again on the corresponding impulsive control and hence it varies usually from period to period.

One repeats this again and again until the system enters the neoclassic era, i.e. balanced growth or dynamic equilibrium and the state of laissez-faire holds. Gómez (1984a, 1986a) deals with some economic and control theoretical aspects of the model. The present paper is mainly concerned with the question how to find an investment-consumption mix-path for the secondary sector along which full employment relative to the corresponding period can be attained given its technology, i.e. given a capital-labour ratio, and its institutions. Moreover, one wishes that this path satisfies certain conditions of optimality, and one looks for conditions that make sure that the state of full employment is not reversible. Hence, our analysis starts right after the corresponding impulse has been applied and ends short before the next impulse becomes necessary, i.e. we deal here with phase 2.) of our control strategy.

2. ON ECONOMIC SURPLUS AND THE ACCUMULATION PROCESS

Let us assume the technology of the productive or advanced sector of the economy is specified by a concave real-valued function $t \to Y(t)$ of class $C^{1,2,2}([0,T] \times \mathbb{R}_+ \times \mathbb{R}_+ \to \mathbb{R}_+)$

$$Y(t) = Y(t,K(t),L(t)) \tag{1*}$$

where $Y(t)$, $K(t)$ and $L(t)$ are flow variables denoting respectively aggregate output, capital and labour services at time t, $T \in \mathbb{R}_+$ arbitrary.

The explicit argument t at the right-hand side of eq. (1^*) represents the process of technical progress. However, in order to keep the exposition simple, we rule out technical progress in the short-run, but its inclusion would not affect our results at least in the case when technical progress is neutral on balance, i.e. when the cost in terms of wage units of capital per unit of output falls at the rate at which output per man-hour rises.

Therefore, eq. (1^*) reduces to the simple form

$$Y(t) = Y(K(t),L(t)). \tag{1}$$

We assume further that the production function, eq. (1), exhibits constant returns to scale, positive marginal productivities which diminish with successive increments of labour L and capital K.

Besides, we assume $\lim_{L\to 0} Y_L = \infty$ and $\lim_{K\to 0} Y_K = \infty$ which exclude some pathological and uninteresting cases. Subscripts denote partial differentiation. To save notation we suppress, unless any confusion may arise, the variable t and simply write Y, K and L, and so on.

Let us recall the following well-known definitional relations for investment I and consumption C:

$$I = s_k(Y-WL) + s_L WL \tag{2}$$

$$C = (1-s_k)(Y-WL) + (1-s_L)WL \tag{3}$$

where W stands for the nominal wage rate. s_L and s_K denote the fraction of wage income WL and capital income $Y-WL$ saved and satisfy the constraints

$$0 \le s_L \le s_K \le 1, \quad s_K, s_L \in \mathbb{R}. \tag{4}$$

The capital accumulation path is given by

$$\dot{K} = I - \delta K \tag{5a}$$

where the dot indicates time derivative, δ denotes the rate of capital depreciation and satisfies

$$\delta > 0, \quad \delta \in \mathbb{R},$$

or, equivalently,

$$\dot{K} = sY - \delta K \tag{5b}$$

where s is the total fraction of output saved and is given by

$$s = s_K\left(\frac{Y-WL}{Y}\right) + s_L\left(\frac{WL}{Y}\right) \tag{6}$$

where the expressions in parentheses on the right-hand side of eq. (6) are respectively the output share of capital and labour. Obviously, s satisfies the condition (4).

Furthermore, one needs the following conditions to hold

$$Y > 0,\ {}^oK > 0,\ {}^oN > 0,\ I > 0,\ C > 0 \tag{7}$$

$$Y,\ {}^{o}K,\ {}^{o}N,\ I,\ C \in \mathbb{R}$$

where ${}^{o}K$ denotes the initial capital endowment, and ${}^{o}N$ the initial labour available to the advanced sector and includes the fraction of idle labour which has to be shifted to the sector and has to be absorbed into production in the time period under consideration. Let us remark that ${}^{o}K$ and ${}^{o}N$ depend upon the corresponding impulsive control. Moreover, let ε denote the population rate of growth and N represent the labour available to the productive sector at time t which is given by

$$N = {}^{o}N\, e^{\varepsilon t} \tag{8a}$$

and ε satisfies

$$\varepsilon > 0, \quad \varepsilon \in \mathbb{R}\ . \tag{8b}$$

The following national income identity holds

$$Y = I + C\ . \tag{1**}$$

A look at eq. (2) covinces ourselves that one of the major problems of the labour-surplus economy is, that employment and wages conflict with investment and through eq. (5) with growth as well. We shall come back to this point.

The relevant range of employment is given by the real interval $[\underline{L},\bar{L}]$, where $\underline{L}$ is the investment maximizing level of employment and $\bar{L}$ is the level of employment at which wage income exhausts total output.

Let us point out that in the conventional world of neoclassic economics the economy would settle at $\underline{L}$, at least under perfect competition. Therefore, in order to increase employment beyond $\underline{L}$ one needs governmental intervention which is brought about indirectly through a system of social preferences and value-judgments which we represent here by the real-valued nonpositive strict concave utility function $c(t) \to U(c(t))$ and the real-valued nonpositive strict concave terminal pay-off function $k(t) \to g(k(t))$, see eq. (12), which we shall specify later. However, one needs to assume that the government or control board is able to require capitalists to save any fraction s_K of profit income as long as s_K remains below the upper bound $\bar{s}_K$, i.e.

$$0 \le s_K \le \bar{s}_K \tag{9}$$

where $\bar{s}_K$ has been negotiated between control board and capitalists.

When L moves towards $\bar{L}$ one has to trade investment, which means future consumption and employment, for present consumption. And this gives rise to a problem of time preference which we mentioned already. For present investment decreases, or, equivalently, present consumption increases, entail future consumption and employment decreases. Consequently, we have to look at the future and present claims associated with a feasible alternative investment-consumption mix, i.e. a pair

(I,C) that would result from an alternative expansion of the level of employment beyond $\underline{L}$ and try to find a logic of balancing these claims.

Thus, we introduce a pair-valued control function $t \to \alpha(t)$ with

$$\alpha(t) = (s_K(t),L(t)) \tag{10}$$

which allows us to rewrite eqs. (2) and (3) in the following more (control) suggestive and simple way

$$I(t) = I(s_K(t),L(t)) = I(\alpha(t)) \tag{2*}$$

$$C(t) = C(s_K(t),L(t)) = C(\alpha(t)). \tag{3*}$$

Concerning the nominal wage W we make the assumption that a minimum wage rate $\bar{W}$ is set exogeneously by the control board which remains constant during the corresponding time period,

$$W \geq \bar{W} . \tag{11}$$

Let us remark that $\bar{W}$ is not necessarily the minimum of subsistence but depends upon the nominal wage rate prevailing in the primary or less productive sector.

We have now enough notation to state our problem a little more precisely.

With this purpose in mind, we introduce the concept of *economic surplus* as the per capita difference between the material production and the consumption necessary to produce it. Necessary consumption comprises the means of production used up which requires replacement and the personal consumption of the direct producers.

Stated in this way, the concept of economic surplus hints at a framework paralleling the theory of social reproduction of the classicists and stressing the continuous renewal nature of production as a material process on the one hand and as a social process on the other. A theory of value or system of adjoint variables, which mimics the equations of motion of classical mechanics for conservative and dissipative systems, provides us with a way of balancing present and future consumption claims associated with the economic policy in course and completes the working framework. See Gómez (1986b).

Concerning the concept of economic surplus there are three points we like to emphasize:

1. We have to deal with *potential* rather than *actual* economic surplus and this is defined as the per capital difference between the material production that could be brought about with the help of employable productive resources and that what might be regarded as essential consumption.

The realization of the potential economic surplus requires substantial institutional changes and a thorough reorganization of production with the market playing a central role as a mechanism of distribution and exchange, but remaining subordinate to the real activities of

production and consumption. This view entails identifying and measuring *excess consumption, output lost* due to the existence of *unproductive workers, irrational* and *wasteful* organization of the existing productive apparatus, and to the existence of under-, unemployment and misemployment of productive resources.
2. The assessment of the magnitude of the potential economic surplus involves difficulties of different kinds, since a) it requires a more or less clear goal visualization, i.e. the picture of a more rationally ordered society, b) it presupposes an adequate knowledge and appraisal of the only partially and indirectly observable performance of the socio-economic organization of the existing social order, c) one has to work with responses of individuals and with goal-depending concepts like rationality, efficiency, etc. which might induce some sort of conceptual circularity.
3. The mode of utilization of the economic surplus, characterized roughly by the path of the investment-consumption mix, constitutes the engine of the process of capital accumulation and determines, together with the issues mentioned in 1. and 2., the pace and direction of the economic development process, and is determined itself by the degree of development of the productive forces, the corresponding structure of socio-economic relations and social welfare views and value judgments forwarded by the government or control board.

Let us proceed to formulate *the process of capital accumulation*. The following notation shall facilitate the argument. Let $c(t)$, $k(t)$, $i(t)$, and $\lambda(t)$ represent consumption, capital, investment, and employment normalized with respect to N

$$c = \frac{C}{N} \quad k = \frac{K}{N} \quad \lambda = \frac{L}{N} \quad i = \frac{I}{N} . \tag{12}$$

Further, let ℓ and y represent respectively the labour-capital and the output-capital ratio:

$$\ell = \frac{L}{K} \quad y = \frac{Y}{K} . \tag{13}$$

Since $Y(K,L)$ is linear and homogeneous, it can be written

$$Y(K,L) = Ky(\frac{L}{K}) \tag{14}$$

where

$$y(\frac{L}{K}) = Y(1,\frac{L}{K}) .$$

Then, eq. (1) obtains the more simple form

$$y(t) = y(\ell(t)) \tag{15}$$

or, simply,

$$y = y(\ell).$$

We get the following explicit expressions for the marginal products Y_L and Y_K by differentiating eq. (14) and applying Euler's theorem for homogeneous functions:

$$Y_L = y'(\ell) \tag{16a}$$

$$Y_K = y-y'\ell\ . \tag{16b}$$

We assume further that changes in the labour supply N and in the labour force participation L are governed by the following stochastic differential equations (SDE) of the diffusion type

$$dN = \varepsilon N\, dt + \pi N\, dB \tag{17}$$

$$dL = a\, L\, dt + b\, L\, dB \tag{18}$$

$$N(0) = {}^{o}N \quad \text{and} \quad L(0) = {}^{o}L \quad \text{given,}$$

where ε, π, a and b are scalars and remain constant within the time period considered, B is a real-valued Brownian motion defined on the abstract probability space $(\Omega, \underline{\underline{F}}, \mathbb{P})$ with the usual properties and endowed with a filtration $\underline{F} = (\underline{\underline{F}}_t)_{t\geq 0}$ of sub-σ-fields of $\underline{\underline{F}}$.

The drift terms εN and aL in eqs. (17) and (18) can be interpreted as the flow of unemployed as well as underemployed labourers (including migrants) at time t searching for a job in the secondary sector and as the flow of jobs openings at time t respectively, and depend in consequence upon the response of individuals, i.e. labourers and entrepreneurs, to changes in the socio-economic environment, attributed to the economic strategy being implemented.

The diffusion coefficients πN and bL in eqs. (17) and (18) are due to unpredictable system fluctuations generated by the effects of a constellation of only partially known decisions and signals of interacting individuals, often pursuing conflicting goals, and by a sort of confusion of aggregate-relative as well as permanent-transitory character typically attached to intertemporal decisional situations under incomplete and asymmetric information, what hinders individuals to ascertain fully the socio-economic state of affairs, its timing as well as their own position within it.

Simple application of Itô's calculus, see Elliott (1982), Liptser and Shiryayev (1977), and Ikeda and Watanabe (1981), delivers the following stochastic differentials

$$d\lambda = \{(a-\varepsilon) + \pi^2 + b\pi\}\lambda dt + (b-\pi)\lambda dB \tag{19}$$

$$d(\ell^{-1}) = \{\frac{I}{L} - (a + \delta - b^2)\ell^{-1}\}dt - b\ell^{-1}dB \tag{20}$$

with $\lambda(0) = {}^o\lambda$ and $\ell(0) = {}^o\ell$ given.

Gómez (1985b) and Davis and Gómez (1986) deal in some detail with eq. (20) which is the stochastic equation of *per worker capital accumulation*. The drift term of eq. (20), i.e. the coefficient of the differential dt, represents the *actual economic surplus*.

The drift term of eq. (19) indicates whether or not the productive sector is on the way to full employment.

Since our main purpose is to design an economic policy which shifts idle labour into more highly productive activities, we direct our attention to the evolution of k, see eq. (12).

Let us recall that ${}^oN$ includes a fraction of surplus labour that depends on the impulsive control and has to be absorbed into the (secondary) production process during the current time period.

Therefore, we consider k as given by

$$k = \left(\frac{K}{L}\right)\left(\frac{L}{N}\right)$$

and after further application of Itô's calculus we obtain

$$\begin{aligned} dk &= d(\ell^{-1}\lambda) \\ &= \ell^{-1}d\lambda + d(\ell^{-1})\lambda + d(\ell^{-1})\cdot d\lambda \end{aligned}$$

which together with eqs. (19) and (20) and some arrangements gives

$$\begin{aligned} dk = {} & [\{(a-\varepsilon) + \pi^2 + b\pi\} + \{\frac{I}{K} - (a + \delta - b^2)\} \\ & - b(b-\pi)]k\, dt - \pi k\, dB \end{aligned} \tag{21}$$

or, equivalently,

$$dk = [\frac{I}{K} - (\varepsilon + \delta - \pi^2) + 2b\pi]k\, dt - \pi k\, dB \tag{21*}$$

We like to point out that the drift term in eq. (21) consists of three stochastic components: the first one being a drift contribution due to changes in the process (of unemployment elimination) λ, eq. (19), the second one due to the actual economic surplus or the actual process of capital formation, eq. (20), and the third one due to the pure stochastic interaction between the processes λ and ℓ^{-1}.

In order to make this point more apparent, we like to make a few comments on eqs. (19), (20), and (21).

Considering λ, ℓ^{-1} and k as functions of time t and of the canonic sample ω, $\omega \in \Omega$, we can interpret the drift term in eqs. (19), (20), and (21) respectively as the mean conditional forward time-derivative of λ, ℓ^{-1} and k with respect to t, i.e.

$$\dot{\lambda}=\frac{\partial\lambda}{\partial t} = \{(a-\varepsilon)+\pi^2+b\pi\}\,\lambda \tag{22a}$$

$$\dot{\ell}^{-1} = \frac{\partial \ell^{-1}}{\partial t} = \frac{I}{L} - (a+\delta-b^2)\,\ell^{-1} \tag{22b}$$

$$\dot{k} = \frac{\partial k}{\partial t} = [\{(a-\varepsilon)+\pi^2+b\pi\} + \{\frac{I}{K} - (a+\delta-b^2)\} -b(b-\pi)]k \tag{22c}$$

or, equivalently,

$$\frac{\partial k}{\partial t} = i - (\varepsilon+\delta)k + (\pi+2b)\pi k \ . \tag{22*c}$$

Let us denote the relative rate of growth of λ, ℓ^{-1} and k respectively, by $\hat{\lambda}$, $\hat{\ell}^{-1}$ and $\hat{k}$, i.e.

$$\hat{\lambda} = \frac{\frac{\partial\lambda}{\partial t}}{\lambda}\ , \quad \hat{\ell}^{-1} = \frac{\frac{\partial \ell^{-1}}{\partial t}}{\ell^{-1}}\ , \quad \hat{k} = \frac{\frac{\partial k}{\partial t}}{k}\ .$$

This enables us forwarding the following compact relationship

$$\hat{k} = \hat{\lambda} + \hat{\ell}^{-1} - b((b-\pi) \tag{23}$$

which suggests that the continuous stochastic control α, eq. (10), i.e. the choice of the investment-consumption mix should be made so that capital accumulation proceeds at a pace that ensures elimination of unemployment. That is, so that

$$\hat{\lambda} \geq 0 \tag{24a}$$

holds.

Eq. (24a) entails that

$$\hat{k} \geq \hat{\ell}^{-1} \tag{24b}$$

provided that

$$\hat{\lambda} \geq b(b-\pi)\ . \tag{24c}$$

That means, provided that uncertainty does not override the employment trend.

Since the inequality

$$N \geq L \tag{24d}$$

holds for any t, the inequalities

$$k \leq \ell^{-1}, \quad 0 \leq \lambda \leq 1 \tag{24e}$$

are trivial.

Since eq. (21) integrates essential characteristics of the process of capital accumulation with respect to the labour supply and of the process of unemployment elimination, we use it here as the *system dynamics*.

In the next section we look at the problem of the social desirability of the feasible investment-consumption mix.

3. FORMULATION OF THE STOCHASTIC CONTROL PROBLEM

In the foregoing section we introduced various concepts in order to characterize the technology of (the productive sector) of the economy as well as institutional and technological constraints and to identify the body of scope of the present enquiry.

Furthermore, we derived a system dynamics which shall enable us to trace out the effect on employment of additional investments and consequently of a faster capital accumulation as prescribed by the path of the alternative investment-consumption mix.

That means that any path $t \rightarrow (I,C)(t)$ determines a path $t \rightarrow \lambda(t)$ and this in turn induces further changes which we shall trace out and evaluate. Therefore, we associate with the given technology and institutions, i.e. k and λ, a path $t \rightarrow {}_*P_K(t)$, where ${}_*P_K(t)$, called the *accounting price of investment,* is given by

$$ {}_*P_K(t) = -\left(\frac{dc}{di}\right) = -\frac{(\frac{\partial c}{\partial \lambda})}{(\frac{\partial i}{\partial \lambda})} \tag{25}$$

which indicated how much consumption $c = c(k,\lambda)$, at any time t and under the given technology and institutions, has the sector to sacrifice in order to forward an additional unit of investment $i = k(k,\lambda)$ along the employment path λ. Furthermore, one associates with the resulting employment path λ, the path of the marginal utility of the current consumption c, $t \rightarrow U_c(t)$, which together with ${}_*P_K$ gives the function, $t \rightarrow ({}_*P_K \cdot U_c)(t)$, the path of the *utility loss*, due to a goal-adequate increment in the rate of investment or, equivalently, due to the resulting decrease in current consumption, which represents a *social supply price of investment*.

And that faces us with the problem of finding an appropriate *social demand price of investment* that correctly reflects the value to the community of a goal-conducive investment increase, i.e. the social utility derived from the future employment and consumption that shall be made possible by this investment increase.

Although we might appeal to investment (asset-) valuation based on the present value of all future utilities on the grounds that this measure is the best estimate of the value of the increments to output (and in turn of the increments to employment and consumption) attributable to the present additions to the capital stock, we rather resort to a direct assessment based on the flow of instantaneous social utility of aggregate consumption per available labour generated by the alternative investment-consumption mix and on the social value of terminal capital.

For that reason and aware of the fact that any path of the investment-consumption mix (I,C) uniquely determines a path of the couple (s_K,L) or, equivalently, any path of the pair (i,c) uniquely determines a path of the pair (s_K,λ) which we call for simplicity α, i.e. $\alpha = (s_K,\lambda)$, we introduce a social welfare criterion J by means of a map $\alpha(\cdot) \to J(\alpha(\cdot))$ with J given by the functional

$$J(\alpha(\cdot)) = \mathbb{M}\left\{\int_0^\tau U(c(t),\alpha(t))dt + g(k(\tau))\right\} \tag{26}$$

$U(\cdot,\alpha(t))$ and $g(\cdot)$ are strict concave functions of class $C^1(\mathbb{R}^n \to \mathbb{R}_-)$. Hence the functional $J(\alpha(\cdot))$ should make possible to measure the social desirability of the control decision $\alpha(\cdot)$. $\mathbb{M}$ denotes mathematical expectation and τ is a stopping time that we consider here as given. τ may become ∞ which would make (26) meaningless. A properly defined social criterion is introduced in (30) below.

In the long range τ, or, more precisely, τ_j, depends on the impulsive control θ_j according to the map $\theta_j \to \tau_j = \tau(\theta_j)$ where j stands for the j-th time period. However, we suppress the index j, since we deal here with the generic time period. Gómez (1984b) deals with the problem of characterizing the stopping time τ.

The path of α, *the continuous control,* determines uniquely the path of c. Nevertheless, for the sake of clarity, we write the continuous control α explicitly as an argument of U, but we shall omit it sometimes, if no confusion may arise. More precisely, the integrand in eq. (26) should read $U(k,\dot{k},\alpha,t)$.

However, a look at eqs. (5a), (1*), (10), (2*), (3*) and (12) convinces ourselves of the notational convenience and correctness of the integrand in eq. (26).

Usually a discounting factor accounts for the explicit appearance of t in the integrand U.

This would have the advantage of rendering the integral in eq. (26) finite, provided the discount factor or rate of time preference is positive.

Finiteness of the integral in eq. (26) is crucial, since we have to state the short-term optimization problem in such a way that it remains consistently formulated when one goes over to long range analysis.

Besides, we do not know a priori whether τ is finite. Relying on results due to Bensoussan and Lions (1978, 1982), we may assume as well finiteness of τ.

The reason why we do not consider t explicitly as an argument of U is because we assume that in the short-term no decisive structural changes occur, the latter taking place only when one goes from the (j-1)-th time period to the next and this as a consequence of the application of the impulsive control θ_j, see Gómez (1984a). However, we shall consider more general cases in forthcoming papers.

In order to guarantee boundness of the integral in eq. (26) we follow the more natural alternative of considering capital saturation.

In a loosely manner we define a level of *capital saturation* as that level beyond which further capital accumulation is no longer worth according to certain rationale. From this point on we may expect the (capital-labour supply) ratio k, eq. (12), to remain constant independently of the prevailing rationale and we may identify it with a particular steady-state growth path

As in the conventional neoclassic growth theory, we associate particular (s_K,λ) paths with particular steady-state growth paths and translate preference for particular growth paths into preference for specific (s_K,λ) paths.

Therefore, in order to identify the level of capital saturation appropriate to us, we look for a steady-state growth path, among all sustainable steady-state growth paths, along which consumption per available labour attains a maximum.

More precisely, we proceed as follows: We consider eqs. (2) and (3) in normalized form

$$c = (1-s_K)(yk-W\lambda) + (1-s_L)W\lambda \tag{27a}$$

$$i = s_K(yk-W\lambda) + s_L W\lambda \tag{27b}$$

and combine them to obtain

$$c = yk - i$$

which together with eq. (22*c) gives

$$c = yk - \frac{\partial k}{\partial t} - [(\varepsilon + \delta) - (\pi + 2b)\pi]k .$$

Hence, setting $\frac{\partial k}{\partial t}$ equal to zero, since capital saturation characterizes our steady-state, we get

$$c = yk - (\varepsilon + \delta)k + (\pi + 2b)\pi k . \tag{28}$$

Before we continue we like to point out once more that $\frac{\partial k}{\partial t}$ (the same holds for eqs. (22a) and (22b)) is to be interpreted as mean conditional forward time-derivative, i.e.

$$\lim_{h \downarrow 0} \frac{\mathbb{M}[k(t+h,\omega)|\underline{\underline{F}}_t] - \mathbb{M}[k(t,\omega)|\underline{\underline{F}}_t]}{h} .$$

However, we stress that for the sake of simplicity and abusing the notation, we write $\dot{k},\dot{\lambda}$ and so on, unless any confusion may arise.

Among the steady-state growth paths that satisfy eq. (28) we are interested precisely in the one that maximizes consumption.

Therefore, we shall look for conditions characterizing the maximum sustainable level of c, where c is to be taken as a function of k and λ with $\lambda \leq 1$, i.e. $c = c(k,\lambda)$.

Differentiating eq. (28) we get

$$\frac{\partial c}{\partial \lambda} = y' \tag{29a}$$

$$\frac{\partial c}{\partial k} = y - y'\ell - (\varepsilon + \delta) + (\pi + 2b)\pi . \tag{29b}$$

Eq. (29a) says that whenever the marginal productivity of labour is positive for all techniques of production, λ must be equal to one which entails the obvious result that maximum consumption requires full employment.

The level of capital compatible with the maximum sustainable level of c, we call it c^*, has to be, according to eq. (29b), precisely the level of capital k^* at which the marginal productivity of capital Y_K, see eq. (16b), renders $\frac{\partial c}{\partial k}$ equal zero, i.e.

$$Y_K = y - y'\ell = (\varepsilon + \delta) - (\pi + 2b)\pi . \tag{29c}$$

Let us denote by U^* the level of social utility corresponding to the maximum sustainable level of consumption c^*, i.e. $U^* = U(c^*)$.

We shall call k^*, c^* and U^* the *golden values* of k, c and U respectively, since they describe a smooth, steady-state growth path with full employment which we call a golden age stressing thereby its *mythical nature,* see Robinson (1856, 1962). Thus, defining k^* by means of eq. (29c), which we shall call the *golden rule of capital accumulation,* we see that whenever ${}^0k \leq k^*$, then it follows that $k(t) \leq k^*$ for all $t \geq 0$.

Hence,

$$c \leq y(\ell^*)k^* \qquad \text{with} \qquad \ell^* = (k^*)^{-1}$$

and

$$U(c) \leq U^* .$$

Thus, we can, following Koopmans (1965), take the golden rule path as a suitable reference path from which to measure social utility by means of the following meaningful social welfare criterion J^*

$$J^*(\alpha(\cdot)) = \mathbb{M} \left\{ \int_0^\tau [U(c,\alpha) - U^*]dt + g(k(\tau)) \right\} . \tag{30}$$

Now, our control problem can be formulated as follows:

Consider the *completely observable control problem:*

$$\sup_{\alpha \in A} J^*(\alpha) \tag{31}$$

subject to eqs. (21) and (27a) and the constraints given by eqs. (9). (11) and (24). $J^*(\alpha)$ is given by eq. (30) and A stands for the set of admissible control α and will be specified in the following section, where we deal in some detail with the solution of (31), prove existence and uniqueness and obtain the adjoint process using a stochastic maximum principle due to Bensoussan (1983a).

We like to close this section recalling a beautiful and enlightening parable due to Marglin (1976) concerning the question why should the maximum sustainable level of consumption per available labour c^* be finite.

"Consider the head of a peasant family to whom the king makes the following offer: 'You can have as much land, ready for the plough, as you like. On one condition. You must clear a sufficient amount of additional land to bequeath to your progeny an equal amount of land, per capita as has been given to you. (In other words, 'give to your progeny as has been given to you'. Whence the name golden rule). If the size of this family remains constant, there need be no bound to the present generation's desire for land. But if the family is growing and it takes time and energy to clear new land, greed for the land's product will inhibit its demand for land. For the larger the initial endowment, the more land it must be cleared to maintain the per capita size of its holding. And the more time it spends in clearing the land, the less time it has to till its original holding. Intuitively, the marginal cost of satisfying the bequest constraint and the marginal benefit of a larger holding must balance at a holding of finite rather than infinite size".

In order to translate the message of Marglin's parable to our context we first substitute the capital ratio k for land and then look at the golden rule as given by eq. (29c) which says that instead of increasing the level of capital to the point that its marginal productivity Y_K fell to zero, the economy would optimally restrict itself to the more modest level k^* at which capital accumulation allows for replacement of the capital stock used up and for endowment of subsequent generations at the prevailing standards, i.e. $(\varepsilon + \delta)$, and for unpredictable fluctuations and informational confusion of the sort mentioned in the remarks following eqs. (17) and (18).

A final comment concerns the social utility U and the terminal pay-off g .

Once again, we follow to some extent Marglin's line of argumentation, see Marglin (1976), when examining the alternative available to a government willing to direct labour force from low productivity employment to more highly productive employment and look at the consequences of such a policy for equilibrium in the market for consumer goods.

If the employment increase makes the marginal productivity of labour fall beyond its marginal cost, the incremental wage income will exceed the incremental output.

Now, the question is, how such an employment policy should be complemented, so that the expansion of the wage bill does not disturb the balance between the demand and supply of consumer goods.

Some policy makers who believe themselves guardians of the status quo, may decide for making up the effect of the employment expansion by keeping down the real wages by means of fiscal or monetary policies, i.e. inflation or taxation. This might be suitable to a goal of aggregate growth but not to one of distributive justice. Others may be inclined to higher taxes on the middle and upper classes which is likely to be opposed, as Marglin (1976) put it, "in the name of anticommunism, preservation of incentives, or some such 'virtue' that is in reality but a mask for special privilege".

The alternative that attracts our attention is that of accompanying any expansion of employment with a shift in the composition of national product from consumption to investment, or, in other words, from claims on current consumption to claims on future consumption, at least equal to the amount by which the employment-induced expansion of consumption exceeds the expansion of output.

But this poses a further problem for a responsible government, for policy makers have now not only to resist the disproportionate influence of those deriving advantages from the established power, but furthermore they have to reshape the prevailing patterns of time preference, social solidarity, motivational behaviour and the like, so that the resulting development policy can be successfully implemented. Therefore, we assume the existence of a *system of social preferences* and *value judgements* by means of which the government forwards, to the limits of political strength and courage, his views and attitudes towards poverty, unemployment, redistribution of consumption, social justice, etc.

Thus, we assume that U describes, according to the government, the social utility flows attained by the *representative man* and g measures the social value attached to the terminal capital, more explicitly, g penalizes terminal deviations from k^*, as we shall see in the next section, i.e. eq. (77).

4. A VARIATIONAL APPROACH TO THE OPTIMAL ALLOCATION OF LABOUR

This section consists of two parts. In the first one we give a rigorous derivation of the stochastic maximum principle relying heavily on results due to Bensoussan (1982b, 1983a).

Bensoussan's powerful and transparent approach relies mainly on variational methods which are very similar to those used in deterministic theory and recovers most of the results which exist in the literature by more elementary methods.

Although stochastic control under partial observations of the state variable would suit better to economic reality and econometric practice, we restrict ourselves to the case of full information. The reason why is because the optimal control of partially observed diffusions is mathematically more involved, needs an infinite dimensional setting and therefore shall be the subject of a forthcoming paper. See Bensoussan (1983b), Fleming (1982), and Fleming and Pardoux (1982).

However, we let the control enter into the diffusion term which represents an improvement against Davis and Gómez (1986) who deal with a similar economic problem.

In the second part of the present section we apply the obtained results to the stochastic control problem formulated in Section 3 and give economic interpretation to several variables and equations with the purpose of stressing the economic meaning of the solution of the control problem and of making ready the ground for a clear treatment of issues concerning attainability and reversibility to be dealt with in Section 5.

4.1. A Rigorous Derivation of the Stochastic Maximum Principle

4.1.1. Notation and setting of the problem. Let $(\Omega, \underline{\underline{F}}, \mathbb{P})$ be a probability space, endowed with a filtration $\underline{F} = (\underline{\underline{F}}_t)_{t\geq 0}$, on which we can construct a standardized $\underline{\underline{F}}_t$-Wiener process $w = (w(t))_{t\geq 0}$ with values in $\mathbb{R}^n$. (In particular, $w(t)$ is a $\underline{\underline{F}}_t$-martingale.)

We assume that

$$\underline{\underline{F}}_t = \sigma(w(s),\ s \leq t)\ . \tag{32}$$

Let $f(\xi,\alpha)$ and $\sigma(\xi,\alpha)$ be such that:

f is continuous from $\mathbb{R}^n \times \mathbb{R}^m \to \mathbb{R}^n$, and continuously differentiable with respect to ξ,α. (33a)

σ is continuous from $\mathbb{R}^n \times \mathbb{R}^m \to L(\mathbb{R}^n;\mathbb{R}^n)$ and continuously differentiable with respect to ξ,α. (33b)

f_ξ, f_α, σ_ξ and σ_α are bounded functions. (33c)

Note that from (33a) - (33c) it follows that:

$$\begin{aligned} |f(\xi,\alpha)| &\leq c(1 + |\xi| + |\alpha|) \\ |\sigma(\xi,\alpha)| &\leq c(1 + |\xi| + |\alpha| \end{aligned} \tag{33d}$$

c is a deterministic constant.

We assume further that

$$A_{ad} = \text{non-empty closed convex subset of } \mathbb{R}^m \tag{34a}$$

and define the space $L_F^2(0,T)$ as

$$L_F(0,T) = \{\alpha \equiv \alpha(t,\omega) \in L^2(\Omega \times (0,T), d\mathbb{P} \otimes dt;\ \mathbb{R}^m):$$
$$\text{a.e. } t,\ \alpha(t) \in L^2(\Omega, \underline{\underline{F}}_t, \mathbb{P}; \mathbb{R}^m)\} \tag{34b}$$

which is a sub-Hilbert space of L^2. We set

$$\mathbb{A} = \{\alpha \in L_F^2(0,T) : \alpha(t) \in A_{ad},\ \text{a.e., a.s.}\} \tag{34c}$$

and $\mathbb{A}$ is a closed convex subset of $L_F^2(0,T)$. An element α of $\mathbb{A}$ shall be called an *admissible control* for any finite $T \in \mathbb{R}_+$. $\mathbb{A}$ is called the space of admissible controls.

4.1.2. The state equation. With any admissible control $\alpha(\cdot) \in \mathbb{A}$, characterized by (34), we associate a process ξ according to the following theorem. $L_F^2(0,T;\mathbb{R}^n)$ denotes the sub-space of $L^2(\Omega \times (0,T), d\mathbb{P} \otimes dt; \mathbb{R}^n)$ consisting of processes z such that a.e.t, $z(t) \in L^2(\Omega, \underline{\underline{F}}_t, \mathbb{P}; \mathbb{R}^n)$.

Theorem 4.1. Under the assumptions (32) and (33), there exists, for any admissible control $\alpha(\cdot)$ characterized by (34) and any T finite, one and only one process $\xi(t)$ satisfying

$$\left.\begin{aligned} d\xi &= f(\xi(t),\alpha(t))\,dt + \sigma(\xi(t),\alpha(t))\,dw(t) \\ \xi(0) &= {}^o\xi \\ \xi &\in L_F^2(0,T;\mathbb{R}^n) \cap L^2(\Omega, \underline{\underline{F}}, \mathbb{P}; C(0,T;\mathbb{R}^n)) \end{aligned}\right\} \tag{35}$$

where ${}^o\xi \in \mathbb{R}^n$.

Proof: See Bensoussan (1982a).

4.1.3. The stochastic control problem. Let the utility functional U and the terminal pay-off functional g be such that:

$$\left.\begin{aligned} U(\xi,\alpha) &: \mathbb{R}^n \times \mathbb{R}^m \longrightarrow \mathbb{R}_- \\ g(\xi) &: \mathbb{R}^n \longrightarrow \mathbb{R}_- \end{aligned}\right\} \tag{36}$$

where U and g are Borel, strict concave, continuously differentiable with respect to (ξ,α) and ξ respectively; U_ξ, U_α and g_ξ hold the following conditions

$$|U_\xi(\xi,\alpha)| \le c(|\xi| + |\alpha| + 1) \tag{37a}$$

$$|U_\alpha(\xi,\alpha)| \le c(|\xi| + |\alpha| + 1) \tag{37b}$$

$$|g_\xi| \le c(|\xi| + 1) \tag{38}$$

$$U(0,0) \in L^{\infty}(0,T) \tag{37c}$$

We define the goal functional

$$J(\alpha(\cdot)) = \mathbb{M}\left\{\int_0^T U(\xi(t),\alpha(t))dt + g(\xi(T))\right\}. \tag{39}$$

Our task is to characterize an admissible control by means of which $J(\cdot)$ attains a supremum, i.e. the control problem can be formulated as: Find an optimal control β satisfying (34), such that

$$J(\beta(\cdot)) = \sup_{\alpha \in A} J(\alpha(\cdot)). \tag{40}$$

4.1.4. The Gâteaux derivative of the functional J. We shall denote by $\beta(\cdot)$ an optimal control satisfying (34) and by $x(\cdot)$ the corresponding state

$$\left.\begin{aligned} &dx = f(x(t),\beta(t))dt + \sigma(x(t),\beta(t))dw(t) \\ &x(0) = {}^{o}x \\ &x \in L^2_F(0,T;\mathbb{R}^n) \cap L^2(\Omega,\underline{\underline{F}},\mathbb{P};C(0,T;\mathbb{R}^n)). \end{aligned}\right\} \tag{41}$$

We obtain the following result:

Lemma 4.1. The functional $J(\alpha(\cdot))$ is Gâteaux differentiable and the following formula holds

$$\begin{aligned} &\frac{d}{d\theta} J(\beta(\cdot) + \theta\alpha(\cdot))\Big|_{\theta=0} = \\ &\quad \mathbb{M}\left\{\int_0^T [U_\xi(x(t),\beta(t))z(t) + U_\alpha(x(t),\beta(t))\alpha(t)]dt \right. \\ &\quad \left. + g'_\xi(x(T))z(T)\right\} \end{aligned} \tag{42}$$

where z is the solution of the linearized version of the state eq. (41), i.e.

$$\left.\begin{aligned} dz = {} & [f_\xi(x(t),\beta(t))z(t) + f_\alpha(x(t),\beta(t))\alpha(t)]dt \\ & + \sum_{j=1}^{m} (\sigma^j_\xi(x(t),\beta(t))z(t) + \sigma^j_\alpha(x(t),\beta(t))\alpha(t))dw^j(t)) \\ z(0) = {} & {}^{o}z = 0 \end{aligned}\right\}$$

$$z \in L^2_F(0,T;\mathbb{R}^n) \cap L^2(\Omega,\underline{\underline{F}},\mathbb{P}\,;\,C(0,T;\mathbb{R}^n))\quad\Big] . \qquad (43)$$

Proof: It is an adaptation of Bensoussan (1983a). ∎

4.1.5. <u>Abstract definition of the adjoint processes p and r.</u> Let φ and $\psi^1,\psi^2,\psi^1,\ldots,\psi^m$ be in $L^2_F(0,T;\mathbb{R}^n)$ with

$$\sum_{j=1}^{m} \lambda^j \,\mathbb{M} \int_0^T |\psi^j(t)|^2 dt < \infty \quad . \qquad (44)$$

Consider ζ to be the solution of

$$\left.\begin{aligned} d\zeta &= (f_\xi(x(t),\beta(t))\zeta(t) + \varphi)dt \\ &\quad + \sum_{j=1}^{m} (\sigma^j_\xi(x(t),\beta(t))\zeta(t) + \psi^j(t))dw^j(t) \\ \zeta(0) &= {}^o\zeta = 0 \\ \zeta &\in L^2_F(0,T;\mathbb{R}^n) \cap L^2(\Omega,\underline{\underline{F}},\mathbb{P}\,;C(0,T);\mathbb{R}^n)) \end{aligned}\right\} . \qquad (45)$$

It is easy to check that the map $\varphi,\psi^1,\psi^2,\ldots,\psi^m \to \zeta$ is linear and continuous from $L^2_F(0,T;\mathbb{R})^{m+1}$ in $L^2_F(0,T;\mathbb{R}^n) \cap L^2(\Omega,\underline{\underline{F}},\mathbb{P}\,,C(0,T;\mathbb{R}^n))$ We consider then the functional

$$\varphi,\psi^1,\psi^2,\ldots,\psi^m \to \mathbb{M}\left\{\int_0^T U_\xi(x(t),\beta(t))\zeta(t)dt + g'_\xi(x(T))\zeta(T)\right\}$$

which in turn is linear and continuous on $L^2_F(0,T;\mathbb{R}^n)^{m+1}$. Therefore, we can define in a unique way stochastic processes

$$\left.\begin{aligned} &p(\cdot) \text{ and } r^1(\cdot),r^2(\cdot),\ldots,r^m(\cdot) \text{ in } L^2_F(0,T;\mathbb{R}^n) \\ &\text{with} \\ &\sum_{j=1}^{m} \lambda^j \,\mathbb{M} \int_0^T |r^j(t)|^2 dt < \infty \end{aligned}\right\} \qquad (46)$$

such that the following relation holds

$$\mathbb{M}\left\{\int_0^T p(t)\varphi(t)dt + \sum_{j=1}^{m} \int_0^T r^j(t)\,\psi^j(t)dt\right\}$$

$$= \mathbb{M} \left\{ \int_0^T U_\xi(x(t),\beta(t))\ \zeta(t)dt + g_\xi'(x(T))\zeta(T) \right\} . \tag{47a}$$

Applying this result to (43), we obtain from (47a)

$$\mathbb{M} \left\{ \int_0^T U_\xi(x,\beta)z\ dt + g_\xi'(x(T))z(T) \right\}$$

$$= \mathbb{M}\left\{ \int_0^T [pf_\alpha(x,\beta)\alpha + \sum_{j=1}^m r^j\sigma^j(x,\beta)\alpha]dt \right\} . \tag{47b}$$

Eq. (47b), together with Lemma 4.1, gives the following lemma.

Lemma 4.2. We have the formula

$$\frac{d}{d\theta} J(\beta(\cdot) + \theta\alpha(\cdot))\Big|_{\theta=0} =$$

$$\mathbb{M} \left\{ \int_0^T (U_\alpha(x,\beta) + p\ f_\alpha(x,\beta) + \sum_{j=1}^m r^j\sigma_\alpha^j(x,\beta))\alpha dt \right\}$$

$$= \mathbb{M} \int_0^T H_\alpha(x,\beta,p,r)\alpha dt \tag{48}$$

where

$$H(\xi,\alpha,p,r) = U(\xi,\alpha) + pf(x,\beta) + \sum_{j=1}^m r^j\sigma^j(\xi,\alpha).$$

4.1.6. The stochastic maximum principle. Theorem 4.2. We assume (32), (33), (34), (36), (37) and (38). Then, for $\beta(\cdot)$ to be a solution of the optimal control problem defined by (39), (40) and (41), and denoting by $x(\cdot)$ the corresponding trajectory, and defining p and r^j in a unique way as in (46) and (47), it is necessary and sufficient that the following condition holds

$$\frac{d}{d\theta} J(\beta(\cdot) + \theta\alpha(\cdot))\Big|_{\theta=0} \le 0. \tag{49}$$

Taking into account Lemma 4.2, condition (48) entails the following two equivalent conditions

$$\text{(a)} \qquad \left. \begin{array}{l} \mathbb{M} \displaystyle\int_0^T (H_\alpha(x,\beta,p,r),\alpha(t) - \beta(t))dt \le 0 \\ \quad \forall\ \alpha(\cdot) \ \text{ satisfying (34)} \end{array} \right\} \tag{50}$$

(b)
$$\left.\begin{aligned}&(H_\alpha(x,\beta,p,r)\alpha - \beta(t)) \le 0\\ &\text{a.e.t.,}\quad dt \otimes d\mathbb{P}\text{-a.s.}\\ &\forall\ \alpha(\cdot)\quad \text{satisfying (34)}\end{aligned}\right\} . \tag{51}$$

Proof: It follows from formula (49) and a classical localization argument. See Bensoussan (1978), for instance. ■

4.1.7. Derivation of the equation for the adjoint processes p and r.

Theorem 4.3. We make the assumptions of Theorem 4.2 and (32). Then the processes p and r^j satisfy

$$\left.\begin{aligned}&-dp = H_\xi(x,\beta,p,r)dt - \sum_{j=1}^m r^j\, dw^j(t)\\ &p(T) = g_\xi(x(T))\end{aligned}\right\} \tag{52}$$

$$\left.\begin{aligned}&p \in L^2_F(0,T;\mathbb{R}^n) \cap L^2(\Omega,\underline{\underline{F}},\mathbb{P};C(0,T;\mathbb{R}^n))\\ &r^j \in L^2_F(0,T;\mathbb{R}^n),\ \sum_{j=1}^m \lambda^j \mathbb{M}\int_0^T |r^j(t)|^2 dt < \infty\end{aligned}\right\} . \tag{53}$$

Proof: Let us denote by Φ, with $\Phi \in L(\mathbb{R}^n,\mathbb{R}^n)$, the fundamental solution of (43).

Then, Φ satisfies

$$\left.\begin{aligned}&d\Phi = f_\xi(x,\beta)\Phi dt + \sum_{j=1}^m \sigma^j_\xi(x,\beta)\Phi dw^j(t)\\ &\Phi(0) = I\end{aligned}\right\} \tag{54}$$

where I denotes the identity matrix.

Further, let us denote by Ψ the adjoint of Φ.

Then, Ψ satisfies

$$\left.\begin{aligned}d\Psi = &[-\Psi f_\xi(x,\beta) + \Psi \sum_{j=1}^m \sigma^j_\xi \sigma^j_\xi]dt\\ &- \sum_{j=1}^m \Psi\sigma^j_\xi(x,\beta)dw^j(t)\\ \Psi(0) = &\ I\end{aligned}\right\} \tag{55}$$

as one easily checks using the Itô rule and the fact that

$$\Phi\Psi = I .$$

$$p(s) = \Lambda(s)\Psi(s). \tag{63}$$

The differential associated with (62a) is

$$d\Lambda = -U_{\alpha}(x,\beta)\Phi(t)dt + \sum_{j=1}^{m} \eta^{j}(t)\, d\, w^{j}(t). \tag{62b}$$

Now we compute the Itô differential of (63) as follows

$$dp = (d\Lambda)\Psi + \Lambda d\Psi + d\Lambda d\Psi . \tag{64}$$

Substituting (62b) and (55) in (64), and recalling that $\Phi\Psi = I$ we get

$$\begin{aligned}-dp = (U_{\xi}(x,\beta) + p(t)f_{\xi}(x,\beta) + \sum_{j=1}^{m} r^{j}\sigma_{\xi}^{j}(x,\beta))dt \\ - \sum_{j=1}^{m} r^{j}\, d\, w^{j}(t)\end{aligned} \tag{65}$$

where

$$r^{j}(t) = \eta^{j}(t)\Psi(t) - p(t)\sigma_{\xi}^{j}(x,\beta)$$

which under the notation introduced in Lemma 4.2 gives

$$\left.\begin{aligned}-dp &= H_{\xi}(x,\beta, p,r)dt - \sum_{j=1}^{m} r^{j}\, d\, w^{j}(t) \\ p(T) &= g_{\xi}(x(T))\end{aligned}\right\} . \tag{66}$$

It remains to see that

$$\mathbb{M} \int_{0}^{T} (p\varphi + \sum_{j=1}^{m} r^{j}\psi^{j})dt = \mathbb{M}\left\{ \int_{0}^{T} U_{\xi}(x,\beta) + g_{\xi}'(x(T))\zeta(T)\right] .$$

To prove that we compute the Itô differential

$$dp\zeta = (dp)\zeta + pd\zeta + dpd\zeta . \tag{67}$$

Then, after substituting (45) and (65) in (67) and some arrangements, we obtain

$$\mathbb{M} g_{\xi}'(x(T))\zeta(T) = \mathbb{M} \int_{0}^{T} (p\varphi - U_{\xi}(x,\beta)\zeta + \sum_{j=1}^{m} r^{j}\psi^{j})dt$$

and this concludes the proof. ■

$$+ \int_0^T g'_\xi(x(T))\Phi(T) \int_0^T \Psi(s)\varphi(s)ds\Big\} . \tag{59b}$$

The right-hand side of (59b) can be rewritten further using the properties of expectation as

$$\mathbb{M}\int_0^t p(t)\varphi(t)dt = \mathbb{M}\Big\{\int_0^T ds \Big(\mathbb{M}_{\underline{\underline{F}}_t}\Big[g'_\xi(x(T))\Phi(T) + \int_s^T U_\xi(x,\beta)\Phi(t)dt\Big]\Psi(s)\Big)\varphi(s)\Big\} . \tag{59c}$$

But (59c) suggests the following representation for $p(s)$

$$p(s) = \Big(\mathbb{M}_{\underline{\underline{F}}_s}\Big[g'_\xi(x(T))\Phi(T) + \int_s^T U_\xi(x,\beta)\Phi(t)dt\Big]\Big)\Psi \tag{60}$$

a.s. $d\mathbb{P}\otimes dt$.

One can easily check that $p(s)$ is a $\underline{\underline{F}}_s$-semi-martingale, local, square integrable and right-continuous.

Now, we set

$$X = g'_\xi(x(T))\Phi(T) + \int_0^T U_\xi(x,\beta)\Phi(t)dt .$$

Therefore, using the Kunita-Watanabe integral representation for local semi-martingales, we obtain

$$\mathbb{M}_{\underline{\underline{F}}_s} X = \mathbb{M}\, X + \int_0^s \sum_{j=1}^m \eta^j(t)dw^j(t) \tag{61a}$$

where the η^j are $\underline{\underline{F}}_s$-processes such that

$$\int_0^s |\eta^j(t)|^2 dt < \infty \qquad \text{a.s.} \tag{61b}$$

Thus, setting

$$\Lambda(t) = \mathbb{M}\, X + \int_0^t \sum_{j=1}^m \eta^j(s)dw^j(t) - \int_0^t U_\alpha(x,\beta)\Phi(s)ds \tag{62a}$$

we get from (60) the following expression for p

Now, let us consider the following particular case of (45)

$$d\zeta = (f_\xi(x,\beta)\zeta + \varphi)dt + \sum_{j=1}^{m} \sigma_\xi^j(x,\beta)\zeta dw^j(t) \qquad (56)$$
$$\zeta(0) = {}^o\zeta = 0.$$

Then, equality (47a) becomes

$$\mathbb{M} \int_0^T p(t)\ \varphi(t)dt$$
$$= \mathbb{M}\left\{ \int_0^T U_\xi(x,\beta)\ \zeta\, dt + g'_\xi(x,T))\zeta(T)\right\} . \qquad (57)$$

Let us compute the Itô differential

$$d\Psi\zeta = (d\Psi)\zeta + \psi d\zeta + d\psi\ d\zeta$$

which together with (55) and (56) gives

$$d\Psi\zeta = \Psi(t)\varphi(t)dt$$

and further

$$\Psi(t)\zeta(t) = \int_0^t \Psi(s)\varphi(s)ds.$$

Hence, we have the following representation for ζ

$$\zeta(t) = \Phi(t) \int_0^t \Psi(s)\varphi(s)ds. \qquad (58)$$

Now, substituting (58) in (57), we obtain

$$\mathbb{M}\int_0^T p(t)\varphi(t)dt = \mathbb{M}\left\{ \int_0^T dt\ U_\xi(x,\beta)\Phi(t) \int_0^t \Psi(s)\varphi(s)ds\right.$$
$$\left. + g'_\xi(x(T))\Phi(T) \int_0^T \Psi(s)\varphi(s)ds\right\} . \qquad (59a)$$

But (59a) can be rewritten as

$$\mathbb{M}\int_0^T p(t)\varphi(t)dt = \mathbb{M}\left\{ \int_0^T ds \int_s^T dt\ U_\xi(x,\beta)\Phi(t)\Psi(s)\varphi(s)\right.$$

4.2. The Optimal Path of Labour Allocation

4.2.1. Some preliminary economic interpretations. We have established the mathematical results we need in order to characterize an optimal control and the corresponding path of labour allocation associated with the stochastic control problem formulated in Section 3, see (31).

To take advantage of the power of the stochastic maximum principle condensed in Theorem 4.2 and Theorem 4.3, we proceed to identify the correspondence between variables and equations in Sections 3 and 4.

We have intentionally used almost the same notation. However, we need some remarks.

Since in Section 3 we deal mainly with capital accumulation in the productive sector, we have to consider the process ξ in (35) as a 1-dimensional real-valued process which corresponds to the capital accumulation process k given by (21). See also (22c*).

Hence, the following correspondence follows

$$f(\xi,\alpha) = i - (\varepsilon + \delta)k + (\pi + 2b)\pi k \tag{68a}$$

$$\sigma(\xi,\alpha) = -\pi k. \tag{68b}$$

Furthermore, $U(\cdot)$ given by (36) in Section 4 corresponds to $U(\cdot)-U^*$, since we like to measure deviation from a Bliss value U^*. See the paragraph following (29c). The social value of investment is described by the process p which evolves according to eq. (52). To understand the rôle played by this equation, we need a little elaboration. First of all, let us have a look at the drift term, i.e. H_k under the present notation. Before we go further, we like to stress the fact that the diffusion coefficients in Section 3 do depend on the control α but that we suppress its use for notational simplicity, e.g. we use k instead of k^α. The Hamiltonian H, obtained in Section 4, under the corresponding notation is given by

$$H(k,\alpha,p,r) = (U(k,\alpha) - U^*) + pf(k,\alpha) - r\pi k \tag{69a}$$

that according to several remarks in Section 3 can be rewritten as

$$H(i,c,p,r) = (U(c) - U^*) + pf(i) - r\pi k. \tag{69b}$$

The Hamiltonian in its form (69b) articiculates, in the form of isoquants, the existing system of social preferences and value-judgements and orders completely alternative combinations of investment i and consumption c. In a loosely way, we may write (69b) simply as $H(i,c)$. Sen (1970, 1982, 1984) deals admirably with questions related to the determination of such social orderings.

On the other hand, the Hamiltonian as given by (69a) measures the total social utility of the representative man associated with the system at any time t, $t \in [0,\tau]$.

We call the first term $U(k,\alpha) - U^*$ and the second $pf(k,\alpha)$ the potential and the kinetic social utility respectively, since they resemble the concepts of potential and kinetic energy in classical physics. The last term $-r\pi k$ accounts for the social utility or social cost due to the system uncertainty or, better, it represents the cost in terms of consumption utility entailed by the risk associated with the control policy at issue.

The logic within the Hamiltonian works as follows: As we have seen, the social utility functional U and therefore also $U - U^*$ is responsible for social preferences, attitudes, value-judgements and so forth. Thus, if the economy control policy is one under which the representative man is expected to postpone certain amount of current consumption, which amounts to giving up certain quantity of current utility, then the second term transforms this amount of current consumption, now in the form of potential future consumption or, equivalently, in the form of investment, into future utility.

For $f(k,\alpha)$ is the drift term of the system dynamics and indicates the ability of the system to generate economic surplus.

Besides, let us recall that $f(k,\alpha)$ shall be interpreted as the mean conditional forward derivative, i.e. $\partial k/\partial t = f(k,\alpha)$.

The last term is unpredictable, it may favour or inhibit the first two terms depending on the sign of π which accounts for unpredictable fluctuations and confusions mostly related to the flow of less productive workers towards the advanced sector. See eqs. (19), (20) and (21).

Summing up, the Hamiltonian as given by eq. (69a), represents the total (net) social utility associated with the control policy α. Thus, eq. (52) gives us, loosely speaking, the total change of the social value of investment due to a change, at time t, in the allocation of output, between consumption and investment, favouring capital formation. The drift term H_k provides the time evolution, see eq. (74a), of the change of the social utility generated by the change just mentioned in the capital endowment and the diffusion term the social disutility, cost or risk, associated with it.

4.2.2. Static characterization of the optimal control policy. Let us denote by α^o the optimal control β of Section 4.1 and by k^o, c^o, i^o and so on the corresponding optimal trajectories of k, c and i respectively. Thereby eq. (69a) becomes

$$H(k^o,\alpha^o,p,r) = (U(c^o,\alpha^o) - U^*) + pf(i^o,\alpha^o) - r\pi k^o. \qquad (69a^*)$$

Here we use the fact that c and i are functionals of k, see (27). Since the Hamiltonian has a maximum at $i^o = i^o(t)$, the control set R^1A is all of $\mathbb{R}_+$ and since H is differentiable in i^o, we must have

$$0 = \frac{dH}{di^o} = -\frac{\partial U}{\partial c^o}\left(-\frac{dc^o}{di^o}\right) + p \quad .$$

Here we use the fact that the Hamiltonian associated with an optimal policy is constant. Now, taking into account (25), (48) and (51), we get

$$p(t) = U'_{c^o}(t)\,{}_*P_K(t) \tag{70}$$

a.e.t, $d\mathbb{P} \otimes dt$ -a.s.

Hence, (70) holds for all $t \in [0,\tau]$ with possible exceptions on $dt \otimes d\mathbb{P}$-null sets. For that reason it is a moment-to-moment relation known in dynamic economics as the *dynamic efficiency condition*. And means that a.s. at any t the social utility of the representative man, derived from the decision of the economy to invest according to i^o, should equate the consumption utility loss of the representative man associated with the consumption she or he has to sacrifice in order to further investments as the control α^o requires.

The dynamic efficiency condition given by (70) amounts to the well-known tangency condition between the investment-consumption transformation functional and the investment-consumption utility substitution functional articulated by means of the family of Hamiltonians (isoquants) $\{H(i,c,p,r)\}$ which in turn defines a *social demand price of investments in terms of current consumption*.

The tangency condition follows easily from (70) which we can rewrite as

$$ {}_*P_K(t) = \frac{p(t)}{U'_{c^o}(t)} = \frac{\frac{\partial H}{\partial i^o}(t)}{\frac{\partial H}{\partial c^o}(t)} \quad . \tag{71}$$

Eq. (71) becomes evident recalling that ${}_*P_K(t)$ is given by (25) and taking into account eqs. (69a) and (68a) on the one hand, and eq. (69b) on the other. From the stochastic maximum principle, Theorem 4.2 and the fact that α^o represents the optimal control, it follows that

$$H(k^o,\alpha^o,p,r) = \max_{\alpha \in A} \{H(k,\alpha,p,r)\} \ . \tag{72}$$

Hence, for fix t and taking into account the differentiability of $H(k,\alpha,p,r)$, Lemma 4.1, one obtains from (72) *static first order conditions of optimality* that fully describe the following three phases the economy undergoes in every time period before entering the neoclassical era:

Phase I $\quad \lambda^o(t) < 1 \qquad s^o_k(t) = \bar{s}_k$

$$\frac{p}{U'_{c^o}} = \frac{(s_K - s_L)W + (1-s_K)y'}{(s_K - s_L)W - s_K\, y'} \tag{73a}$$

a.e.t, $dt \otimes d\mathbb{P}$-a.s.

Phase II $\quad \lambda^o(t) = 1 \qquad s^o_k(t) = \bar{s}_k$

$$1 \le \frac{p}{U'_{c^o}} \le \frac{(s_K - s_L)W + (1-s_K)y'}{(s_K - s_L)W - s_K\, y'} \tag{73b}$$

a.e.t, $dt \otimes d\mathbb{P}$-a.s.

Phase III $\lambda^o(t) = 1$

$$s_k^o(t) < \bar{s}_k \qquad \frac{p}{U'_{c^o}} = 1 \tag{73c}$$

a.e.t, $dt \otimes d\mathbb{P}$-a.s.

The Hamiltonian, together with the initial conditions and constraints, determines whether the economy finds itself in phase I, II or III.

As we shall see in Section 5, the economy optimally develops by moving from phase I to phase II and from phase II to phase III when it starts from a capital intensity ${}^ok$ which is low enough. However, the economy not always has to begin with phase I. A sufficiently large initial endowment of capital ${}^ok$ may put the economy also in phase II or even in phase III.

The phases are to be interpreted as follows: If optimality dictates full employment, i.e. $\lambda^o(t) = 1$, or unemployment, i.e. $\lambda^o(t) < 1$, and binding investment policy, i.e. $s_K^o(t) = \bar{s}_K$, or not, i.e. $s_K^o(t) < \bar{s}_K$, which case holds is indicated at the left-hand side under the corresponding phase heading, then the relative social desirability of the couple $(s_K^o(t),\lambda(t))$ or alternatively of the resulting investment-consumption mix $(i^o(t),c^o(t))$ has to be measured by means of the corresponding weight ${}_*P_K(t)$ resulting from the first-order conditions, see eqs. (73).

4.2.3. Dynamic characterization of the optimal control policy. At this stage we like to point out that the Maximum Principle condensed in Theorems 4.2 and 4.3 enables us splitting the intertemporal optimization problem (40) into a static, i.e. (72), and a dynamic one, i.e. (76) and (77). In other words, the Maximum Principle allows a time decentralization of the decision process and this is extremely convenient for applications. Having considered the static features in the foregoing sub-section, let us look at the dynamic aspects of optimality.

From (52) in Theorem 4.3 we obtain under the notation of the present section

$$-\frac{\partial p}{\partial t} = H_k \tag{74a}$$

where the left-hand side in (74a) has to be interpreted as the mean forward conditional derivative with respect to time, i,e. $\partial p/\partial t$ is given by

$$\lim_{h\downarrow 0} \frac{\mathbb{M}[p(t+h,\omega)|\underline{\underline{F}}_t] - \mathbb{M}[p(t,\omega)|\underline{\underline{F}}_t]}{h} .$$

Taking into account (27), (68), and (69), and after a few arrangements, (74a) becomes

$$\frac{\frac{\partial p}{\partial t}}{p} + [(\varepsilon + \delta - \pi^2) - 2b\pi] + r\pi$$

$$= \frac{(1-s_K) + s_K(\frac{p}{U'_c})}{(\frac{p}{U'_c})}\;(y - y'\ell)$$

$$= \frac{{}_*P_\pi (y - y'\ell)}{{}_*P_K} \tag{74b}$$

where ${}_*P_\pi$ is the accounting price of a unit of profits, i.e.

$${}_*P_\pi = (1 - s_K) + s_K \cdot {}_*P_K \ .$$

${}_*P_\pi$ measures the social value of a unit of profit which is allocated among consumption, i.e. the fraction $1-s_K$, and investment, i.e. the fraction s_K .

Recall that the social value of consumption equals 1, while the social value of investments in terms of consumption is ${}_*P_K$, see (25) and (73a), which is greater than 1 . The reason why investment is valued higher than consumption is because investment generates future employment. This higher social valuation is a consequence of the institutional constraints introduced in Section 3.

The expression ${}_*P_\pi(y-y'\ell)$ in (74b) measures the social value of marginal productivity of capital Y_K, see (16b). Then, since ${}_*P_K$ is the social value of investment, the ratio in the right-hand side of (74b) stands for the investment rate of return which we denote as customary by ${}_*r_K$.

Then, (74b) acquires the more simple form

$$-\frac{\frac{\partial p}{\partial t}}{p} + [(\varepsilon + \delta - \pi^2) - 2b\pi] + r\pi = {}_*r_K \ . \tag{74c}$$

On the other hand, the relative rate of decay of the premium attached to investment is given by the logarithmic derivative of ${}_*P_K = (p/U'_c)$

$$\frac{{}_*\dot{P}_K}{{}_*P_K} = \frac{\frac{\partial p}{\partial t}}{p} - \frac{\dot{U}'_c}{U'_c} - \frac{s_K(1-s_K)(y')^2 b\ell^2 \pi\eta}{{}_*P_K^2} \ . \tag{75a}$$

where η stands for the elasticity of marginal utility, i.e. $\eta = -d(\log U')/d(\log C)$.

Similarly, the relative rate of decay of the marginal utility of (aggregate) consumption $\partial U/\partial C$, that we shall denote by ${}_*i_C$ and call the *consumption rate of discount,* is given by the following relation

$$_{*}i_C = - \frac{\frac{d}{dt}\left(\frac{\partial U}{\partial C}\right)}{\frac{\partial U}{\partial C}} .$$

Due to the fact that the argument of the utility function is $c = \frac{C}{N}$, it results

$$_{*}i_C = - \frac{\dot{U}'_c}{U'_c} + \varepsilon . \tag{75b}$$

Therefore, combining (74c), (75a) and (75b) it follows that

$$_{*}i_C + [(\delta - \pi^2) - 2b\pi + r\pi] = {}_{*}r_C \tag{76}$$

where $_{*}r_C$, called the *consumption rate of return,* is given by

$$_{*}r_C = {}_{*}r_K + \frac{_{*}\dot{P}_K}{_{*}P_K} .$$

The compact relation (76) reveals the economic content of Theorem 4.3 and in particular of eq. (52). It says that in order to motivate postponing claims on current consumption, the consumption rate of return $_{*}r_C$ has to make up first for the relative decay of the utility of consumption due to the passing of time, and also for the decay due to capital depreciation and uncertainty, the second term in the left-hand side of (76). Thus, we shall call the optimality requirement stated by (76), the *instantaneous intertemporal consistency condition.*

The *transversality condition* is given by

$$g(k(T)) = - p(T)(k(T) - k^{*})^{-} . \tag{77}$$

Here z^{-} means the negative part of z, i.e. $z^{-} = -\min(0,z)$.

Summing up, we shall call a path of investment and employment, $(s_K(t),\lambda(t))$, optimal if and only if it satisfies: the static first-order i.e. conditions (73), the instantaneous intertemporal consistency condition (76), the transversality condition (77) and the dynamic efficiency condition (70).

Our results in Section 4.2, as well as in Section 5, depend heavily on Marglin (1976). Burmeister and Dobell (1970), Neumann (1982), and Chakravarty (1969) have been also very useful throughout the economic sections.

In the next section we deal with questions of attainability and reversibility of optimal paths.

5. ATTAINABILITY AND REVERSIBILITY OF THE NEOCLASSIC PHASE

As we have seen in Subsection 4.2.2., the static first-order conditions (73) entail that, with a sufficiently low labour-capital ratio ℓ, the economy optimally develops moving from phase I, characterized by unemployment and an investment binding constraint, to phase II at which full employment holds but the investment binding requirement still remains. Furthermore, that after bringing $_*P_K$ down to unity in phase II, the economy enters the neoclassic phase III and reaps the rewards for postponing consumption and furthering accumulation. Now, we shall look for conditions that guarantee attainability of these three phases and that prevent the economy to move from higher-numbered phases, once they have been attained, to lower-numbered ones.

Intuitively, the rationale behind this issue is as follows: We look first for a capital endowment to be given to any worker in the production process as well as to those who are supposed to enter into production during the current planning period. However, we shall keep the amount of labour per unit of capital endowment, i.e. ℓ, low enough as to allow capital accumulation. Throughout phase I ℓ shall remain constant while λ and k grow up. Recall that

$$k = \ell^{-1} \lambda . \tag{78a}$$

That means that in phase I, we are mainly eliminating unemployment and widening capital.

In phase II, we shall hold λ constant, remind that in phase II $\lambda = 1$, and let k grow, so that labour productivity goes beyond wages. That means, phase II is one of capital deepening under saving binding constraints. The fact that $_*P_K$ falls down to unity makes sure that capital deepening comes to an end.

In phase III, the savings constraint is no longer binding, consumption increases and the economy moves towards the golden values.

5.1. Attainability of the Golden Value k^*

Let $\bar{s}$ denote the ratio s given by eq. (6) when $s_k = \bar{s}_k$ and y^* the process $y(\ell)$ when $\ell = \ell^*$. Further, let $\underline{\ell}$ denote the level of ℓ at which the economic surplus given by eq. (22c*) or, alternatively, by eq. (80a) attains a maximum.

A simple computation shows that at $\underline{\ell}$ the marginal product y' is given by

$$y' = \left(\frac{s_K - s_L}{s_K}\right) W \tag{78b}$$

which we denote $\underline{y'}$.

<u>Lemma 5.1</u>. A necessary and sufficient condition for the attainability of k^* is that the following inequality holds:

$$\bar{s}y^* > (\varepsilon + \delta) - (\pi + 2b)\pi \quad . \tag{79}$$

Proof: Let us recall that the mean forward conditional-derivative $\frac{\partial k}{\partial t}$ is given by

$$\frac{\partial k}{\partial t} = i - (\varepsilon + \delta) + (\pi + 2b)\pi \quad . \tag{22c*}$$

Then, substituting (27b) in (22c*) we get

$$\frac{\frac{\partial k}{\partial t}}{k} = s_K(y - W\ell) + s_L W\ell - (\varepsilon + \delta) + (\pi + 2b)\pi \tag{80a}$$

or, equivalently, see eq. (6),

$$\frac{\frac{\partial k}{\partial t}}{k} = s\,y - (\varepsilon + \delta) + (\pi + 2b)\pi \ . \tag{80b}$$

Hence, positive capital accumulation amounts to positivity of the right-hand side of eq. (80b). Thus, setting $s = \bar{s}$ in (80b) gives the accumulation feasibility condition (79) at the golden-rule values $\ell = \ell^*$ and $y = y^*$.

Now, the lemma follows almost immediately by means of a continuity argument, see Marglin (1976).

Indeed, the continuity of $y = y(\ell)$ and (79) entail positive accumulation of capital, regardless of the level of k, in a neighbourhood of ℓ^* as well as at ℓ^* itself. Therefore, ℓ can be brought sufficiently close to ℓ^* within this neighbourhood by sacrificing employment. Furthermore, observe that attaining ℓ^* and correspondingly c^* is tantamount to attaining k^*.

Finally, knowing the feasibility of attaining and sustaining k^* this must optimally happen in finite time. Otherwise, from (30) it follows easily that such a path would be dominated by any path attaining c^* in finite time, and therefore cannot be optimal. ■

Lemma 5.2. The golden-rule level of the capital per available labour ratio k^* associated with ℓ^* is necessarily higher than the level $\underline{k}$ associated with the level $\underline{\ell}$ of the labour-capital ratio at which the economic surplus, right-hand side of (80), attains a maximum, i.e. the inequalities

$$\ell^* < \underline{\ell} \tag{81a}$$

$$\underline{k} < k^* \tag{81b}$$

hold.

Proof: Let us recall that k^* by definition, i.e. (29c), satifies

$$y^* - y^{*\prime}\ell^* = (\varepsilon + \delta) - (\pi + 2b)\pi \tag{29c}$$

where $y^{*\prime}$ stands for y' at $\ell = \ell^*$.

Eq. (29c) together with the attainability condition (79) gets

$$\bar{s}\, y^* > y^* - y^{*\prime}\ell^* \quad . \tag{82a}$$

Rewriting the left-hand side of (82a), see eqs. (6) and (80a), we obtain

$$\bar{s}_K \left[y^* - \left(\frac{s_K - s_L}{s_K} \right) W\ell^* \right] > y^* - y^{*\prime}\ell^*$$

and, since $\bar{s}_K \leq 1$, we obtain further

$$y^* - \left(\frac{s_K - s_L}{s_K} \right) W\ell^* > y^* - y^{*\prime}\ell^* \ . \tag{82b}$$

From eq. (82b) one obtains easily

$$\underline{y}' = \left(\frac{s_K - s_L}{s_K} \right) W < y^{*\prime}$$

and this inequality delivers (81a). Since at the golden-rule level $\lambda = 1$, (78a) provides (81b). ∎

<u>Lemma 5.3.</u> The relative rate of growth of the aggregate consumption is given by the following relation

$$\frac{\frac{\partial C}{\partial t}}{C} = sy - \delta + \frac{(1-s_K)y' + (s_K - s_L)W}{(1-s_K)y + (s_K - s_L)W\ell} \frac{\partial \ell}{\partial t}$$

$$- \frac{(y - W\ell)}{(1-s_K)y + (s_K - s_L)W\ell} \dot{s}_K + \frac{\frac{1}{2}(1-s_K)y'' b^2\ell^2}{(1-s_K)(y-W\ell) + (1-s_L)W\ell} \ . \tag{83a}$$

<u>Proof</u>: It follows easily from (3), some arrangements and Itô's calculus. ∎

<u>Lemma 5.4.</u> The relative rate of growth of consumption per available labour c is given by

$$\frac{\frac{\partial c}{\partial t}}{c} = \frac{\frac{\partial C}{\partial t}}{C} - \varepsilon + (\pi + 2b)\pi \ . \tag{83b}$$

Proof: The computation of (83b) follows via Itô's calculus and is similar to that of (21) and (22).

Lemma 5.5. Similarly, the following relation for $_*P_K$ holds

$$\frac{\frac{\partial_* P_K}{t}}{_*P_K} = \frac{(1-s_K)y''\frac{\partial \ell}{\partial t} + (W-y')\dot{s}_K - \frac{1}{2}(1-s_K)y''' \, b^2\ell^2}{(s_K-s_L)W + (1-s_K)y'}$$

$$+ \frac{s_K y''\frac{\partial \ell}{\partial t} - (W-y')\dot{s}_K - s_K y''' b^2\ell^2[(s_K-s_L)W-s_K y']^{-1}}{(s_K-s_L)W - s_K\, y'} \ . \tag{83c}$$

Let us remind that the time derivative of random processes are to be interpreted as mean forward conditional-derivative as we did in Sections 3 and 4.

Before we go further to analyze the question related to the once-and-for-all transition from phase I to II, and from II to III, we would like to make a few remarks intended to facilitate the argumentation.

First of all, we like to observe that (75b) can be written as

$$_*i_c = \eta\frac{\dot{c}}{c} + \varepsilon \tag{84a}$$

where η is the elasticity of the marginal utility U_c' and therefore measures the relative rate of change of marginal utility with an increase in consumption c of one percent. The strict concavity of U entails that $\eta > 1$.

Now, combining (83a), (83b) and (84b) one obtains (84b) that we state in the following lemma.

Lemma 5.6. We have the formula

$$_*i_c = \eta\left\{sy - \delta + \frac{(1-s_K)y' + (s_K-s_L)W}{(1-s_K)y + (s_K-s_L)W\ell}\,\frac{\partial \ell}{\partial t}\right.$$

$$- \frac{(y-W\ell)}{(1-s_K)y + (s_K-s_L)W\ell}\,\dot{s}_K + \frac{\frac{1}{2}(1-s_K)y''\, b^2\ell^2}{(1-s_K)(y-W\ell) + (1-s_L)W\ell}$$

$$\left. - \varepsilon + (\pi + 2b)\pi\right\} + \varepsilon \ . \tag{84b}$$

Next, we list properties and requirements which characterize the three phases.

<u>Phase I</u>: $\lambda < 1$ $\quad s_K = \bar{s}_K$ $\quad {}_*P_K \geq 1$

$\dot{\lambda} > 0$ $\quad \dot{k} > 0$ $\quad \left(\frac{s_K - s_L}{s_K}\right)W \geq y'$

$\dot{\ell} = 0$ $\quad \frac{\partial H}{\partial \lambda} = 0$ $\quad s_K = 0$

<u>Phase II</u>: $\lambda = 1$ $\quad s_K = \bar{s}_K$ $\quad {}_*P_K \geq 1$

$\dot{\lambda} = 0$ $\quad \dot{s}_K = 0$ $\quad {}_*\dot{P}_K < 0$

$\dot{k} > 0$ $\quad \frac{\partial H}{\partial \lambda} \geq 0$ $\quad \left(\frac{s_K - s_L}{s_K}\right) W \leq y'$

$\ell = k^{-1}$ $\quad \ddot{k} > 0$

<u>Phase III</u>: $\lambda = 1$ $\quad s_K < \bar{s}_K$ $\quad {}_*P_K = 1$

$\dot{\lambda} = 0$ $\quad \dot{s}_K < 0$ $\quad \dot{P}_K = 0$

$\dot{k} > 0$ $\quad \ddot{k} \gtrless 0$ $\quad \left(\frac{s_K - s_L}{s_K}\right) W < y'$

$\ell = k^{-1}$

We have relaxed the notation with respect to the time derivative of stochastic processes.

The following lemma will be needed.

<u>Lemma 5.7</u>. If the path variable ℓ is such that (85b) and (85c) below are simultaneously satisfied, then it must follow that $\ell \leq \ell_\varepsilon$ where ℓ_ε is defined by

$$\ell_\varepsilon = \inf \{\ell > \underline{\ell} \mid sy = \varepsilon + \delta - (\pi + 2b)\pi\} \tag{85a}$$

$$\frac{{}_*P_\pi (y-y')}{{}_*P_K} = \eta \{sy - (\varepsilon+\delta) + (\pi + 2b)\pi\} + \varepsilon \tag{85b}$$

$$\left(\frac{s_K - s_L}{s_K}\right)W > y' \quad . \tag{85c}$$

<u>Proof</u>: Our proof goes indirectly. Therefore, let us assume ℓ is such

that (85b) and (85c) are satisfied but that $\ell > \ell_\varepsilon$. From (85a) follows that

$$sy \le (\varepsilon + \delta) - (\pi + 2b)\pi$$

and since $\eta > 1$, we obtain from (85b)

$$\frac{{}_*P_\pi (y - y'\ell)}{{}_*P_K} \le sy \; . \tag{86a}$$

On the other hand,

$$\frac{{}_*P_\pi}{{}_*P_K} = \frac{(s_K - s_L)W}{(s_K - s_L)W + (1 - s_K)y'}$$

which together with (85c) gives

$$\frac{{}_*P_\pi}{{}_*P_K} > s_K \; . \tag{86b}$$

Now, using (86b) in (86a) provides the inequality

$$s_K(y - y'\ell) < sy \tag{86c}$$

or, equivalently,

$$s_K(y - y'\ell) < s_K\left[y - \left(\frac{s_K - s_L}{s_K}\right)W\ell\right] . \tag{86d}$$

But this inequality contradicts (85c). ■

Theorem 5.1. Let us denote the optimal value of ℓ_o by ℓ^o, i.e. the value of ℓ associated with the optimal control α^o. Then, it follows that in phase I, ℓ^o satisfies the inequality

$$\ell^o < \ell_\varepsilon \; . \tag{87}$$

Proof: We proceed here again in an indirect way. Hence, we assume that $\ell^o > \ell_\varepsilon$ and show that this leads to a contradiction. We consider the cases:

$$\dot{\ell}^o = 0 \qquad \dot{\ell}^o < 0 \qquad \dot{\ell}^o > 0 \; .$$

Then, taking into account the intertemporal consistency condition (76) as well as the lemmas 5.3. - 5.6., and the characterization of phase I, one proves that for the first two cases (85b) holds.

Thus, since in phase I $\left(\frac{s_K - s_L}{s_K}\right) W > y'$, then (85c) holds also and application of Lemma 5.7. delivers the contradiction.

The last case, i.e. $\ell^o > \ell_\varepsilon$ and $\dot{\ell}^o > 0$, can be reduced to one of the first two cases or lead to a contradiction of the attainability of the golden-rule values.

Therefore, we conclude that in phase I $\ell^o < \ell_\varepsilon$ which entails that phase I must come to an end in finite time. ∎

We like to point out that optimality dictates unemployment whenever the social value of the net contribution of the marginal worker falls to zero before full employment can be reached. And further that $\partial H/\partial\lambda = 0$ is a necessary condition to be in phase I.

We like to state the following lemma.

Lemma 5.8. Once the economy reaches full employment with $\dot{k} > 0$, then $\dot{k}$ must remain positive as long as the economy remains in phase II and maintains full employment.

Proof: The fact that in phase II $\ell = k^{-1}$ and a small calculation deliver the lemma. ∎

Lemma 5.9. We have the formula

$$\frac{\partial H}{\partial \lambda} = p\, s_K \left[y' - \left(\frac{s_K - s_L}{s_K}\right)W\right] +$$

$$U_c'(c) \left[y' - s_K\left(y' - \left(\frac{s_K - s_L}{s_K}\right)W\right)\right] . \tag{88}$$

Proof: The lemma follows from eqs. (69a), (68a), and (27) after a small calculation and a few arrangements. ∎

Lemma 5.10. We have also the formula

$$\frac{d}{dt}\left(\frac{\partial H}{\partial \lambda}\right) = U_c'(c)\Big\{(1-s_K)(y-y'\ell)\dot{k} + \left[y'-s_K\left(y' - \left(\frac{s_K-s_L}{s_K}\right) W\right)\right]$$

$$\dot{\lambda} \left[y' - s_K\left(y' - \left(\frac{s_K-s_L}{s_K}\right) W\right)\right] +$$

$$U_c'(c)\,[\,(1-s_K)y''\,\ell] + p\, s_K\, y''\, \dot{\ell}$$

$$-\{U_c'(c)(1-s_K)(y-y'\ell) + p[s_K(y-y'\ell) - (\varepsilon+\delta) + (\pi+2b)\pi]$$

$$-r\pi\}\left[\,y' - \left(\frac{s_K-s_L}{s_K}\right) W\,\right]$$

$$+ \frac{1}{2}(1-s_K)U_c''(c)\left\{\left[y'-s_K(y' - (\frac{s_K-s_L}{s_K})w\right]\left[(y''-y')\lambda\pi^2 + y''\,\ell\lambda(b-\pi)^2 + y'''\,b^2\ell^2\right]\right\} . \qquad (89)$$

Proof: One applies Itô's formula to $\frac{\partial H}{\partial \lambda}$. Recall that $H = H(k,\alpha,p,r)$ which is given by eq. (69a) and that $\alpha = \alpha(\lambda, s_K)$. Furthermore, the processes k,λ,p are random and their differentials are given respectively by eqs. (21), (19) and (52), see also (74a). ∎

It remains to show that when the economy enters phase II and phase III, that phase I, and phase II respectively, can never recur.

The following theorems hold unambiguously only for the fairly restrictive case $\bar{s}_K = 1$. However, it is likely that the theorems also hold without this assumption.

Theorem 5.2. Whenever the economy has entered phase II, phase I can never recur.

Sketch of the proof: One has to prove that once phase II has lasted for a positive interval of time, $\partial H/\partial\lambda$, given by Lemma 5.9, can never again become zero.

Hence, one has to prove that for small values of $\partial H/\partial\lambda$, its time derivative, given by Lemma 5.10,

$$\frac{d}{dt}\left(\frac{\partial H}{\partial \lambda}\right)$$

must be positive.
Recall that phase II is characterized by conditions like $\lambda = 1$, $s_K = \bar{s}_K$, $\dot{\lambda} = 0$, and so on, see the corresponding table. ∎

Phase II comes to an end when $_*P_K$ falls down to unity. In phase III the following lemma is valid.

Lemma 5.11. In phase III the intertemporal consistency condition (76) becomes simply

$$_*i_c + [(\delta - \pi^2) - 2b\pi + r\pi] = y - y'\ell . \qquad (90)$$

Proof: It follows from (74b), (74c), (76) and the fact that in phase III $_*P_K = {}_*P_\pi = 1$. ∎

Theorem 5.3. Show that the transition from phase II to phase III is irreversible.

Sketch of the proof: Combining (90) and (84b) one gets a formula for $\dot{s}_K$ and proves that for $_*P_K = 1$ and s_K close to $\bar{s}_K$ its time derivative $\dot{s}_K$ has to be negative. See the characteristic features of phase III in the corresponding table.

Hence, once the economy has lived in phase III for a while, then s_K can never become equal to $\bar{s}_K$. Therefore, the economy would remain by itself in phase III for ever.

■

6. CONCLUSIONS

We have modelled the process of capital accumulation as the interplay of the processes of unemployment elimination and economic surplus formation and we have thereby gained some insights into qualitative and quantitative aspects of their interaction as the economy moves towards full employment. Furthermore, we have obtained conditions for the attainability and reversibility of the three phases the system undergoes before entering its optimum state.

However, we still have to learn more about the time the economy spends in the different phases in order to be able to use the model in economic policy issues. The semi-martingale calculus, stochastic control and its various maximum principles are powerful tools which may be helpful shaping a stochastic theory of value from analogies with certain quantities occurring in quantum mechanics and thermodynamics.

7. ACKNOWLEDGEMENTS

The main part of this paper has been written during my visit at the Research Center BiBoS of the University of Bielefeld in 1985. Therefore, I wish to thank this institution for its kind hospitality and financial support. I would especially like to thank S. Alberverio for fruitful and stimulating discussions, valuable suggestions and his kind guidance.
I owe a tremendous thank to A. Bensoussan who patiently explained to me his maximum principle and made clear to me some aspects of his fascinating and powerful theory.
J.L. Menaldi and H. Nagai read the manuscript and made useful comments.
I am very indebted to H. Bauer and M. Neumann for encouragements, advice and a continuous interest in my work that crystalizes into a joint research project supported by funds from the DFG which I gratefully acknowledge. Finally, to Mrs. Jegerlehner goes my gratitude for the skill and care she brings to the task of typing a difficult manuscript.

8. REFERENCES

Baran, P.A. (1957). The Political Economy of Growth. Monthly Review, New York.

Bensoussan, A. (1978). Control of Stochastic Partial Differential Equations. In Distributed Parameter Systems, edited by W.H. Ray and D.G. Lainiotis. Marcel Dekker, N.Y.

Bensoussan, A. (1982a). Stochastic Control by Functional Analysis Methods. North-Holland, Amsterdam.

Bensoussan, A. (1982b). Lectures on stochastic control. In Lecture Notes in Mathematics, Nr. 972, Springer Verlag, Berlin.

Bensoussan, A. (1983a). Stochastic Maximum Principle for Distributed Parameter Systems. Journal of the Franklin Institute, Vol. 315, No. 5/6, pp. 387-406.

Bensoussan, A. (1983b). Maximum principle and dynamic programming approaches of the optimal control of partially observed diffusions. Stochastics, Vol. 9, pp. 169-222.

Bensoussan, A. and J. Lions (1978). Applications des Inéquations Variationnelles en Contrôle Stochastique. Dunod, Paris.

Bensoussan, A. and J. Lions (1982). Contôle Impulsionnel et Inéquations Quasi-Variationnelles. Dunod, Paris.

Burmeister, E. and A.R. Dobell (1970). Mathematical Theories of Economic Growths. The Macmillan Company, London.

Burmeister, E. (1980). Capital Theory and Dynamics. Cambridge University Press, Cambridge.

Chakravarty, S. (1969). Capital and Development Planning. The MIT-Press, Cambridge, Mass.

Davis, M.H.A. and G. Gómez (1986). The semi-martingale approach to the optimal resource allocation in the controlled labour-surplus economy. In Lecture Notes in Mathematics, Nr. Ed. by S. Albeverio et al., Springer Verlag, Berlin

Elliott, R.J. (1982). Stochastic Calculus and Applications. Springer Verlag, Berlin.

Fleming, W. (1982). Non-linear semigroups for controlled partially observed diffusions. SIAM Journal of Control and Optimization, Vol. 20.

Fleming, W. and E. Pardoux (1982). Optimal control of partially observed diffusions. SIAM Journal of Control and Optimization, Vol. 20.

Gómez, G. (1984a). Modelling the economic development by means of impulsive control techniques. Mathematical Modelling in Sciences and Technology, pp. 802-806, Eds. X.J. Avula and R.E. Kalman. Pergamon-Press, N.Y.

Gómez, G. (1984b). On the Markov Stopping Rule Associated with the Problem of Controlling a Dual Economy. Dynamic Modelling and Control of National Economies, 1983, pp. 197-204. Eds. T. Basar and L.F. Pau. Pergamon-Press, N.Y.

Gómez, G. (1985a). The intertemporal labour allocation inherent in the optimal stopping of the dual economy: the dynamic case. Methods of Operations Research, Vol. 49, pp. 523-543.

Gómez, G. (1985b). Labour allocation policies based on semi-martingale methods. To be presented at the 5th IFAC/IFORS Conference on Dynamic Modelling of National Economies, Budapest, June 17-20 (1986).

Gómez, G. (1986a). A mathematical dynamic model of the dual economy emphasizing unemployment, migration and structural change. Institute of Mathematics, University of Erlangen-Nürnberg, Erlangen, FRG.

Gómez, G. (1986b). Equations of motion in economics and mechanics: towards a stochastic theory of value. Institute of mathematics, University of Erlangen-Nürnberg, Erlangen, FRG.

Ikeda, N. and S. Watanabe (1981). Stochastic Differential Equations and Diffusion Processes, North-Holland, Amsterdam.

Krylov, N.V. (1980). Controlled Diffusion Processes, Springer Verlag, Berlin.

Liptser, R.S. and A.N. Shiryayev (1977). Statistics of Random Processes I. Springer Verlag, Berlin.

Koopmans, T. (1966). On the Concept of Optimal Economic Growth. In The Econometric Approach to Planning. Rand McNally, Chicago.

Marglin, S. (1976). Value and Price in the Labour-Surplus Economy. Clarendon Press, Oxford.

Neumann, M. (1982). Theoretische Wolkswirtschaftslehre III. Verlag Vahlen, München.

Robinson, J. (1962). Essays in the Theory of Economic Growth. The Macmillan Press, London.

Robinson, J. (1956). The Accumulation of Capital.The Macmillan Press, London.

Sen, A. (1970). Collective Choice and Social Welfare. North-Holland, Amsterdam.

Sen, A. (1982). Choice, Welfare and Measurement. Basil Blackwell, Oxford.

Sen, A. (1984). Resources, Values and Development. Harvard University Press, Cambridge, Mass.

WAVELET TRANSFORMS AND EDGE DETECTION

A. Grossmann

Centre de Physique Théorique, Section II

CNRS – Luminy – Case 907

F-13288 MARSEILLE CEDEX 9 (France)

Abstract : A wavelet transform of a function is, roughly speaking, a description of this function across a range of scales. We use the technique of wavelet transforms to detect discontinuities in the n-th derivative of a function of one variable.

The term "edge detection" appears in the study of vision (see e.g. [1], [2]) ; it refers to the problem of localizing rapid changes of intensity in a signal.

"Wavelet transform" (see e.g. [3], [4]) describes a procedure of harmonic analysis ; it means the decomposition of a rather arbitrary function into contributions that are all "of the same shape" i.e. can be obtained from each other by translations and by changes of scale.

In this talk I shall describe a simple example involving the two subjects. The problem will be the detection of discontinuities of the n-th derivative of a function of one variable.

S. Albeverio et al. (eds.), Stochastic Processes in Physics and Engineering, 149–157.

1. WAVELET TRANSFORMS

I start by describing wavelet transforms.

Let s(t) be the function to be analyzed. Assume for the moment that s(t) is square integrable over $\mathbb{R}$.

The "signal" s will be analyzed in terms of a function g(t), the "analyzing wavelet". The minimal assumptions on g are :

(i) g is square integrable over $\mathbb{R}$

and

(ii) The Fourier transform of g, denoted by $g^\wedge$, is square integrable with respect to the dilation invariant measure $d\omega/|\omega|$. We write

$$c_g = 2\pi \int |g^\wedge(\omega)|^2 |\omega|^{-1} d\omega \qquad (1)$$

Intuitively, the first condition insures some "concentration" of g and of $g^\wedge$. In the examples that we shall study, g will satisfy much stronger conditions ; namely it is infinitely differentiable, and its simultaneous concentration in the "time" and "frequency" variable will be close to the limit allowed by the Heisenberg uncertainty principle.

The second condition, which is central to the argument, is, in practice, equivalent to the requirement $g^\wedge(0) = 0$, i.e. to the condition that g should have no zero-frequency component. This is in turn equivalent to the condition $\int g(t)dt = 0$, i.e. , to the requirement that g should have zero mean.

Our aim is to express the "signal" s(t) as a superposition of contributions each of which is of the same shape as the analyzing wavelet, i.e. obtained from the analyzing wavelet by a shift and a dilation of its argument. The "elementary contributions" will thus form a two-parameter family of functions $g^{(x,y)}$ labelled by a shift parameter x, and a dilation parameter y. We define

$$g^{(x,y)}(t) = |y|^{-1/2} g((t-x)/y) \qquad (x \in \mathbb{R},\ y \in \mathbb{R},\ y \neq 0) \tag{2}$$

The functions $g^{(x,y)}$ are not, in general, mutually orthogonal. Nevertheless, one has the following result :

Fix g, and associate to any square integrable s(t) the function

$$S(x,y) = c_g^{-1/2} |y|^{-1/2} \int \overline{g((t-x)/y)}\ s(t)dt \tag{3}$$

which is, except for the normalization coefficient, the inner product of $g^{(x,y)}$ and of s in $L^2(\mathbb{R})$. Then s(t) can be reconstructed from S(x,y) through the formula

$$s(t) = c_g^{-1/2} \iint S(x,y)\ g^{(x,y)}(t)\ y^{-2}\ dxdy \tag{4}$$

One has

$$\iint |S(x,y)|^2\ y^{-2}\ dxdy = \int |s(t)|^2\ dt \tag{5}$$

which shows that the inversion formula (4) is quite "stable" ; the correspondence $s \to S$ is an isometry between $L^2(\mathbb{R})$ and a closed subspace of $L^2(\mathbb{R}^2, y^{-2}\ dxdy)$, which can be explicitly characterized.

We shall say that S is the wavelet transform of s, with respect to the analyzing wavelet g.

It is useful to express S also in terms of the Fourier transforms $g^\wedge$ and $s^\wedge$: one has

$$S(x,y) = c_g^{-1/2}\ |y|^{1/2} \int e^{ix\omega}\ \overline{g^\wedge(\omega y)}\ s^\wedge(\omega)d\omega \qquad (6)$$

Let us now discuss the intuitive meaning of S(x,y) ; this will be relevant for edge detection. The main point to notice is that, for fixed y, S(x,y) is essentially the convolution of the signal s with the wavelet g scaled by the dilation parameter y. For instance

$$S(1,x) = c_g^{-1/2} \int \overline{g(t-x)}\, s(t)\ \ dt \qquad (7)$$

In some very rough sense, S(1,x) is a picture of s on a scale given by the "spread" of g.

A comparison of (6) with (3) shows that (3) contains information about s at all scales, the scale being governed by the dilation parameter y. If y is small, then S(x,y), considered as a function of x, shows fine detail, while S(x,y) for large y describes gross features of s.

This picture helps in understanding the reason for the existence of the stable inversion formula (4) ; the formula (4) incorporates information over all scales, and this makes it stable. On the other hand, the reconstitution of s from the values $S(x,y_0)$ for one fixed y_0 is deconvolution, which is a notoriously ill-posed problem.

2. THE WAVELETS g_n

Let us now make a specific choice of analyzing wavelet.

Fix an integer $n \geqslant 1$, and consider the function $\hat{g}_n(\omega) = \omega^n \exp(-\omega^2/2)$

It is the Fourier transform of

$$g_n(t) = (-i)^n (d/dt)^n \exp(-t^2/2) = i^n He_n(t) \exp(-t^2/2) \quad (8)$$

where

$$He_n(t) = 2^{-n/2} H_n(t/\sqrt{2}) \quad (9)$$

and H_n denotes the usual Hermite polynomial.

The function (8) satisfies (i) and (ii). (Notice that (ii) does not hold for $n = 0$).

The wavelets (8) are well known to people in edge detection : for $n = 2$, $g_n(t)$ is referred to as the "Mexican hat" ; it appears in discussions of vision.

Now look at the transform of a signal s with respect to the analyzing wavelet (8). One obtains :

$$S(x,y) = c_g^{-1/2} y^n |y|^{-1/2} \int \exp((t-x)^2/2y^2) s^{(n)}(t)dt \quad (10)$$

where $s^{(n)}$ is the n-th derivative of s. In other words :
Except for normalization factors, the transform (10), considered as a function of x for fixed y, is the n-th derivative of the signal smoothed with a gaussian filter of width y.

In order to discuss discontinuities we shall now extend the transform (3), (10) , to spaces larger than $L^2(\mathbb{R})$.

The smoothness and fast decrease of the functions (8) allow us to extend the definition of wavelet transform with respect to (8) to signals s that belong to a class of distributions including the Schwartz space $\mathcal{S}'$. In particular, (3) is now defined for all polynomials and all derivatives of δ-functions.

One should notice a difference between the transform $s \to S$ on $L^2(\mathbb{R})$ and its extension to $\mathcal{S}'$. On L^2, the transform is isometric ; it follows in particular that the image of a nonzero square integrable signal cannot be zero. On the other hand, the wavelet transform, under g_n , of any polynomial of order $\leqslant n$ is identically zero by (6).

Indeed the Fourier transform of any such polynomial is a combination of derivatives of order $\leqslant n$ of the δ-function positioned at $\omega=0$, and $g_n(\omega)$ has a zero of order n at $\omega = 0$.

The above argument does not apply, of course, to piecewise polynomial functions that cannot be expressed as a single polynomial. The Fourier transform of such a function involves necessarily negative powers of ω , and cannot be supported on $\omega = 0$.

3. DISCONTINUITIES

Let us look at the simplest case, with

$$s(t) = \begin{cases} 0 \text{ for } t < 0 \\ 1 \text{ for } t \geq 0 \end{cases} \tag{11}$$

The derivative of this s is the δ-function at t = 0 By (8), we immediately obtain :

$$S_n(x,y)=(2\pi(n-1)!)^{-1/2}(-i)|y|^{1/2}g_{n-1}(-x/y) \tag{12}$$

$$= (2\pi(n-1)!)^{-1/2}\, i^{n-1}\, He_{n-1}\,(-x/y)\exp(-x^2/2y^2)$$

What is the behaviour of this $S_n(x,y)$ as a function of x, for fixed y ? This function is practically zero except on an interval around the origin, of size proportional to y. On this interval, the function oscillates ; it has n-1 zeros.

In the open upper x,y plane, the zeros of $S_n(x,y)$ lie on n-1 straight lines which all point toward the point x = 0 , y = 0, i.e. towards the discontinuity (if we identify the t-axis with the line y = 0). The ratios of these "zero-crossings" of S are independent of y. This result has straightforward generalizations to piecewise polynomial signals s : the wavelet transform with respect to (8) can be made to display discontinuities of high-order derivatives in splines. Moreover the resulting pattern of zeros is quite stable under perturbations ; it is just a description of the fact that a discontinuity has no intrinsic scale.

The above results can be extended in many respects. They allow us to draw the following conclusions.

1) The use of wavelet transforms enables one to circumvent , in a certain sense, the ill-posedness of the problem of numerical differentiation.

2) The concepts and results developped in the study of edge detection are useful in harmonic analysis.

Acknowledgements :

The topics discussed here are a partial snapshot of work done in collaboration with I. Daubechies, R. Kronland, Y. Meyer, J. Morlet, T. Paul and P. Tchamitchian. The crucial first information on the subject of vision came from R. Balian, and essential help was provided by M. Bertero, Ch. de Mol, E. de Micheli and G. Sandini. This work was performed within the framework of the RCP 820, "Ondelettes" of the CNRS.

REFERENCES :

[1] T. Poggio, H. Voorhes and A. Yuille : 'A regularized solution to edge detection' ; Massachusetts Institute of Technology, Artificial Intelligence Laboratory. A.I. Memo 833 (1985)

[2] D. Marr. 'Vision. A computational investigation into the human representation and processing of visual information.' W.H. Freeman and Co, San Francisco 1982.

[3] Y. Meyer : 'Principe d'incertitutde, bases hilbertiennes et algèbres d'opérateurs' : Séminaire Bourbaki, 1985-86, n° 662.

[4] A. Grossmann and J. Morlet : 'Decomposition of Hardy functions into square integrable functions of constant shape' : SIAM J. Math. Anal. **15**, 4 (1984).

LIE ALGEBRAIC METHOD IN FILTERING AND IDENTIFICATION

M. Hazewinkel
Centre for Mathematics and Computer Science
P.O.Box 4079, 1009 AB Amsterdam
The Netherlands

ABSTRACT. These lectures concern (nonlinear) filtering. Very roughly the art of obtaining best estimates for some stochastic time-varying variable x on the basis of observations of another process y. The more concrete object under consideration being a stochastic dynamical system $dx=f(x)dt+G(x)dw$, where w is Wiener noise, with observations $dy=h(x)dt+dv$, corrupted by further noise. The subject as presented here involves ideas and techniques from Lie algebra theory, stochastics, differential topology, approximation theory and partial differential equations and has relations with quantum theory and stochastic physics. The lectures are adressed to practitioners in any one of these areas assuming that as a rule they are not experts in the other ones.

1. INTRODUCTION

Filtering is concerned with making estimates of quantities associated with a stochastic process $\{x_t\}$ on the basis of information gleaned from a related process $\{y_t\}$. The process $\{x_t\}$ is called the *signal* or *state* process and $\{y_t\}$ is the *observation* process. In this paper the following more concrete realization will be considered

$$dx_t = f(x_t)dt + G(x_t)dw_t, \quad x_t \in \mathbf{R}^n,\ w_t \in \mathbf{R}^m \tag{1.1}$$

$$dy_t = h(x_t)dt + dv_t,\ y_t \in \mathbf{R}^p,\ v_t \in \mathbf{R}^p \tag{1.2}$$

Here f is a function $\mathbf{R}^n \to \mathbf{R}^n$; G is an $n \times m$ matrix valued function on $\mathbf{R}^n$, h is a function $\mathbf{R}^n \to \mathbf{R}^p$ and w_t and v_t are Wiener processes, assumed independent of each other and also independent of the initial random variable x_0. More precisely these equations can be written

$$x_t = x_0 + \int_0^t f(x_s)ds + \int_0^t G(x_s)dw_s \tag{1.3}$$

$$y_t = \int_0^t h(x_s)ds + v_t \tag{1.4}$$

where the last term of (1.3) is a stochastic integral in the sense of *Ito*.

Much more loosely one can look at equations (1.1) and (1.2) as

$$\dot{x} = f(x) + G(x)\dot{w} \tag{1.5}$$

$$\dot{y} = h(x) + \dot{v} \tag{1.6}$$

with $\dot{w}$ and $\dot{v}$ white noise. Thus we have a differential equation $\dot{x} = f(x)$ on $\mathbf{R}^n$ which is subject to continuous random shocks whose intensity and direction (distribution) is state dependant and as observations we have an integral of some function of x and these observations are corrupted by more noise.

The general filtering problem for the state process $\{x_t\}$ with observation process $\{y_t\}$ is now to calculate for (interesting) functions ϕ of the state the conditional expectation

$$E[\phi(x_t)\,|\,y_s,\ 0 \leq s \leq t] = \widehat{\phi(x_t)}, \tag{1.7}$$

S. Albeverio et al. (eds.), Stochastic Processes in Physics and Engineering, 159–176.

i.e. the best (least squares) estimate of $\phi(x_t)$ given the observations y_s up to time t. That is we are interested in calculation procedures for $\widehat{\phi(x_t)}$. In many (engineering) applications the data come in sequentially and one does not really want a calculating procedure which needs all the data y_s, $0 \leq s \leq t$, every time t that it is desired to find $\widehat{\phi(x_t)}$; rather we would like to have a procedure which uses a statistic m_t which can be updated using only the new observations y_s, $t \leq s \leq t'$ to its value $m_{t'}$ i.e.

$$m_{t'} = a(m_t, t', t, \{y_s : t \leq s \leq t'\}) \tag{1.8}$$

and from which the desired conditional expectation can be calculated directly, i.e.

$$\widehat{\phi(x_t)} = E[\phi(x_t) | y_s, 0 \leq s \leq t] = b(t, y_t, m_t). \tag{1.9}$$

Finally to actually implement the filter it would be nice if m_t were a finite dimensional quantity. All this leads to the (ideal) notion of a *finite dimensional recursive filter.* By definition such a filter is a system

$$d\xi_t = \alpha(\xi_t)dt + \sum_{i=1}^{p} \beta_i(\xi_t)dy_{it} \tag{1.10}$$

driven by the observations y_{it}; y_{it} is the i-th component of y_t, $i = 1,...,p$; together with an output map

$$\widehat{\phi(x_t)} = \gamma(\xi_t) \tag{1.11}$$

More precisely formulated our problem is now the following: given a system (1.1)-(1.2) and a function ϕ on $\mathbf{R}^n$, how can we decide whether for these data there exists a finite dimensional recursive filter (1.10)-(1.11) which calculates $\widehat{\phi(x_t)}$, the best least squares estimate, and how do we find the functions (vectorfields) $\alpha, \beta_1, \cdots, \beta_p, \gamma$ of (1.10)-(1.11).

Now this may of course be a totally unreasonable question to ask. It could be that such nice filters virtually never exist. That is not the case though. In the case of linear systems

$$dx_t = Ax_t dt + Bdw_t \tag{1.12}$$

$$dy_t = Cx_t dt + dv_t \tag{1.13}$$

where now A, B, C are matrices of the appropriate sizes (which may be time varying), the well known Kalman-Bucy filter is precisely such a filter as (1.10)-(1.11). The equations are as follows. The statistic ξ_t is a pair (m_t, P_t) consisting of an n-vector and a symmetric $n \times n$ matrix P_t. These evolve according to

$$dP_t = (AP_t + P_t A^T + BB^T - P_t C^T C P_t)dt \tag{1.14}$$

$$dm_t = Am_t dt + P_t C^T (dy_t - Cm_t dt). \tag{1.15}$$

Here X^T denotes the transpose of a matrix X. This filter was discovered in 1961 and it is hard to overestimate its importance: whole books are devoted to its applications into single specialized fields and substantial companies can make a good living doing little more than Kalman-Bucy filtering. Naturally, efforts immediately started to find similar filters for more general systems than (1.12)-(1.13). This turned out to be unexpectedly difficult and this is still the case though there exists hosts of approximate filters of various kinds which (seem to) work well in a variety of situations; there is very little systematically known about how to construct approximate filters or about how to predict that a given one or class will work well when applied to a given collection of systems.

The approach based on Lie-algebraic considerations which I will try to discuss and explain below seems to hold great promise both in understanding the difficulties involved and in providing some kind of systematic foothold in the area of constructing approximate filters. For, as will become clear below, the existence of finite dimensional recursive filters for a nontrivial statistic will be a rare event.

Let me pause at this point to point out that identification problems can easily be construed as filtering problems. By way of illustrating this point consider again a linear system

$$dx_t = Ax_t + Bdw_t, \quad dy_t = Cx_t dt + dv_t \tag{1.16}$$

where now the matrices A,B,C are (partially) unknown. By adding to (1.16) the stochastic equations

$$da_{ij} = 0, \quad db_{kl} = 0, \quad dc_{qx} = 0 \tag{1.17}$$

for all unknown a_{ij}, b_{kl}, c_{qr}, one obtains a system (1.16)-(1.17) (of much larger state space dimension). And solving the filtering problem for the functions which project the vector $(x, (a_{ij}), (b_{kt}), (c_{qr}))$ onto a suitable component means identifying that particular coefficient.

2. THE DMZ-EQUATION AND THE ESTIMATION ALGEBRA

Let $\{x_t\}$ be a diffusion process as in (1.1)-(1.2) above. Given sufficient regularity of f,G,h the conditional expectation $\hat{x}_t$ will have a density $\pi(x,t)$.

Theorem 2.1. Under appropriate regularity conditions there exists an unnormalized version $\rho(x,t)$ of $\pi(x,t)$ which satisfies an equation

$$d\rho = \mathfrak{L}\rho dt + \sum_{i=1}^{p} h_i(x)\rho dy_{it} \tag{2.2}$$

where $\mathfrak{L}$ is the second order differential operator given by

$$(\mathfrak{L}\psi) = \frac{1}{2}\sum_{i,j=1}^{n} \frac{\partial^2}{\partial x_i \partial x_j}((GG^T)_{ij}\psi) - \sum_{i=1}^{n}\frac{\partial}{\partial x_i}(f_i\psi) - \frac{1}{2}\sum_{j=1}^{p} h_j^2\psi \tag{2.3}$$

Here $(GG^T)_{ij}$ is the (i,j)-th component of the $n \times n$ matrix $G(x)G(x)^T$ and f_i,h_j are the i-th and j-th component respectively of f and h.

Several comments are in order. First of all equation (2.2) is in Fisk-Stratonovic form. The corresponding Ito equation looks the same with $\mathfrak{L}$ changed by removing the $-\frac{1}{2}\sum h_j^2\psi$ term. The word "unnormalized" means that $\rho(x,t) = \sigma(t)\pi(x,t)$ where $\sigma(t)$ is an unknown function of time. Under appropriate reachability conditions on (1.1) $\rho(x,t)$ is a positive function. That ρ is unnormalized does not hurt much as $\rho(x,t)$ still suffices to calculate such things as $\widehat{\phi(x_t)}$. Indeed

$$\widehat{\phi(x_t)} = \left(\int \rho(x,t)dx\right)^{-1} \int \rho(x,t)\phi(x)dx \tag{2.4}$$

Theorem 2.1 was proved by Duncan [13], Mortensen [28] and Zakai [36] and the corresponding equation 2.1 is often refered to as the Duncan-Mortensen-Zakai or DMZ equation.

It is a stochastic partial differential equation being driven by the stochastic processes $y_1,...,y_p$.

It is important to note (Brockett [5]), that equations (2.2), (2.4) together constitute in fact a recursive filter in the sense of (1.10)-(1.11). The role of ξ_t is played by $\rho(x,t)$ so that instead of a point ξ evolving on a finite dimensional M we have an evolving density, i.e. a point ρ in an infinite dimensional space of positive functions evolving with time.

The simplest nontrivial example of a system (1.1)-(1.2) is

$$dx = dw, \quad dy = xdt + dv \tag{2.5}$$

i.e. one dimensional Wiener noise linearly observed corrupted by further noise. In this case the DMZ-equation becomes

$$d\rho = \left(\frac{1}{2}\frac{\partial^2}{\partial x^2} - \frac{1}{2}x^2\right)\rho dt + x\rho dy_t, \quad \left(\frac{\partial \rho}{\partial t} = \frac{1}{2}\frac{\partial^2 \rho}{\partial x^2} - \frac{1}{2}x^2\rho + x\rho\dot{y}\right) \tag{2.6}$$

i.e. we are dealing with the Euclidean Schrödinger equation with an extra forcing term. This is not an accident but part of a general pattern of which we shall see a further manifestation below in section 8, cf. also Mitter [26,27] for other remarks on this theme. I do not know wether the use Bismut makes of the filtering equations when dealing with a stochastic approach to index theorems and the Dirac operator can also be fitted into this framework.

3. ROBUSTNESS AND NUMERICAL MATTERS

As it stands equation (2.2) is not a very useful object for applications. It is a stochastic partial differential equation (with as probability space a space of paths $\{y\}$) and as such a solution is in principle only defined apart from a set of measure zero. On the other hand actual observations will always consist of piecewise smooth $y(t)$ and the class of all such is of measure zero. Thus there arises the question whether there exist a version of (2.2) which can be interpreted pathwise for all $y(t)$ and for which the solutions of (2.2) for piecewise smooth $(y(t)$ carry (approximative) information, cf. Clark [9] and Davis [11]. Fortunatedly the time dependent gauge transformation

$$\tilde{\rho}(x,t) = \exp(-h_1(x)y_1(t) - \cdots - h_p(x)y_p(t))\rho(t,x) \tag{3.1}$$

transforms (2.2) into an equation

$$\frac{\partial\tilde{\rho}}{\partial t} = \mathcal{L}\tilde{\rho} - \sum_{i=1}^{p} y_i(t)\mathcal{L}_i\tilde{\rho} + \sum_{i,j=1}^{p} y_i(t)y_j(t)\mathcal{L}_{ij}\tilde{\rho} \tag{3.2}$$

where $\mathcal{L}_i = [h_i,\mathcal{L}] := h_i\mathcal{L}-\mathcal{L}h_i$ and $\mathcal{L}_{ij} = \mathcal{L}_{ji} = \frac{1}{2}[h_i,[h_j,\mathcal{L}]]$, and this equation, which does not anymore involve derivatives of y, can simply be interpreted as a family of partial differential equations parametrized by the possible observation paths $y(t)$.

Equation (3.2) can of course be verified directly (remembering that (2.2) is a Fisk-Stratonovic integral so that the ordinary rules of calculus apply; removing the term $-\frac{1}{2}\sum h_i^2$ from $\mathcal{L}$ gives the corresponding Ito equation and then Ito calculus of course also gives (3.2). An easier way of obtaining (3.2) is to observe that (3.1) in (2.2) gives $d\tilde{\rho} = \exp(-\sum h_i y_i)\mathcal{L}\exp(\sum h_i y_i)\tilde{\rho}$ and to use the version of the Baker-Campbell-Hausdorff formula which says

$$\exp(-rA)B\exp(rA) = \sum_{k=0}^{\infty}(-1)^k\frac{r^k}{k!}ad_A^k(B) \tag{3.3}$$

where $ad_A(B) = [A,B] = AB-BA$, $ad_A^k(B) = ad_A(ad_A^{k-1}(B))$ for linear operators A,B. In our case the contributions of (3.3) for $k\geqslant 2$ disappear because then A is a function, $B=\mathcal{L}$ is a second order differential operator, so $[A,B]$ is first order, $[A,[A,B]]$ is a function and $[A,[A,[A,B]]] = 0$.

Also of course there still remains the question of how to use equation (3.2) or (2.2) effectively to calculate certain desired conditional expectations. A direct numerical discretization approach is out of the question. Typically x is a fairly large dimensional object; for example around 27 for certain problems involving helicopters. Taking three data points per coordinate axis (which is ridiculous) then gives $3^{27}\approx 2.10^{14}$ space grid points! So other methods must be tried. It seems likely that the Lie-algebraic considerations to be discussed below will help. Other promising work into the numerics of the nonlinear filtering equations has been started by Pardoux-Talay [29].

4. WEI-NORMAN THEORY

It is important to note that the filtering equation (3.2) (or (2.2)) is of the general form

$$\dot{x} = (A_1x)u_1 + \cdots + (A_kx)u_k \tag{4.1}$$

where the A_i are linear operators and the u_i known functions of time. Of course in (3.2) the role of x is played by ρ, an infinite dimensional object. Here for the moment lets consider (4.1) as a finite dimensional object. Let us also assume that the $A_1,\ldots,A_k$ who are now, say, $n\times n$ matrices, form the basis of a Lie algebra. (By adding a few more terms with corresponding u_i equal to zero this can of course always be assured.) Let us look for solutions of the form (Wei-Norman [35]).

$$x(t) = e^{g_1A_1g_2A_2}...e^{g_kA_k}x(0) \tag{4.2}$$

Differentiating this gives

$$\dot{x} = \dot{g}_1A_1e^{g_1A_1}e^{g_2A_2}...e^{g_kA_k}x(0) + e^{g_1A_1}\dot{g}_2A_2e^{g_2A_2}...e^{g_kA_k}x(0) + ... \tag{4.3}$$

and inserting

$$e^{-g_1A_1}e^{-g_2A_2}...e^{-g_iA_i}e^{g_iA_i}...e^{g_1A_1}$$

just after $\dot{g}_iA_i$ in the i-th term equation (1.1) can be rewritten

$$\dot{x} = \sum_{i=1}^{k}\dot{g}_i(A_i + \sum_{r=1}^{i-1}\sum_{\substack{j_1,...,j_{i-1}\\ j_1+...+j_{i-1}>0}}\frac{g_1^{j_1}...g_{i-1}^{j_{i-1}}}{j_1!...j_{i-1}!}ad_{A_1}^{j_1}...ad_{A_{i-1}}^{j_{i-1}}(A_i))x \tag{4.4}$$

$$= \sum_{i=1}^{k}\dot{g}_i(A_i + h_{ij}(g_1,...,g_k)A_j)$$

with $h_{ij}(0,...,0) = 0$, where, again, the Campbell-Baker-Hausdorff formula (3.3) has been used. Note that h_{ij} are universal functions which only depend on the Lie algebra and the chosen basis. Thus it remains to solve (equating the coefficients of the basic elements A_i in (4.4) and (4.1))

$$\begin{aligned}
&\dot{g}_1 + \dot{g}_2h_{11}(g_1,...,g_k) + \dot{g}_2h_{21}(g_1,...,g_k) + ... + \dot{g}_kh_{k1}(g_1,...,g_k) = u_1\\
&\dot{g}_2 + \dot{g}_1h_{12}(g_1,...,g_k) + \dot{g}_2h_{22}(g_1,...,g_k) + ... + \dot{g}_kh_{k2}(g_1,...,g_k) = u_2\\
&\qquad ...\\
&\dot{g}_k + \dot{g}_1h_{1k}(g_1,...,g_k) + \dot{g}_2h_{2k}(g_1,...,g_k) + ... + \dot{g}_kh_{kk}(g_1,...,g_k) = u_k
\end{aligned} \tag{4.5}$$

which can be done for small t and $g_1(0) = ... = g_k(0) = 0$ because $h_{ij}(0,...,0) = 0$. In general a representation (4.2) for the solution is only possible for small t. However things change if the Lie-algebra in question is solvable, then ([35]) there is such a representation for all t. More precisely there is a suitable basis such that there is such a representation for all t. How this comes about is easy to see in the case that the Lie algebra L is nilpotent. Indeed let

$$L \underset{\neq}{\supset} L^{(1)} = [L,L] \underset{\neq}{\supset} L^{(2)} = [L,L^{(1)}] \underset{\neq}{\supset} ... \underset{\neq}{\supset} L^{(m)} = [L,L^{(m-1)}] = 0 \tag{4.6}$$

be a basis such that $A_1,...,A_{k_1}$, $A_{k_1+1},...,$ $A_{k_2},...,A_{k_{m-1}+1},...,$ $A_{k_m} = A_k$, $k_1<k_2<...<k_m$ such that $A_{k_i+1},...,A_{k_m}$ is a basis for $L^{(i)}$, $i=0,...,m-1$ $(k_0=1, k_m=k)$. Then it immediately follows from (4.4) that $h_{ij} = 0$ for $j \leq i$ and the set of equations (4.5) gets a nice triangular structure. Moreover $h_{ij}(g_1,...,g_k)$ involves only $g_1,...,g_{i-1}$ (this is always the case, cf.(4.5), so the h_{1j} in (4.5) are always all zero) and the resulting equations (4.5) for the nilpotent case are therefore of the form

$$\begin{aligned}
&\dot{g}_1 = u_1,...,\dot{g}_{k_1} = u_{k_1}\\
&\dot{g}_{k_1+1} = u_{k_1+1} + \alpha_{k_1+1}(u_1,...,u_{k_1};g_1,...,g_{k_1}),...,\dot{g}_{k_2} = u_{k_2} + \alpha_{k_2}(u_1,...,u_{k_1};g_1,...,g_{k_1})\\
&\dot{g}_{k_2+1} = u_{k_2+1} + \alpha_{k_2+1}(u_1,...,u_{k_2};g_1,...,g_{k_2}),...,\dot{g}_{k_3} = u_{k_3} + \alpha_{k_3}(u_1,...,u_{k_2};g_1,...,g_{k_2})\\
&\qquad ...
\end{aligned} \tag{4.7}$$

where the α_j are known (universal) functions of the u's and g's.

These considerations are not limited to Lie-algebras of matrices. Indeed the left hand sides of equations (4.5) only depend on the abstract structure of the Lie algebra in question and the choice of basis. Thus all this equally applies to Lie-algebras of say differential operators (given suitable definitions of $\exp(tA)$), though in order to have a finite set of equations (4.5) one needs of course a finite dimensional algebra. It also follows from (4.4) that the Wei-Norman equations are compatible with homomorphisms of Lie algebras, more precisely quotients. Indeed if $\mathfrak{A}\subset L$ is an ideal and $A_1,...,A_{k_1},A_{k_1+1},...,A_k$ is a basis of L such that $A_{k_1+1},...,A_k$ is a basis of $\mathfrak{A}$ then the h_{ij} are zero for $j\in\{1,...,k_1\}$ and $i\in\{k_1+1,...,k\}$. So in the case of a topologically nilpotent algebra L, or more generally one with a chain of ideals $\mathfrak{A}_1\supset\mathfrak{A}_2\supset\mathfrak{A}_3\supset...$ such that $\cap\mathfrak{A}_i=0$ and $L/\mathfrak{A}_i$ finite dimensional for all i one can in principle still do Wei-Norman theory with now infinite ordered product expressions $x=e^{g_1A_1}e^{g_2A_2}...e^{g_kA_k}...x_0$ in the sense that the equations for the g_i belonging to a quotient $L/\mathfrak{A}_j$ involve only those same g_i. Of course now questions of convergence arise.

5. THE ESTIMATION LIE ALGEBRA

The considerations of the previous section already make it clear that the Lie algebra generated by the operators which occur in equation (2.2) or (3.2) contains important information concerning the filtering problem. One therefore defines the estimation Lie algebra $EL(\Sigma)$ of a system Σ given by (1.1)-(1.2) as the Lie algebra of differential operators generated by the 2-nd order differential operator $\mathfrak{L}$ and the multiplication operators $h_1,...,h_p$.

$$EL(\Sigma) = Lie(\mathfrak{L},h_1,...,h_p). \tag{5.1}$$

Note that the Lie algebra generated by the operators which occur in (3.2) is in any case a subalgebra of $EL(\Sigma)$. Often it is equal.

Example 5.2. Consider again the simplest nonzero linear system (2.5). Then $p=1$ and $\mathfrak{L}=\frac{1}{2}d^2/dx^2-\frac{1}{2}x^2$. So we have in this case the Lie algebra $Lie(\frac{1}{2}d^2/dx^2-\frac{1}{2}x^2,x)$. Now $[\frac{1}{2}d^2/dx^2-\frac{1}{2}x^2,x]=d/dx$ (as operators on functions), $[\frac{1}{2}d^2/dx^2-\frac{1}{2}x^2,\frac{d}{dx}]=x$, $[\frac{d}{dx},x]=1$ and $[?,1]=0$. So in this case we obtain the well-known oscillator Lie algebra, which is four dimensional with basis $\frac{1}{2}d^2/dx^2-\frac{1}{2}x^2$, x, d/dx, 1. It is solvable (but not nilpotent) with as derived algebra the nilpotent Heisenberg algebra with basis x, d/dx, 1.
$EL(\Sigma)$ is (of course) an invariant of Σ meaning that a change of coordinates in Σ (a diffeomorphism $x\to x'$ taking Σ to Σ') will yield isomorphic estimation Lie algebras. The algebra also has a gauge transformation invariance. A gauge transformation $\rho(x,t)\to\psi(x)\rho(x,t)$, where $\psi(x)\neq0$ for all x, transforms the DMZ-equation in such a way that the operators in the new equation generate an isomorphic Lie algebra.

The new equation may again have the form of a DMZ-equation, and in this way systems which are definitely not equivalent as systems may have equivalent filtering problems associated to them. An example are the 1-dimensional Benes systems (cf. various contributions in [19]).

In a way which will (hopefully) become clearer below the estimation Lie algebra $EL(\Sigma)$ encodes information about how difficult the filtering problem for Σ is. For example if it is finite dimensional (a very rare case) Wei-Norman theory does the job for small time; if it is also solvable one thus gets a filter. If it is infinite dimensional but solvable things become more difficult but asymptotic expansions are possible, cf. below; etc.

6. THE BC PRINCIPLE

Let me now describe a second reason why the Lie algebra $EL(\Sigma)$ of a system Σ is important for filtering problems. I like to call it the *BC principle*, not because it is very old, though it could have been maybe, nor is it named after Johny Hart's chartoon character; the BC stand for Brockett and Clark [6] who first enunciated it.

Suppose we have a filter (1.10)-(1.11) on a finite dimensional manifold M for a statistic $\widehat{\phi(x_t)}$. We may as well assume that it is minimal, i.e. has minimal dim(M). The α and $\beta_1,...,\beta_p$ in (1.10) are vectorfields on M. Let $V(M)$ denote the Lie algebra of smooth vectorfields on M. Then the BC principle states the following

6.1. BC Principle
If (1.10)-(1.11) is a minimal filter for a statistic then $\mathcal{L} \mapsto \alpha$, $h_1 \mapsto \beta_1,...,h_p \mapsto \beta_p$ defines an antihomomorphism of Lie algebras from $EL(\Sigma)$ into $V(M)$.

Here "anti" means the following: if $\phi: L_1 \to L_2$ is a map of vectorspaces from the Lie-algebra L_1 to the Lie-algebra L_2, it is called an antihomomorphism of Lie-algebras if $\phi([A,B]) = -[\phi(A),\phi(B)]$ for all $A,B \in L_1$.

Example 6.2. Consider again the simplest nonzero linear system (2.5). It is linear so there is the Kalman-Bucy filter for the conditional state $\hat{x}$. This filter is

$$dP_t = (1-P_t^2)dt, \quad dm_t = P_t(dy_t - m_t dt). \tag{6.3}$$

So the two vectorfields α and β of the filter are respectively

$$\alpha = (1-P^2)\frac{\partial}{\partial P} - Pm\frac{\partial}{\partial m}, \quad \beta = P\frac{\partial}{\partial m}. \tag{6.4}$$

A simple calculation shows $[\alpha,\beta] = \frac{\partial}{\partial m}$, and it is now indeed a simple exercise to show that $\frac{1}{2}\frac{d^2}{dx^2} - \frac{1}{2}x^2 \mapsto \alpha$, $x \mapsto \beta$, induces an antimorphism of Lie-algebras. (It also induces a homomorphism, but that is an accident which happens for linear systems (1.12)-(1.13) if the drift term Ax is absent).

A feeling of why the BC principle should be true can be generated as follows. Think for the moment of two automata with given initial state and with outputs (Moore automata), which, when fed the same string of input data, produce exactly the same string of output data. Suppose the second automaton is minimal. Then it is wellknown (and easy to prove by constructing the minimal automaton from the input-output data) that there is a homomorphism of the subautomaton of the first consisting of the states reachable from the initial state to the second automaton; this homomorphism so to speak makes visible that the two machines do the same job. A similar theorem holds for initialized finite dimensional systems [Sussmann [34]), in particular for systems of the form

$$\dot{x} = \alpha(x) + \sum_{i=1}^{m} \beta_i(x)u_i, \quad y = \gamma(x) \tag{6.5}$$

Here the picture produced by theorem is the following commutative diagram

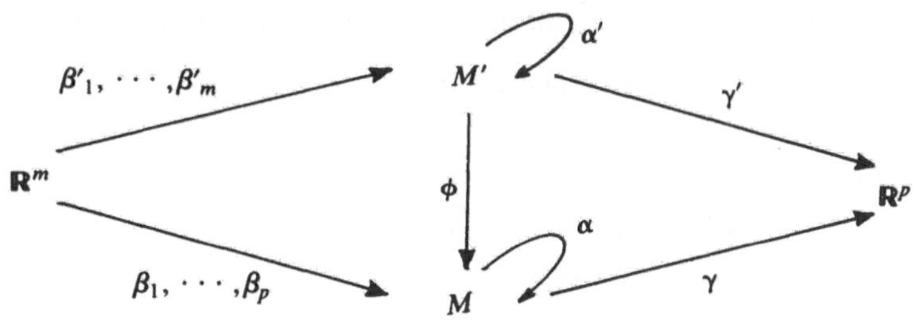

(The theorem asserts the existence of a differentiable map ϕ defined on the reachable from x'_0 subset of M' which makes the diagram commutative. This in particular implies that $d\phi$ takes the vectorfields

$\alpha',\beta'_1,\ldots,\beta'_m$ into $\alpha,\beta_1,\ldots,\beta_m$ respectively, and, ϕ being a differentiable map, $d\phi$ induces a homomorphism from the Lie algebra generated by $\alpha',\beta'_1,\ldots,\beta'_m$ to $V(M)$.

In the case of the BC principle we also have two "machines" which do the same job: one is the postulated minimal filter, the other is the infinite dimensional machinge given by the DMZ-equation (2.2) and the ouput map (2.4). So we are in a similar situation as above but with M' infinite dimensional. A proof in this case follows from considerations of Hijab [20].

The fact that in the case of the BC-principle we get an antihomomorphism arises from the following. Given a linear space V and an operator A on it we can define a (linear) vectorfield on V by assigning to $v \in V$ the tangent vector Av. (So we are considering the equation $\dot{v} = Av$.) This defines an anti-isomorphism of the Lie algebra of operators on V to the Lie algebra of linear vectorfields on V.

What about a converse to the BC principle? I.e. suppose that we have given an antihomomorphism of Lie-algebras $EL(\Sigma) \to V(M)$ into the vectorfields of some finite dimensional manifold. Does there correspond a filter for some statistic of Σ. Just having the homomorphism is clearly insufficient. There are also explicit counterexamples. This is understandable for in any case we completely ignored the output aspect when making the BC-principle plausible. This is not trivial contrary to what the diagram above may suggest. It is not true that given ϕ and any γ one can take $\gamma' = \gamma \circ \phi$. The problem is that γ' as a function on M' = space of unnormalized densities is of a very specific type cf. (2.4).

Even apart from that things are not guaranteed. What we need of course is a ϕ making the left half of the diagram above commutative. Then, if $m' \in M'$ is going to the mapped on $m \in M$, obviously the isotropy subalgebra of $EL(\Sigma)$ at m' will go into the isotropy subalgebra of $V(M)$ at m.

For the case of finite dynamical systems there are positive results of Krener [21] stating that in such a case this extra condition is also sufficient to guarantee the existence of ϕ locally.

The whole clearly relates to seeing to what extend a manifold can be recovered from its Lie algebra of vectorfields (via its maximal subalgebras of finite codimension) and whether differentiable maps can be recovered from the map between Lie-algebras they induce. This question has been examined by Pursell-Shanks [30].

A more representation theoretic way of looking at things is as follows. Both $EL(\Sigma)$ and $V(M)$ come with a natural representation on the space of functionals on M' and the space of functions on M respectively. If there were a ϕ as in the diagram above ϕ would also induce a map between these representation spaces compatible with the homomorphism of Lie algebras. That therefore is clearly a necessary condition. This way of looking at things contains the isotropy subalgebra condition and also contains output function aspects. Thus the total picture regarding a converse to the BC-principle is not unpromising but nothing is established.

Except for one quite positive aspect. If $EL(\Sigma)$ is finite dimensional, the Wei-Norman equations practically define the filter, for small time in the general case, for arbitrary time in the solvable case.

7. EXAMPLES OF ESTIMATION ALGEBRAS

7.1. The cubic sensor

This is the one dimensional system

$$dx_t = dw_t, \;\; dy_t = x_t^3 dt + dv_t \tag{7.2}$$

and it is about the simplext nonlinear system imaginable. Its estimation Lie algebra is generated by $\frac{1}{2}\frac{d^2}{dx^2} - \frac{1}{2}x^6,\; x^3$.

Theorem 7.3. (Hazewinkel-Marcus [17]). EL(cubic sensor) = W_1, where $W_1 = \mathbf{R}<x, \frac{d}{dx}>$ is the Lie algebra of the differential operators (any order, zero included) with polynomial coefficients.

Example 7.4. $dx_1 = dw$, $dx_2 = x_1^2 dt$; $dy_1 = x_1 dt + dv_1$, $dy_2 = x_2 dt + dv_2$. In this case the estimation Lie algebra is $W_2 = \mathbf{R}<x_1, x_2, \frac{\partial}{\partial x_1}, \frac{\partial}{\partial x_2}>$, the Lie algebra of all differential operators in two variables with polynomial coefficients.

Example 7.5. $dx_1 = dw$, $dx_2 = x_1^2 dt$, $dy = x_1 dt + dv_1$. In this case the estimation Lie algebra has a basis A, B_i, C_i, D_i $i = 1,2,...$ with the commutation relations $[A, B_i] = C_i$, $[A, C_i] = B_1 + 2B_{i+1}$, $[B_i, C_j] = -D_{i+j}$ and all other commutation relations between basis elements are zero. Note that in this case the Lie algebra is infinite dimensional but has many ideals $\mathfrak{A}_i$ such that $L / \mathfrak{A}_i$ is finite dimensional.

Example 7.6. $dx = dw_1 + \epsilon x dw_2$, $dy = x dt + dv$. Here again $EL = W_1$.

It has become clear that as a rule estimation algebras tend to be infinite dimensional (except in the linear case: then EL(linear system) has dimension $2n+2$ if the linear system is completely reachable and observable); it has also become noticeable that the Weyl-Heisenberg algebras or Weyl algebras W_n have a tendency to appear very often.

Conjecture 7.7. Consider systems (1.1)-(1.2) with polynomial f, G, h. Then generically, i.e. for almost all f, G, h, the estimation algebra will be W_n.

8. THE SEGAL-SHALE-WEIL REPRESENTATION AND ALL KALMAN-BUCY FILTERS

8.1. The linear systems Lie-algebra ls_n

Consider all differential operators in n indeterminates with polynomial coefficients

$$D = \Sigma c_{\alpha\beta} x^{\alpha} \frac{\partial^{\beta}}{\partial x^{\beta}} \tag{8.2}$$

where $\alpha = (\alpha_1, ..., \alpha_n)$, $\beta = (\beta_1, ..., \beta_n)$ are multiindices $\alpha_i, \beta_j \in \mathbf{N} \cup \{0\}$. Consider those D which are of total degree ≤ 2; i.e. such that $|\alpha| + |\beta| > 2 \Rightarrow c_{\alpha\beta} = 0$ where $|\alpha| = \alpha_1 + \cdots + \alpha_n$. As is readily verified these form a finite dimensional Lie algebra (under the commutator product $[D_1, D_2] = D_1 D_2 - D_2 D_1$) of dimension $2n^2 + 3n + 1$. A basis is

$$1; x_1, ..., x_n; \frac{\partial}{\partial x_1}, ..., \frac{\partial}{\partial x_n}; \frac{\partial}{\partial x_i \partial x_j}, i,j = 1,...,n; x_i x_j, i,j = 1,...,n; x_i \frac{\partial}{\partial x_j}, i,j = 1,...,n. \tag{8.3}$$

The operators of total degree ≤ 1 form a subalgebra h_n (basis: $1; x_1, ..., x_n$; $\partial / \partial x_1, ..., \partial / \partial x_n$) which is in fact an ideal. The quotient is isomorphic to the symplectic algebra sp_n of all real $2n \times 2n$ matrices M such that

$$MJ + JM^T = 0, \quad J = \begin{pmatrix} 0 & I_n \\ -I_n & 0 \end{pmatrix}. \tag{8.4}$$

The isomorphism is given by

$$E_{i,n+j} + E_{j,n+i} \mapsto x_i x_j; \; E_{n+i,j} - E_{n+j,i} \mapsto \frac{\partial^2}{\partial x_i \partial x_j}; \tag{8.5}$$

$$E_{i,j} - E_{n+j,n+i} \mapsto x_i \frac{\partial}{\partial x_j} + \tfrac{1}{2}\delta_{ij}; \; i,j = 1,...,n.$$

Here $E_{i,j}$ is the matrix with a 1 at spot (i,j) and 0 everywhere else; these linear combinations of the $E_{i,j}$ form a basis of sp_n; this isomorphism exhibits sp_n as a subalgebra complementary to h_n; i.e. as a Levi-factor for the short exact sequence $0 \to h_n \to ls_n \to sp_n \to 0$).

8.2. The oscillator representation

There is a famous representation of sp_n which occurs in the framework of symmetries of boson fields (Shale, Segal), in algebraic number theory (Weil), and a multitude of other places, known variously as the Segal-Shale-Weil representation or the oscillator representation. One way to obtain it is as follows. Let H_n denote the Heisenberg group, $H_n = \mathbf{R}^n \times \mathbf{R}^n \times S^1$ with multiplication

$$(x,y,z)(x',y',z') = (x+x',y+y',e^{-2\pi i<x,y'>}zz') \tag{8.7}$$

where $< , >$ denotes the standard scalar product on $\mathbf{R}^n$. The Lie algebra of H_n is of course $h_n = \mathbf{R}^n \times \mathbf{R}^n \times \mathbf{R}$. And the Lie-bracket of h_n can be interpreted as giving (and given by) a bilinear form $\mathbf{R}^{2n} \times \mathbf{R}^{2n} \to \mathbf{R}$ defined by the matrix J, cf. (8.4) above. Thus the Lie group Sp_n of sp_n can be seen as a group of automorphism of h_n and H_n which is moreover the identity on the centre $S^1 \subset H_n$. Let ρ be the standard Schrödinger representation of H_n in $L^2(\mathbf{R}^n)$

$$\begin{aligned}&(x,0,0) \to M_x,\ (M_x f)(x') = e^{2\pi i<x,x'>} f(x'),\ \ f \in L^2(\mathbf{R}^n)\\ &(0,y,0) \to T_y,\ (T_y f(x') = f(x'-y),\ \ f \in L^2(\mathbf{R}^n)\\ &(0,0,z) \to S_z,\ (S_z f)(x') = z f(x'),\ \ f \in L^2(\mathbf{R}^n)\end{aligned} \tag{8.8}$$

Now let $g \in Sp_n$ be seen as a group of automorphisms of H_n. Then $h \mapsto \rho(g(h))$ is another irreducible representation of H_n with the same central character. So by the Stone-von Neumann theorem there is an $\omega(g)$ intertwining them, i.e. such that

$$\omega(g)\rho(h)\omega(g)^{-1} = \rho(g(h)). \tag{8.9}$$

These $\omega(g)$ are unique up to scalar factors and therefore define a projective representation of Sp_n. The factors can be fixed up to define a representation of the two-fold covering $\tilde{S}p_n$ of Sp_n. This is the Segal-Shale-Weil representation.

8.3. All Kalman-Bucy filters

Now consider something apparently totally unrelated, namely the DMZ-filtering-equation (2.2) for a linear dynamical system

$$dx = Axdt + Bdw,\ dy = Cxdt + dv,\ x \in \mathbf{R}^n,\ w \in \mathbf{R}^m,\ y,v \in \mathbf{R}^p. \tag{8.11}$$

It is a trivial remark that the operators occurring in (2.2) are all in ls_n in this case. And in fact the Lie algebra generated by them will consist of the second order operator $\mathfrak{L}$ and a subalgebra of h_n stable under $\mathfrak{L}$. In most cases, to be precise in the case that the system (A,B,C) is completely reachable and completely observable, this will be all of h_n, giving us generically an estimation algebra of dimension $2n+2$ which is a subalgebra of ls_n, which has dimension $2n^2+3n+1$.

The Kalman-Bucy filter defines by BC principle an antihomomorphism of this Estimation Lie algebra $EL(A,B,C)$ into the vector Lie algebra of vector fields $V(\mathbf{R}^N)$, $N = n + \frac{1}{2}n(n+1)$.

Theorem [16] 8.12. For varying (A,B,C) these anti-homomorphisms fit together to define a antihomomorphism of all of ls_n into $V(\mathbf{R}^N)$ with as kernel the centre $\mathbf{R}1$. This representation can be lifted to one on $V(\mathbf{R}^{N+1})$ which is faithful.

The explicit formulas are as follows. Interpret a point $x \in \mathbf{R}^{N+1}$ as a triple $x = (c,m,P)$ consisting of a scalar c, an n-vector m, and a symmetric $n \times n$ matrix P. The antihomomorphism is then given by

$$1 \to \frac{\partial}{\partial c} \tag{8.13}$$

$$x \to m_i \frac{\partial}{\partial c} + \sum_t P_{it} \frac{\partial}{\partial m_t} \tag{8.14}$$

$$\frac{\partial}{\partial x_i} \to -\frac{\partial}{\partial x_i} \tag{8.15}$$

$$x_i x_j \to (m_i m_j + P_{ij})\frac{\partial}{\partial c} + \sum_t (m_i P_{jt} + m_j P_{it})\frac{\partial}{\partial m_t} \tag{8.16}$$

$$+ \sum_{s,t} P_{is} P_{jt} \frac{\partial}{\partial P_{st}} + \sum_t P_{it} P_{jt} \frac{\partial}{\partial P_{tt}}$$

$$x_i \frac{\partial}{\partial x_j} \to -m_i \frac{\partial}{\partial m_j} - \delta_{ij}\frac{\partial}{\partial c} - P_{ij}\frac{\partial}{\partial P_{jj}} - \sum_t P_{it}\frac{\partial}{\partial P_{jt}} \tag{8.17}$$

$$\frac{\partial^2}{\partial x_i \partial x_j} \to \frac{\partial}{\partial P_{ij}} \text{ if } i \neq j, \quad \frac{\partial^2}{\partial x_i^2} \to 2\frac{\partial}{\partial P_{ii}}. \tag{8.18}$$

Inversely these formulas can be checked directly to give an antihomomorphism of Lie algebras and this thus verifies the BC principle for linear dynamical systems and also for families of such depending on a parameter.

Changing all the minus signs in (8.17) and (8.15) into plus signs gives a faithful homomorphism of ls_n into $V(\mathbf{R}^{N+1})$.

Restricting this homomorphism to sp_n as given by (8.5) then defines a homomorphism of sp_n into $V(\mathbf{R}^{N+1})$.

The final remark is that this realization of sp_n as a Lie algebra of vector fields on $\mathbf{R}^{N+1}$ has much to do with the Segal-Shale-Weil representation. The precise statement is that the mapping

$$(c,m,p) \to \exp(c + <2\pi i m, x> - 2\pi^2 P(x)) \in L^2(\mathbf{R}^n) \tag{8.19}$$

(where $P(x)$ is the quadratic form defined by the symmetric matrix P; note that apart from a scaling factor this is the normal distribution with mean m and covariance P) linearizes the vectorfields in the image of) ls_n in $V(\mathbf{R}^{N+1})$ and switching from a linear vectorfield to the operator which defines it then defines a representation of $ls_n \supset sp_n$. This is another real form of the Segal-Shale-Weil representation meaning that after tensoring with $\mathbf{C}$ (=extending scalars to the complexes), they become isomorphic.

9. W_n AND $V(M)$

We have seen that the Weyl-Heisenberg algebra $W_n = \mathbf{R}<x_1,...,x_n; \partial/\partial x_1,...,\partial/\partial x_n>$ of all differential operators with polynomial coefficients often occurs in filtering problems, i.e. as an Estimation Lie algebra. Given the BC-principle it is therefore of interest to know something about its relations with another class of infinite dimensional Lie algebras, viz the Lie algebras $V(M)$ of smooth vectorfields on a finite dimensional manifold. The algebra W_n has a one-dimensional centre $\mathbf{R}.1$ consisting of the scalar multiples of the identity operator.

Theorem 9.1. (Hazewinkel-Marcus [17]). Let α: $W_n \to V(M)$ or $W_n/\mathbf{R}.1 \to V(M)$ be a homomorphism or antihomomorphism of Lie algebras, where M is a finite dimensional manifold. Then $\alpha = 0$.

The original proof of this result ([17]) was long and computational. Another much shorter proof based on the nonexistence of finite dimensional representations of h_n for which 1 gets mapped onto the unit operator has more recently been given by Toby Stafford.

10. THE CUBIC SENSOR

Consider again the cubic sensor, i.e. the one-dimensional system

$$dx = dw, \quad dy = x^3 dt + dv \tag{10.1}$$

consisting of Wiener noise, cubically observed with further independent noise corrupting the observations. As noted before (theorem 7.3)

$$EL(\text{cubic sensor}) = W_1. \tag{10.2}$$

Now suppose that we have a finite dimensional filter for some conditional statistic $\widehat{\phi(x_t)}$ of the cubic sensor. By the BC-principle (6.1) if follows that there is an antihomomorphism of Lie-algebras $W_1 = EL$ (cubic sensor) $\xrightarrow{\alpha} V(M)$. By theorem 9.1 it follows that $\alpha=0$ and from this it is not hard to see that the only statistics of the cubic sensor for which there exists a finite dimensional exact recursive filter are the constants.

A direct proof of this, which sort of proves the BC-principle in this particular case along the way, is contained in Hazewinkel-Marcus-Sussmann [18].

11. PERTURBATIONS AND APPROXIMATIONS

Let us start with an example. Consider the weak cubic sensor

$$dx = dw, \quad dy = xdt + \epsilon x^3 dt + dv \quad (\Sigma_\epsilon) \tag{11.1}$$

If $\epsilon = 0$, this the simplest nontrivial linear system for which there is the Kalman filter. For $\epsilon \neq 0$ one can prove that $EL(\Sigma_\epsilon) = W_1$ again, [15]. So for all $\epsilon \neq 0$ there is no recursive exact filter for any nonconstant statistic. Yet it is hard to believe that for small ϵ the Kalman-Bucy filter would not do a good job of first approximation. A question thus arises whether the estimation Lie algebra also has things to say about approximate filters. In this section and the following ones I shall argue that it does.

The first observation is as follows. If one actually commutes the two generators $\frac{1}{2}d^2/dx^2 - \frac{1}{2}(x+\epsilon x^3)^2$, $(x+\epsilon x^3)$ repeatedly of course eventually all the basis elements of W_1 appear. But they appear with higher and higher powers of ϵ and the ϵ-degree grows faster than the degree in (α) and (β) of the $x^\alpha \partial^\beta / \partial x^\beta$.

A precise version of this is as follows. Consider the two generators just listed as operators over the ring $\mathbb{R}[\epsilon]$ (or $\mathbb{R}[[\epsilon]]$), i.e. consider ϵ as an extra variable. Then it makes sense to consider

$$EL(\Sigma_\epsilon) \otimes_{\mathbb{R}[\epsilon]} \mathbb{R}[\epsilon] / \epsilon^n =: EL(\Sigma_\epsilon) \mathrm{mod}(\epsilon^n). \tag{11.2}$$

This simply amounts to setting $\epsilon^m = 0$ for $m \geqslant n$ whenever it appears. The set of all $\epsilon^i x^j d^k / dx^k$ with $\epsilon \geqslant n$ form an ideal in $\mathbb{R}[\epsilon]\langle x, d/dx \rangle$, so this makes sense. Now observe

Proposition 11.3. [15]. The Lie-algebras $EL(\Sigma_\epsilon)$ mod ϵ^n are finite-dimensional for all n.

As an example $EL(\Sigma_\epsilon)$ mod ϵ^2 turns out to be 14 dimensional with basis

$$\frac{1}{2}\frac{d^2}{dx^2} - \frac{1}{2}x^2 - \epsilon x^4,\ x,\ \epsilon x^3,\ \frac{d}{dx},\ 1,\ \epsilon,\ \epsilon x^2 \frac{d}{dx},\ \epsilon x,$$

$$\epsilon x \frac{d}{dx},\ \epsilon \frac{d^2}{dx^2},\ \epsilon \frac{d}{dx},\ \epsilon \frac{d^3}{dx^3},\ \epsilon x \frac{d^2}{dx^2},\ \epsilon x^2.$$

This is a general phenomenon.

Theorem 11.4. [15]. Let Σ_ϵ be a system of the form

$$dx = (Ax + \epsilon P_A(x))dt + (B + \epsilon P_B(x))dw, \quad dy = (Cx + \epsilon P_C(x))dt + dv \tag{11.5}$$

where P_A, P_B, P_C are polynomial vector and matrix valued functions of the approximate dimensions. Then $EL(\Sigma_\epsilon)$ mod (ϵ^n) is finite dimensional for all n. It is also solvable.

In [15] this is proved for the case $P_B = P_A = 0$. The proof generalizes immediately (simply give ϵ a negative enough degree to make degree decreasing all terms in the generators of $EL(\Sigma_\epsilon)$ in which ϵ appears (both $x_i, \partial / \partial x_j$ are given degree 1 in this argument).)

The next obvious question is: do these "finite dimensional quotients of $EL(\Sigma_\epsilon)$" actually compute anything, do they correspond to filters for some statistic? In the case of the weak cubic sensor this is (11.1) easy to answer. Consider the unnormalized conditional density $\rho(x,t,\epsilon)$ and (formally) expand it as a power series in ϵ

$$\rho(x,t,\epsilon) = \rho_0(x,t)+\epsilon\rho_1(x,t)+\epsilon^2\rho_2(x,t)+... \tag{11.6}$$

Then $EL(\Sigma_\epsilon)$ mod (ϵ^n) corresponds to the first n coefficients $\rho_0(x,t),...,\rho_{n-1}(x,t)$, and via Wei-Norman theory actually computes them. This is generally true, also in the setting of theorem 11.4. In the case of the weak cubic sensor (11.6) actually converges (for small ϵ). That, it appears, is not generally true. But it is still true that (11.6) gives an asymptotic expansion (Blankenschip-Liu-Marcus [4]). The Lie algebras being solvable one can of course implement these approximate filters, using the Wei-Norman technique. This was done in [4] and also the results were compared with the extended Kalman filter (EKF). The zero-th order approximation (of course) performed worse than EKF but the first order approximation performed better!

These Lie algebras $EL(\Sigma_\epsilon)$ tend to become large rapidly and to actually produce the, say FORTRAN, code is a long, but mechanical, job, prone to errors. Even the simplest nontrivial case needs several pages of densely written code. It is thus natural to try to let the computer do the job itself and in this way these ideas and techniques are being implemented in an expert system which is a joint effort of INRIA and the Department of Electrical Engineering of the University of Maryland (cf. Blankenschip [3]; the system also contains many other facets of stochastic control, filtering and optimization).

From the point of view developed in section 4 above the fact that calculating $\rho_0(x,t),...,\rho_{n-1}(x,t)$ corresponds to $EL(\Sigma_\epsilon)$ mod (ϵ^n) can be understood as follows. Choosing a basis suitably the remarks made in section 4 about the compatibility of Wei-Norman theory with quotients say that $\rho(x,t,\epsilon)$ admits an "expansion"

$$\rho(x,t,\epsilon) = e^{g_1(t)A_1}e^{g_2(t)A_2}...e^{g_n(t)A_n}...\rho_0(x) \tag{11.7}$$

with $g_1,...,g_{m(n)}$ where $m(n)=\dim(EL(\Sigma_\epsilon)$ mod $(\epsilon^n))$ depending only on $EL(\Sigma_\epsilon)$ mod (ϵ^n). The operators A_i in (11.7) involve higher and higher powers of ϵ. Writing out the exponentials one recovers (11.6). (And this point of view also strongly suggests (because also higher derivatives appear in the A_i) that the best one can hope for in general is an asymptotic expansion.)

12. THE PROFINITE DIMENSIONAL CASE

A Lie-algebra L is said to be profinite dimensional if there is a sequence of ideals $L\supset\mathfrak{A}_1\supset\mathfrak{A}_2\supset\mathfrak{A}_3\supset...$ such that

$$\dim L\,/\,\mathfrak{A}_i < \infty \text{ for all } i \tag{12.1}$$

$$\bigcap_i \mathfrak{A}_i = \{0\}. \tag{12.2}$$

Suppose the estimation Lie algebra $EL(\Sigma)$ has this property. Then again, as in the previous section, one can write an expansion

$$\rho(x,t) = e^{g_1A_1}e^{g_2A_2}...e^{g_nA_n}...\rho_0(x) \tag{12.3}$$

and consider the possible approximants

$$\rho^{(n)}(x,t) = e^{g_1A_1}e^{g_2A_2}...e^{g_nA_n}\rho_0(x). \tag{12.4}$$

Using, again, that the equations for $g_1,...,g_n$; $n=n(m)=\dim L\,/\,\mathfrak{A}_m$ do not depend on $g_{n+1},...$. Abstractly, there is no immediate reason to expect the higher $e^{g_kA_k}$ to be small, though one would expect this to the case in the majority of the interesting cases, even in more general cases than this, as I shall argue below in section 14.

Profinite dimensional estimation Lie algebras occur frequently. Consider systems

$$dx = f(x)dt + G(x)dw_t, \quad dy = h(x)dt + dv \tag{12.5}$$

with the additional assumptions that f,G and h are analytic (totally around zero) and that $f(0)=G(0)=0$.

Theorem 12.6. [17]. Under the assumptions made immediately above $EL(\Sigma)$ is profinite dimensional.

If one adds the condition that $h(0)=0$ (which surely does no harm; removing a known constant from the observation equation is a triviality) the resulting estimation Lie algebra is even solvable (meaning that all the quotients $L/\mathfrak{A}_i$ are solvable).

Another case of a profinite dimensional estimation Lie algebra (different from the class of theorem 12.6, the identification case to be treated below, and the perturbation case of section 11 above) is example 7.5. As a rule one should probably not expect that "the statistic calculated by $L/\mathfrak{A}_i$" of a system whose estimation Lie algebra happens to be profinite dimensional, is easily interpretable (recognizable) as the statistic of an interesting quantity. In the case of example 7.5 this is however the case, (Liu-Marcus [23]).

13. IDENTIFICATION OF LINEAR DYNAMICAL SYSTEMS

Suppose now that we are faced with a somewhat different problem. Namely suppose one has reason to believe, or simply does not know anything better to do, that a given phenomenon, say a time series, is modeled by a linear dynamical system

$$dx = Axdt + Bdw, \quad dy = Cxdt + dv \tag{13.1}$$

Now, however, the coefficients in A,B,C are unknown and also have to be estimated from the observation $y(t)$. That is the system (13.1) has to be identified. It is easy to turn this into a filtering problm by adding the (stochastic) equations

$$dA = 0, \quad dB = 0, \quad dC = 0 \tag{13.2}$$

(or just $dr_{ij}=0$ whether the r_{ij} run through the coefficients which are unknown, if A,B,C are partly known; for example because of structural considerations). The resulting filtering problem is nonlinear.

13.1. Observation

The estimation Lie algebra of the system (13.1)-(13.2) is a sub-Lie-algebra of the current Lie algebra $ls_n \otimes \mathbf{R}[A,B,C]$ where $\mathbf{R}[A,B,C]$ stands for the ring of polynomials in the indeterminates a_{ij}, b_{kl}, c_{rs}.

A corollary is that these estimation algebras are profinite dimensional. And looking a bit more closely at them, they are solvable [37]. Thus the ideas and considerations of the previous two sections can be brought into play and one can try to do infinite dimensional Wei-Norman theory etc. This is attempted in Krishnaprasad-Marcus-Hazewinkel [37]. In this rather special case it turns out that the higher approximations (the zero-th approximation is simply the family of Kalman-Bucy filters parametrized by A,B,C also discussed in section 8 above) have to do with sensitivity equations: sensitivities of the ouput $y(t)$ with respect to changes in the parameters A,B,C.

As stated above, though, the problem is degenerate and likely to cause all kind of difficulties. The problem is that the conditional density $\rho(x,A,B,C,t)$ will be degenerate because the A,B,C are not uniquely determined by the observations. Indeed if S is an invertible $n \times n$ matrix then the system (13.1) given by the matrices SAS^{-1}, SB, CS^{-1} instead of A,B,C gives exactly the same input-output behaviour. Thus we should really be considering this problem on a suitable quotient space $\{(A,B,C)\}/GL_n$. These quotient spaces as a rule are not diffeomorphic to open sets in some $\mathbf{R}^n$. This is one way in which stochastic systems like (1.1)-(1.2) on nontrivial manifolds naturally arise and it leads to the necessity of finding a DMZ-equation in this more general context. Work in this direction has been done by Ji Dunmu and T.E. Duncan.

Let me add one observation. For the filters giving $\hat{x}, \hat{A}, \hat{B}, \hat{C}$ for problem (13.1)-(13.2) one expects $\hat{x}$

to move fast relative $\hat{A},\hat{B},\hat{C}$. Thus it would make sense to consider a system

$$dx = (A_0+\epsilon A_1)xdt+(B_0+\epsilon B_1)dw,\ \ dy = (C_0+\epsilon C_1)dt+dv \tag{13.4}$$

$$dA_1 = 0,\ \ dB_1 = 0,\ \ dC_1 = 0$$

(where A_0,B_0,C_0 are assumed known) and apply the ideas of section 11 above to find optimal directions of change (i.e. the A_1,B_1,C_1).

14. ASYMPTOTIC EXPANSIONS AND APPROXIMATE HOMOMORPHISMS

The ideas to be outlined below in this section are still speculative but there are quite a number of positive signs.

First however let me point out that the procedures based on Wei-Norman techniques as described in sections 11 and 12 above clearly indicate that existence, uniqueness and regularity results for solutions of the DMZ-equation have a lot to do with the existence of asymptotic expansions ([2,4]). For regularity results etc. cf. e.g. work of D. Michel, J.-M. Bismut, E. Pardoux, M. Chaleyat-Maurel, D. Ocone, Th. Kurtz, W.E. Hopkins Jr., H.J. Sussmann a.o. ([25,8,22,2] and references in these papers).

Let us consider a control system of the form

$$\dot{x} = f(x)+\Sigma u_i g_i(x) \tag{14.1}$$

where the f and g_i are vectorfields. To make thinking easier assume that 0 is a stable and asymptotically stable equilibrium for the unforced equation. A system like (14.1) is intended as a model of something and as such one can argue that say the values of $f(x),g_i(x)$ are relatively well known, the values of their (partial) derivatives (w.r.t. the x_i) will be less well known, the second partial derivations are still less well determined etc..

Thus, intuitively, for systems which represent or model real (stable) things one would expect that in many cases the behaviour of (14.1) will depend primarily on the first few terms which appear in the Lie algebra generated by f and the g_i. The higher brackets should matter less and less.

That means that instead of looking at $Lie\{f,g_1,...,g_m\}$, the Lie algebra generated by $f,g_1,...,g_m$ as a Lie algebra without further structure, we should look at it as a Lie algebra with a given set of generators and sort of keep track of how often these generators are used to generate further elements of the algebra. (For each time a bracket is taken a differentiation is applied, and thus the higher brackets of the $f,g_1,...,g_m$ depend only on the deeper parts of the Taylor expansions of $f,g_1,...,g_m$.)

Personally I would also say that having noises rather than precise deterministic controls u_i would enhance this type of (structural?) stability.

A precise way to keep track of how often the generators are used is to introduce one extra counting indeterminate z and to consider instead of $L = Lie\{f,g_1,...,g_m\}$ the Lie algebra generated by the vectorfields $\{zf,zg_1,...,zg_m\}$. This Lie algebra L_z is topologically nilpotent, i.e. if $L_z^{(n)}=[L_z,L_z^{(n-1)}]$, $L_z^{(0)}=L_z$, then $\cap L_z^{(m)}=\{0\}$. And a homomorphism $L_z \to V(M)$ into the vectorfields on M with kernel $L_z^{(n)}$ precisely means "respecting the structure of the Lie algebra L up to brackets of order n". All this is very much related to the ideas of nilpotent approximation as introduced by Stein, Rothschild, Goodman and Rockland, [32,14,31] in the study of hypoellipticity and taken up by Crouch in system theory [10].

Thus in filtering theory it would seem natural to look at the Lie algebra of operators $EL_z(\Sigma)$ generated by the operators

$$z_0\mathfrak{L},\ z_1h_1,...,z_ph_p$$

where the $z_0,z_1,...,z_p$ are additional variables (so as to give, if desired, certain observations more weight than others and to be able to set certain of them, especially z_0, equal to 1). The idea would be then to study the filters produced by Wei-Norman type techniques for the various finite dimensional quotients and to see whether this produces viable expansions.

15. REMOVING OUTLIERS

A final idea in much the same spirit as before is the following. Suppose we are again dealing with a system

$$dx = f(x)dt + G(x)dw, \quad dy = h(x)dt + dv. \tag{15.1}$$

Suppose also to make thinking easier that the thing is more or less stable so that x tends to remain in some bounded partion of $\mathbb{R}^n$ (f asymptotically stable) and maybe suppose also that h is proper, so that large y observations are exceedingly rare and should probably be discounted. Suppose that $e^{-\|x\|^2}$ is differential algebraically independent of f,G,h. This is for example this case if f,G,h are polynomial and also if they are of compact support. In other cases other functions with similar properties can presumably be found. Now instead of (15.1) consider the modified system

$$dx = f(x)dt + G(x)dw, \quad dy = e^{-a\|x\|^2}h(x)dt + dv \tag{15.2}$$

where $a>0$ is a small parameter. Note that the only thing which (15.2) does with respect to (15.1) is to discount large y observations.

Now consider the estimation Lie algebra of the sytem (15.2).

Theorem 15.3. If $e^{-a\|x\|^2}$ is differentially algebraically independent of f,G,h then the estimation Lie algebra of (15.2) is pro-finite dimensional and solvable. To be more precise it is finite dimensional and solvable mod $(a^i e^{-ja\|x\|^2}, i+j \geq n)$ for all n.

Thus the yoga of the previous sections can again be applied and the behaviour of the resulting filters as a goes to zero could be studied.

REFERENCES

[1] J.S. BARAS, Group invariance methods in nonlinear filtering of diffusion processes, In: [19], 565-572.

[2] J.S. BARAS, G.L. BLANKENSCHIP, W.E. HOPKINS Jr., Existence, uniqueness and asymptotic behaviour of solutions to a class of Zakai equations with unbounded coefficients, IEEE Trans. AC-28 (1983), 203-214.

[3] G.L. BLANKENSCHIP, Lecture at MTNS'85, Stockholm, To appear North Holland Publ. Cy.

[4] G.L. BLANKENSCHIP, C.-H. LIU, S.I. MARCUS, Asymptotic expansions and Lie algebras for some nonlinear filter problems, IEEE Trans. AC-28 (1983), 787-797.

[5] R.W. BROCKETT, Remarks on finite dimensional nonlinear estimation, In: C. Lobry (ed.), Analyse des systèmes, Astérisque 76 (1980), Soc. Math. de France.

[6] R.W. BROCKETT, J.M.C. CLARK, The geometry of the conditional density equation In: O.L.R. Jacobs et al. (eds), Analysis and optimization of stochastic systems, New York, 1980, 299-309.

[7] R.S. BUCY, J.M.F. MOURA (eds), Nonlinear stochastic problems, Reidel, 1983.

[8] J.M.C. CLARK, The design of robust approximations to the stochastic differential equations of nonlinear filtering, In: J.K. Skwirzynski (ed.), Communication systems and random process theory, Sijthoff & Noordhoff, 1978.

[9] M. CHALEYAT-MAUREL, D. MICHEL, Hypoellipticity theorems and conditional laws, Z. Wahrsch. und verw. Geb. 65 (1984), 573-597.

[10] P.E. CROUCH, Solvable approximations to control systems, SIAM J. Control and Opt. 32 (1984), 40-54.

[11] M.H.A. DAVIS, Pathwise nonlinear filtering, In [19], 505-528.

[12] M.H.A. Dsavis, S.I. MARCUS, An introduction to nonlinear filtering, In [19], 565-572.

[13] T.E. DUNCAN, Probability densities for diffusion processes with applications to nonllinear filtering theory, Ph. D. thesis, Stanford, 1967.

[14] R.W. GOODMAN, Nilpotent Lie groups, structure and applications to analysis, LNM 562, Springer, 1976.

[15] M. HAZEWINKEL, On deformations, approximations and nonlinear filtering, Systems Control Lett. 1 (1982), 29-62.

[16] M. HAZEWINKEL, The linear systems Lie algebra, the Segal-Shale-Weil representation and all Kalman-Bucy filters, J. Syst. Sci. & Math. Sci. 5 (1985), 94-106.

[17] M. HAZEWINKEL, S.I. MARCUS, On Lie algebras and finite dimensional filtering, Stochastics 7 (1982), 29-62.

[18] M.HAZEWINKEL, S.I. MARCUS, H.J. SUSSMANN, Nonexistence of finite dimensional filters for conditional statistics of the cubic sensor problem, Systems Control Lett. 3 (1983), 331-340.

[19] M. HAZEWINKEL, J.C. WILLEMS (eds), Stochastic systems: the mathematics of filtering and identification and applications, Reidel, 1981.

[20] O.B. HIJAB, Finite dimensional causal functionals of brownian motion, In [6], 425-436.

[21] A.J. KRENER, On the equivalence of control systems and the linearization of nonlinear systems, SIAM J. Control 11 (1973), 670-676.

[22] Th.G. KURTZ, D. OCONE, A martingale problem for conditional distributions and uniqueness for the nonlinear filtering equations, Lect. Notes Control and Inf. Sci. 69 (1985), 224-235.

[23] C.-H. LIU, S.I. MARCUS, The Lie algebraic structure of a class of finite dimensional nonlinear filters, In: C.I. Byrnes, C.F. Martin (eds), Algebraic and geometric methods in linear systems theory, Amer. Math. Soc., 1980, 277-297.

[24] S.I. MARCUS, Algebraic and geometric methods in nonlinear filtering, SIAM J. Control Opt. 22 (1984), 817-844.

[25] D. MICHEL, Régularité des lois conditionelles en théorie du filtrage non-linéaire et calcul des variations stochastiques, J. Funct. Anal. 14 (1981), 8-36.

[26] S.K. MITTER, On the analogy between mathematical problems of non-linear filtering and quantum phisics, Ric. di Automatica 10 (1980), 163-216.

[27] S.K. MITTER, Nonlinear filtering and stochastic mechanics, In: [19], 479-504.

[28] R.E. MORTENSEN, Optimal control of continuous time stochastic systems, Ph.D. thesis, Berkeley, 1966.

[29] E. PARDOUX, D. TALAY, Discretization and simulation of stochastic differential equations, Acta Appl. Math. 3 (1982), 182-203.

[30] L.E. PURSELL, M.E. SHANKS, The Lie algebra of a smooth manifold, Proc. Amer. Math. Soc. 5 (1954), 468-472.

[31] Ch. ROCKLAND, Intrinsic nilpotent approximation, preprint MIT, LIDS-R-1482, 1985, to appear Acta Appl. Math.

[32] L.P. ROTHSCHILD, E.M. STEIN, Hypoelliptic differential operators and nilpotent groups, Acta Math. 137 (1976), 247-320.

[33] H.J. SUSSMANN, Approximate finite dimensional filters for some nonlinear problems, Stochastics 7 (1982), 183-203.

[34] H.J. SUSSMANN, Existence and uniqueness of minimal realizations of nonlinear systems, Math.

Syst. Theory 10 (1977), 349-356.

[35] J. WEI, E. NORMAN, On the global representation of the solutions of linear differential equations as a product of exponentials, Proc. Amer. Math. Soc. 15 (1964), 327-334.

[36] M. ZAKAI, On the optimal filtering of diffusion processes, Z. Wahrsch. verw. Geb. 11 (1969), 230-243.

[37] P.S. KRISHNAPRASAD, S.I. MARCUS, M. HAZEWINKEL, Current algebras and the identification problem, Stochastics 11 (1983), 65-101.

WHITE NOISE ANALYSIS AND THE LEVY LAPLACIAN

Takeyuki HIDA and Kimiaki SAITO
Department of Mathematics
Faculty of Science, Nagoya University
Chikusa-ku, Nagoya 464
JAPAN

ABSTRACT. In line with the harmonic analysis on the space $(L^2)^-$ of generalized Brownian functionals we are given the Lévy's Laplacian Δ_L and discuss its roles in the causal calculus on $(L^2)^-$. There we can find interesting relations to the Lévy group as well as to the Fourier transform introduced by H.-H. Kuo.

0. INTRODUCTION

The white noise analysis which discusses the analysis of Brownian functionals has an aspect of the harmonic analysis arising from the infinite dimensional rotation group. There the Laplacian plays a key role as it does in the finite dimensional analysis.

There are, however, many operators that may be viewed as infinite dimensional analogue of the ordinary Laplacian. Each of them has its own specific character and plays some roles in the white noise analysis. Among them, the Lévy Laplacian, discussed in [1] is just fitting for our present purpose to carry out the differential and integral calculus of generalized Brownian functionals.

As will be seen in Section 3, one can find intimate relation between the Lévy Laplacian and the Lévy group $\mathcal{G}$ (introduced in [1] and we consider it in Section 2) that comes from the permutation of the coordinate axes. Indeed, we shall see in Theorem 1 that Δ_L commutes with the group $\mathcal{G}$ and it is a quadratic form of the differential operators. We can even observe that Δ_L is like an operator obtained by taking the limit of

$$\frac{1}{N}\sum_1^N \frac{\partial^2}{\partial \xi_n^2}$$

as $N \to \infty$. It is our hope to prove that the group $\mathcal{G}$ would characterize the operator Δ_L.

Also in the last section, we shall show some other important characteristic properties of Δ_L such as the relation to the Fourier transform on $(L^2)^-$ introduced by H.-H. Kuo [8] and an evaluation of the

S. Albeverio et al. (eds.), Stochastic Processes in Physics and Engineering, 177–184.

norm $\|\Delta_L\phi\|_{(L^2)}$- for ϕ in the domain of Δ_L. Further relations between the Lévy Laplacian and the infinite dimensional rotation group will be discussed in another paper.

1. BACKGROUND

Let the time parameter set T be taken to be the unit interval $[0,1]$, and let H be the Hilbert space involving real, square integrable functions on T. We start with a Gel'fand triple

$$E \subset H \subset E^*,$$

where E is a nuclear space and E^* is the dual space of E. The standard Gaussian measure μ is introduced on E^* by the characteristic functional $C(\xi) = \exp(-\|\xi\|^2/2)$:

$$C(\xi) = \int \exp(i\langle x,\xi\rangle)d\mu(x), \quad \xi \in E.$$

With this measure μ almost every x may be viewed as a sample function of white noise $\dot{B}(t)$ which is the time derivative of Brownian motion $B(t)$.

A complex Hilbert space $(L^2) = (L^2(E^*,\mu)$ is formed, and we are given the Wiener-Itô decomposition :

(1) $$(L^2) = \oplus_0^\infty H_n \qquad \text{(Fock space)}.$$

A member of (L^2) is a Brownian functional.

The integral representation of a Brownian functional is obtained by the T-transform:

(2) $$(T\phi)(\xi) = \int \exp(i\langle x,\xi\rangle)\phi(x)d\mu(x) \qquad \phi \in (L^2).$$

If ϕ is restricted to the subspace H_n, then we have

(3) $$\begin{aligned}(T\phi)(\xi) &= i^n C(\xi)U(\xi)\\ U(\xi) &= \int_{T^n} F(u_1,\cdots,u_n)\xi(u_1)\cdots\xi(u_n)du^n\end{aligned}$$

and establish an isomorphism :

(4) $$\begin{aligned}&\phi \to F \in \widehat{L^2}(T^n) \quad (= \text{symmetric } L^2(T^n) \text{ space}),\\ &\|\phi\|_{(L^2)} = \sqrt{n!}\,\|F\|_{L^2(T^n)}.\end{aligned}$$

The function F is called the kernel of the representation of ϕ.

In order to carry out the causal calculus (see [3]), we are led to introduce a class of generalized Brownian functionals. First, we have to define the space $H_n^{(n)}$ $(\subset H_n)$ of test functionals :

(5) $$\begin{aligned}&H_n^{(n)} = \{\phi \in H_n;\ \text{kernel of } \phi \text{ is in } H^{(n+1)/2}(T^n) \cap \widehat{L^2}(T^n)\}.\\ &\|\phi\|_n = \sqrt{n!}\cdot H^{(n+1)/2}(T^n)\text{-norm of the kernel.}\end{aligned}$$

The space $\mathcal{H}_n^{(-n)}$ is defined to be the dual space of $\mathcal{H}_n^{(n)}$. Thus we obtain a triple :

$$\mathcal{H}_n^{(n)} \subset \mathcal{H}_n \subset \mathcal{H}_n^{(-n)}.$$

A member in $\mathcal{H}_n^{(-n)}$ is called a generalized (Brownian) functional. The $\mathcal{T}$-transform is extended to the space $\mathcal{H}_n^{(-n)}$ and we can define U-functional of a generalized Brownian functional. The explicit form of the U-functional is given by the $\mathcal{S}$-transform (see [6]) :

$$(\mathcal{S}\phi)(\xi) = \int \phi(x+\xi)d\mu(x).$$

A typical example of a generalized Brownian functional ϕ is given by the following U-functional

$$U(\xi) = \int_{T^k} G(u_1,\cdots,u_k)\xi(u_1)^{n_1}\cdots\xi(u_k)^{n_k}du^k, \quad G \in L^2(T^k). \tag{6}$$

Applying the transform $\mathcal{S}^{-1}$ to this U we get a generalized Brownian functional expressed in the form

$$(\mathcal{S}^{-1}U)(x) = \int_{T^k} G(u_1,\cdots,u_k):x(u_1)^{n_1}\cdots x(u_k)^{n_k}:du^k, \tag{7}$$

where : : means the renormalization (for details, see [5]). After P. Lévy, we call it a normal functional. Note that the functional is a realization of a polynomial, indeed, a Hermite polynomial in $\dot{B}(t)$'s.

A weighted sum $(L^2)^- \equiv (L^2)^-_{\{c_n\}}$ is defined as the dual space of the space $(L^2)^+ \equiv (L^2)^+_{\{c_n\}}$ of test functionals:

$$(L^2)^+_{\{c_n\}} = \{\phi = \Sigma\,\phi_n;\ \phi_n \in \mathcal{H}_n^{(n)},\ \Sigma\, c_n\|\phi\|_n^2 < \infty\},$$

where $\|\ \|_n$ is the $\mathcal{H}_n^{(n)}$-norm, and $0 < c_n\uparrow$.

Example. An exponential function of the form

$$\phi(x) = :\exp(c\int x(t)^2dt):, \quad c < 1/2, \tag{8}$$

is a generalized functional living in $(L^2)^-$, the U-functional of which is given by

$$\exp(c'\int \xi(t)^2dt), \quad c' = c/(1-2c).$$

The renormalization in (8) is multiplicative, since the functional is exponential.

We are now ready to introduce a differential operator ∂_t, which is, formally speaking, the partial differential operator with respect to the variable $x(t)$, or to the $\dot{B}(t)$.

Definition. Let $U(\xi)$ be the U-functional associated with an (L^2)-functional ϕ. If $U(\xi)$ is Fréchet differentiable and if the derivative $\frac{\partial U}{\partial \xi(t)}$ is a U-functional associated with some ϕ_t' in the space $(L^2)^-$, then ϕ is said to be ∂_t-differentiable and we define

$$\phi_t' = \partial_t \phi.$$

By using the duality between $(L^2)^+$ and $(L^2)^-$ we can define the adjoint operator ∂_t^* for ∂_t.

There are many interesting operators acting on $(L^2)^-$ that are quadratic forms of ∂_t and ∂_t^*; among others

$$N = \int \partial_t^* \partial_t dt$$

is well known as the number operator. In fact, N acts in such a way that

$$N\phi = n\phi, \quad \text{for } \phi \in H_n.$$

The operator

$$\Delta_\infty = -N$$

is the infinite dimensional Laplace-Beltrami operator, and it is often called Ornstein-Uhlenbeck operator.

2. THE LEVY LAPLACIAN

We are interested in an infinite dimensional analogue of the finite dimensional Laplacian. Originally it was proposed by P. Lévy (see, for example, [1]). We expect that such an operator would be a quadratic form of the differential operators ∂_t, $t \in T$.

Before we give the definition of the Lévy Laplacian, let us remind the definition of the second order Fréchet derivative.

Let $U(\xi)$ be a functional on E, and let $U'(t)$ be the Fréchet derivative of U. If the variation $\delta U'(t)$ is expressed in the form

(10) $$\delta U'(t) = U''_{\xi\xi}(t)\delta\xi(t) + \int U''_{\xi\eta}(t,s)\delta\xi(s)ds,$$

then $U''_{\xi\xi}$ and $U''_{\xi\eta}$ are called the second order Fréchet derivatives.

Let ϕ be an (L^2)-functional with U-functional $U(\xi)$. Assume that $U''_{\xi\xi}(t)$ is a U-functional associated with some generalized functional ϕ_t''. Then the Lévy Laplacian Δ_L is defined to be an operator acting in the form

(11) $$\Delta_L = \int \phi_t'' dt.$$

We are now in a position to discuss the domain of the Laplacian. Let $\mathcal{D}$ be the vector space spanned by the normal functionals. We introduce a norm $\| \ \|_e$ to $\mathcal{D}$:

(12) $$\|\phi\|_e^2 = \|\phi\|_-^2 + \int \|\partial_t\phi\|_-^2 dt + \int \|\partial_t^2 \phi dt\|_-^2 dt, \quad \| \ \|_- : (L^2)^-\text{-norm}.$$

Obviously $\|\phi\|_e < \infty$ for any $\phi \in \mathcal{D}$. The completion of $\mathcal{D}$ with respect to the norm $\| \ \|_e$ will be denoted by $\mathcal{D}$ and is taken to be the domain of the Lévy Laplacian.

Obviously, we have

Proposition. For ϕ in $\mathcal{D}$

$$\|\Delta_L\phi\|_- \leq \|\phi\|_e.$$

Now one may ask if there is any connection between the Lévy Laplacian and the limit

$$\lim \frac{1}{N}\Sigma_1^N \frac{\partial^2}{\partial\xi^2} \equiv \tilde{\Delta}$$

which was the motivation of the Lévy's work. We are actually able to give it an interpretation in the following manner (see [1], Part III). Let ξ_n be a complete orthonormal system in H which is "equally dense", that is,

$$\frac{1}{N}\Sigma_1^N \xi_n(t)^2 \to 1 \quad \text{in} \quad L^1(T), \quad \text{as} \quad N \to \infty.$$

Then, if $U(\xi)$ is the U-functional associated with a functional in $\mathcal{D}$, then

$$\int U''_{\xi\xi}(t)dt = \tilde{\Delta}U,$$

as is expected.

There is an interesting observation on the Lévy Laplacian. H.-H. Kuo proposes to express Δ_L by the formula

$$\Delta_L = \int \partial_t^2 (dt)^2. \tag{13}$$

In fact, we are able to give a plausible explanation to this expression, since the first term of the right hand side of (9) may be written as

$$\int U''_{\xi\xi}(s)\delta_t(s)\delta\xi(s)ds$$

and since $\delta_t(t)$ may be written as $1/dt$ like in the non-standard analysis.

Example. For $\phi(x) = \int f(t) :x(t)^2: dt$ we have

$$\Delta_L\phi(x) = 2\int f(t)\, dt.$$

Remark. A functional $\int x(t)x(1-t)\, dt$ is not in the domain of the Laplacian.

3. SOME FURTHER PROPERTIES OF THE LEVY LAPLACIAN

Before we discuss some more interesting properties of the Lévy Laplacian, we have to prepare a notion of the Lévy group which is an important subgroup of the infinite dimensional rotation group. Let E be as in Section 1. Set

$$O(E) = \{g;\ g \text{ is a linear homeomorphism of } E \text{ and } \|g\xi\| = \|\xi\| \text{ for every } \xi \in E.\}$$

It forms a group (see e.g. 2, Chapt.5). Let $\{\xi_n\}$ be the complete orthonormal system given in the last section. Take a finite, say n, dimensional subspace E_n of E spanned by ξ_k, $1 \le k \le n$, and consider the rotations on E_n. They can define members in $O(E)$ and the collection of them is a subgroup G isomorphic to $SO(n)$. The inductive limit is also a subgroup denoted by G_∞ which we shall call the finite dimensional rotation group. We then come to the Lévy group (see [1] Part III). Let π be a permutation of positive integers and define g_π by

$$g_\pi \xi = \Sigma a_n \xi_{\pi(n)} \quad \text{for} \quad \xi = \Sigma a_n \xi_n .$$

The collection

$$\mathcal{G} = \left\{ g \in O(E);\ g = g_\pi,\ \forall \varepsilon > 0, \exists N(\varepsilon),\ n > N(\varepsilon) \atop \#\{i \le n;\ \pi(i) > n\} < \varepsilon n \right\}$$

forms a subgroup of $O(E)$, and it is called the Lévy group.

Incidentally, we note that G_∞ is a normal subgroup of the group G generated by G_∞ and $\mathcal{G}$.

Let g^* be the adjoint of g. Then g^* is a linear transformation acting on E^* and it keeps the measure invariant. We can therefore define an operator U_g acting on (L^2):

$$(U_g \phi)(x) = \phi(g^* x).$$

If U_g is restricted to (L^2), then U_g is a unitary operator.

Theorem 1. The Lêvy Laplacian Δ_L enjoys the following properties.

i) Δ_L annihilates the space (L^2),

ii) it is a quadratic form of ∂_t, $t \in T$, in a generalized sense,

iii) it commutes with the Lévy group in the sense that

$$\Delta_L U_g = U_g \Delta_L \quad \text{for any} \quad g$$

Proof. i) The integral representation tells us that any functional in $\mathcal{H}_n$ is expressed in the form (7) with $n_1 = \cdots = n_k = 1$. Hence the associated U-functional has no term $U''(t)$. This means that the Lévy Laplacian vanishes.

ii) is obvious if we see the Kuo's expression (12).

iii) is proved if we see the expression $\tilde{\Delta}$.

The last topic is related to the Kuo's Fourier transform. He takes two different white noises with the symbols x and y. For ϕ in $(L^2)_x^-$ define $\hat{\phi}$ by

$$(14) \qquad \hat{\phi}(y) = \int :\exp(-i \int_T y(u)x(u)du):_y \phi(x) d\mu(x).$$

Theorem 2. For ϕ in $\mathcal{D}$, we have

(15) $$(\Delta_L+1)\hat{\phi} = -\widehat{\Delta_L\phi}.$$

Proof. It suffices for us to prove (15) for a normal functional $\phi(x)$ given by

$$\phi(x) = \int_{T^k} G(u_1,\cdots u_k) :x(u_1)^{n_1}\cdots x(u_k)^{n_k}: du^k, \quad \Sigma n_j = n.$$

By Theorem 6 in Kuo [8] we have

$$\hat{\phi}(y) = (-i)^n \Sigma_0^\infty \frac{(-1)^N}{N!2^N} \times \int_{T^N}\int_{T^k} G(u_1,\cdots,u_k):$$

$$y(u_1)^{n_1}\cdots y(u_k)^{n_k} y(u_{k+1})^2\cdots y(u_{K+N})^2: dt^{k+N}.$$

While

$$(\Delta_L+1)\hat{\phi}(y) = (-i)^n \Sigma_0^\infty \frac{(-1)^N}{N!2^N} \sum_{j=1}^k n_j(n_j-1) \int_{T^N}\int_{T^k}$$

$$G(u_1,\cdots,u_k):y(u_1)^{n_1}\cdots y(u_{j-1})^{n_j-1} y(u_j)^{n_j-2}$$

$$\times\ y(u_{j+1})^{n_j+1}\cdots y(u_k)^{n_k}\ y(u_{k+1})^2\cdots y(u_{k+N})^2: dt^{k+N}.$$

In a similar manner we can see that $-\widehat{\Delta_L\phi}(y)$ has also the same expression as above. Thus we have proved the theorem.

Before closing this paper, we would like to state a proposition, although we do not have so deep insight.

Proposition. Exponential functions given by (8) are all eigenfunctionals of the operator Δ_L with eigenvalues $2c'$.

REFERENCES

[1] P. Lévy, Problemes concrets d'analyse fonctionnelle, Gauthier-Villars, 1951.

[2] T. Hida, Brownian motion (English translation), Springer-Verlag, New York, 1980.

[3] ______, Causal nanlysis in terms of white noise, in Quatum Fields -Algebra, Processes, ed. L Streit, Springer-Verlag, New York, 1980, 1-19.

[4] ______, Generalized Brownian functionals and stochastic integrals. Appl. Math. Optimization 12, 1984, 115-123.

[5] ______, Analysis of Brownian functionals. Lecture Notes, IMA University of Minnesota, 1986.

[6] I. Kubo and S. Takenaka, Calculus of Gaussian white noise, Proc. Japan Acad. 56A: 376-380, 411-416; 57A:433-437; 58A:186-

189, 1980, 1981, 1982.

[7] H.-H. Kuo, Brownian functionals and applications, Acta Applicandae Mathematicae 1, 175-188, 1983.

[8] ________, On Fourier transform of generalized functionals. J. of Multivariate Analysis, 12, 415-431, 1982.

[9] N. Obata, A note on certain permutation groups in the infinite dimensional rotation group. to appear.

[10] L. Streit and T. Hida, Generalized Brownian functionals and the Feynman integrals, Stochastic Processes and their Applications. 16, 55-69, 1983.

AN APPLICATION OF ADAPTIVE OPTIMAL CONTROL

R. Hut G.J. Olsder
Dept. of Mathematics and Informatics
Delft University of Technology
P.O. Box 356, 2600 AJ Delft
The Netherlands

ABSTRACT. A combined estimation and optimal control problem is considered. The application deals with two ships in which one receives passive sonar measurements of the other. Three numerical schemes are given and their results are compared.

1. INTRODUCTION

In this paper a problem of combined optimal control and estimation has been considered. The theory of estimation is well developed, see for instance [1], [2], as is the field of optimal control [3]. In practical problems in which some parameters may not be known, adaptive control is usually employed. In an adaptive control problem estimates of parameters are updated regularly and the control is adjusted to these new estimates (as if they were completely known), see for instance [4]. Adaptive control is used in order to maintain a certain performance for the controlled system, such as for instance that stability should be assured. In the current paper we do not only want to control the system, but also want to minimize a criterion function.

The application considered deals with two ships. One of these ships (S_1) can manoeuvre, whereas the other ship (S_2) will follow a rectilinear course. S_1 would like to know the distance between the two ships, but does not measure this distance directly. The only measurements available are the direction in which S_2 is seen from S_1 with respect to a fixed reference axis. In nautical terms one would speak about bearings-only measurements obtained by means of passive sonar. The accuracy with which the distance between S_1 and S_2 can be calculated from the bearings-only measurements depends on the manoeuvre chosen by ship S_1. It is for instance known, that if S_1 also performs a rectilinear course with constant speed, then this distance cannot be determined at all [5]. The optimal manoeuvre of S_1 (i.e. it minimizes the variance of the distance at a given time instant) depends on the actual position of S_2, which is exactly what S_1 would like to know. Due to this and also to the nonlinearities in the problem, the problem we want to solve is a very difficult one.

S. Albeverio et al. (eds.), Stochastic Processes in Physics and Engineering, 185–192.

In this paper two numerical approaches have been described and they are compared to the one in [5]. These approaches deal with on-line optimization, which in principle could be implemented in practical situations.

In section 2 the model is given. It consists of systems equations (a set of three scalar nonlinear differential equations) and an observation equation. In section 3 the extended Kalman filter is used for estimation purposes, given a fixed manoeuvre. Lastly, in section 4, numerical schemes for (sub)optimal manoeuvres are given together with results.

2. THE MODEL

We are given two ships, ship S_1 and ship S_2. Quantities describing the characteristic features of S_i will be provided with an index i. Both ships move in an undisturbed, flat, sea. Ship S_1 measures the angle in which it "hears" (by means of sonar) the other ship S_2. This angle, called δ, is measured with respect to a fixed reference axis e.g. the North direction, and the half line in which direction S_2 is spotted by S_1 is called the line of sight (l.o.s.). The speeds v_i, $i = 1,2$, are assumed to be constant. Ship S_1 can change the direction in which it moves. This direction is indicated by u and is measured with respect to the l.o.s. Though not very realistic it is assumed that S_1 can change its course instantaneously, i.e. the function u which is the control variable, is not necessarily continuous with respect to the time t. It is assumed that ship S_2 moves according to a straight line and hence does not change its course. Interesting variables to S_1 are the distance r to S_2 and the angle γ between the l.o.s. and the direction in which S_2 moves. The distance r is time dependent, as is γ. The latter quantity is time dependent since the l.o.s. is not constant; it will change because the ships move.

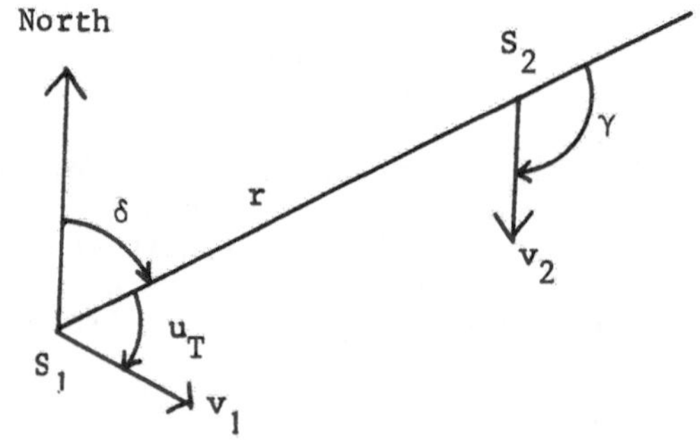

Fig. 1

The quantities introduced so far have been indicated in Fig. 1. It is possible to describe the configuration of S_2 relative to S_1 with respect to different quantities (e.g. the direction in which S_2 moves could be determined w.r.t. the North direction). The choice for the quantities used in this paper has been described in [5].

It is not difficult to derive the model for the time dependent behaviour of the various variables;

$$\dot{r} = - v_1 \cos u + v_2 \cos\gamma \tag{1a}$$

$$\dot{\gamma} = \frac{1}{r} (v_1 \sin u - v_2 \sin\gamma) \tag{1b}$$

$$\dot{\delta} = - \frac{1}{r} (v_1 \sin u - v_2 \sin\gamma) \tag{1c}$$

$$\dot{v}_2 = 0 \tag{1d}$$

The differential equations are easily derived by considering two subsequent relative positions with a difference of Δt time units and then by letting $\Delta t \downarrow 0$. Equation (1d), $\dot{v}_2 = 0$, has been added to the system equations since v_2 is not known to ship S_1; it is a parameter which is to be determined by S_1. A possible way of determining such a parameter is to add it to the system equations (as done) and then using a filter technique (to be described in section 3) such as to estimate not only v_2 but also r and γ. The information which S_1 gets in order to (try to) estimate these variables is the angle δ as a function of time.

By defining $x_1 = \frac{1}{r}$, $x_2 = \gamma$, $x_3 = v_2$ and $y = \dot{\delta}$ we can write the equations formally as

$$\dot{x} = f(x,u) \tag{2}$$

$$y = h(x,u) \tag{3}$$

or, written out in scalar equations, as

$$\begin{aligned} \dot{x}_1 &= - x_1^2(x_3 \cos x_2 - v_1 \cos u) \\ \dot{x}_2 &= - x_1(x_3 \sin x_2 - v_1 \sin u) \\ \dot{x}_3 &= 0 \end{aligned} \tag{4}$$

$$y = - x_1(v_1 \sin u - x_3 \sin x_2) \tag{5}$$

The first three equations are the system equations; the latter equation is called the measurement equation. In practice, not $y = \dot{\delta}$ is measured, but δ itself. The motivation for this has been given in [5]. Essentially we try to keep the dimension of the state vector as small as possible. In (4) this dimension is three whereas in (1) it was four.

The measurement y will be corrupted by noise and therefore to the right-hand sides of (3) and (5) a term w(t) will be added; w(t) is the measurement noise which is assumed to be white and have zero mean. Please note that no system noise has been assumed.

3. FILTER APPROACH

The system equations and measurement equation are

$$\dot{x} = f(x,u),\ x(0) = x_0;\quad y = h(x,u) + w, \tag{6}$$

where x_0 is the initial condition. Since these equations are nonlinear, the Kalman filter cannot be used in order to obtain estimates for the statevector x. Instead, the extended Kalman filter (EKF) will be used [2]. For that reason (6) must be linearized about an estimate of x(t), called $\hat{x}(t)$. Define

$$A = \frac{\partial f}{\partial x} = \begin{pmatrix} 2\hat{x}_1(v_0\cos u-\hat{x}_3\cos\hat{x}_2) & \hat{x}_1{}^2\hat{x}_3\sin\hat{x}_2 & -\hat{x}_1{}^2\cos\hat{x}_2 \\ V_0\sin u-\hat{x}_3\sin\hat{x}_2 & -\hat{x}_1\hat{x}_3\cos\hat{x}_2 & -\hat{x}_1\sin\hat{x}_2 \\ 0 & 0 & 0 \end{pmatrix} \tag{7}$$

$$H = \frac{\partial h}{\partial x} = -(V_0\sin u-\hat{x}_3\sin\hat{x}_2 \quad -\hat{x}_1\hat{x}_3\cos\hat{x}_2 \quad -\hat{x}_1\sin\hat{x}_2) \tag{8}$$

The 3 * 3 covariance matrix P of the state x satisfies the Riccati equation belonging to the EKF;

$$\dot{P} = AP + PA' - PH'R^{-1}HP \tag{9}$$

where R is the covariance of the measurement noise and ' denotes the transposed. The inverse of P, to be denoted by $Q = P^{-1}$, satisfies

$$\dot{Q} = -QA - A'Q + H'R^{-1}H;\quad Q(0) = 0. \tag{10}$$

The filter equation becomes

$$\dot{\hat{x}} = f(\hat{x},u) + P\,H'R^{-1}(y - h(\hat{x},u)) \tag{11}$$

When using these filter equations, the control u(.) and an initial estimate $\hat{x}(0)$ together with an initial covariance matrix P(0) must be known. If these quantities are given, (9) and (11) can be integrated with respect to time t; together there are 9 scalar equations to be integrated (3 from (11) and (9) has 6 essentially different equations and not 9, due to symmetry of P). If we want to compare the solution to the real evolution of the state, (6) must also be integrated.

4. (SUB-)OPTIMAL MANOEUVRING AND SOME RESULTS.

The extended Kalman filter discussed in section 3 does not only give an estimate of the state x but also yields an approximation of the covariance matrix P of the state x. It is clear that this matrix P(t) depends on the choice of the control function u(.). We would like to choose u(.) in such a way as to minimize some criterion related to the norm of P at a given final time t = T. For this criterion we could choose trace P(T) or the maximal eigenvalue of P(T). Since we are interested explicitly in the distance r(T), the criterion chosen in this paper will be $p_{11}(T)$, the (1,1) st entry of P(T). The relation between p_{11}, the variance of x_1, and r is given via $r = 1/x_1$.

It is very difficult to calculate the optimal control, see [5] or [2] chapter 11. In the latter reference the dynamic programming approach is discussed. We will resort to suboptimal solutions for u(.). In [5] such a suboptimal control has been calculated by using a first order gradient algorithm applied to the necessary conditions as formulated by means of the maximum principle of Pontryagin. This gradient algorithm finds, given a certain u(.), a better one. The algorithm converged if the initial u(.) was chosen in the neighbourhood of the (sub-)optimal solution, which turned out to have a zig-zag character. In a specific example the initial choise of u(.) was

$$u(t) = \begin{cases} \pi/3 , & 0 \le t \le .05; \\ -\pi/3 , & .05 \le t \le .1 = T. \end{cases} \tag{12}$$

Two comments can be made with respect to the solution approach of [5]:
1. the method is off-line and cannot be used on-line.
2. the linearization as described in (7) and (8) takes place about the real trajectory x(t). This real x(t) is of course not known in practice, but the results give a lower bound for the criterion function (i.e. $p_{11}(T)$).

In this paper a method will be presented in which the optimization procedure is on-line. Instead of using the maximum principle for finding the minimum of $p_{11}(T)$ subject to the differential equations (9) and (11) we now construct u(.) by minimization of $\dot{p}_{11}(t)$ as a function of u(t) at each instant of time. The functional relationship between $\dot{p}_{11}(t)$ and u(t) is given as one of the scalar equations of (9). While this method is on-line, the result will in general be suboptimal. Obvious advantages of this approach are that it can be applied in real situations and the numerical procedure is simpler than the one used in the off-line situation. We will now give numerical results. The following parameters are the same for all results to follow;

final time T = .1 hours
(noise covariance $R)^{-1}$ = 2500 rad^{-2}
speed ship S_1; v_1 = 6 knots.

For reason of comparison we start with some results obtained in [5];

I Initial values needed to start numerical procedure;

$$Q(0) = P^{-1}(0) = 0 \text{ rad}^2$$

u(t) of eq. (12)

$$x_1(0) = .5 \text{ (sea mile)}^{-1}$$

$$x_2(0) = \pi/3 \text{ rad.}$$

$$x_3(0) = 10 \text{ knots}$$

The program gave the following results when the optimal control (also calculated, but not repeated here) is applied;

$$x_1(T) = .43; \quad x_2(T) = .68; \quad x_3(T) = 10$$

$$P(T) = \begin{pmatrix} .00019 & -.00074 & .00023 \\ -.00074 & .0095 & -.037 \\ .00023 & -.037 & .22 \end{pmatrix}$$

$$r_{mean}(T) = 2.3; \quad var(r(T)) = .0053 \tag{13}$$

II The following results have been obtained by minimizing $\dot{p}_{11}(t)$ as a function of u(t), as described above. This minimization was done each .01 seconds (in between u was kept constant). The linearization was again taken with respect to the real trajectory. For the initial value of P(0) was chosen

$$P(0) = \begin{pmatrix} 10^4 & \frac{1}{2}\cdot 10^4 & \frac{1}{2}10^4 \\ \frac{1}{2}10^4 & 10^4 & 0 \\ \frac{1}{2}10^4 & 0 & 10^4 \end{pmatrix}$$

The results of the integration procedure are as follows:

t	u	x_1	x_2	x_3	p_{11}	p_{22}	p_{33}
0	4.71	0.5	1.05	10.0	10000	10000	10000
0.01	1.57	0.487	0.976	.	6.57	3.66	6280
2	4.74	74	0.962	.	2.9E-3	5.72E-1	29.9
3	4.71	61	0.897	.	4.8E-4	4.39E-2	2.4
4	1.57	48	36	.	3.5E	1.78E	1.1
5	4.71	34	0.828	.	3.2E	1.42E	1.1
6	4.71	22	0.775	.	2.1E	7.09E-3	5.94E-1
7	4.71	0.409	0.722	.	1.8E	4.3E	4.1E
8	1.57	0.397	0.672	.	1.5E	2.9E	3.1E
9	4.71	85	59	.	1.3E	2.4E	2.9E
0.10	4.71	0.373	0.622	10.0	1.2E-4	1.8E-3	2.4E-1

To compare these results with those of I, the final values of r and var(r) can be calculated:

$$r_{mean}(T) = 2.68; \quad var(r(T)) = .0059 \tag{14}$$

III The next results have been obtained by again instantaneously minimizing $\dot{p}_{11}(t)$ as above, but this time the linearization took place along an estimate of the real trajectory. This estimate was obtained as follows. The estimate $\hat{x}$ was obtained from (11) in which $y=h(x,u)+w$ was substituted. For the noise the specific sample $w(t) \equiv 0$ was chosen. To obtain the results in the next table, equations (6), (9) and (11) had to be integrated simultaneously. The initial values for $\hat{x}(0)$ can be read in this table. For P(0) we choose $P(0) = 10^4\ I$, I being the identity matrix.

t	u	$\hat{x}_1$	$\hat{x}_2$	$\hat{x}_3$	x_1	x_2	x_3	p_{11}	p_{22}	p_{33}
0.00	4.71	0.55	1.15	9.0	0.5	1.1	10.0	10000	10000	10000
1	1.57	49	1.00	71	49	0.976	.	1.3	5.8	688
2	4.70	48	0.996	9	7	65	.	2.1E-3	0.31	13.3
3	4.60	46	13	9.93	6	0.902	.	4.4E-4	3.32E-2	1.53
4	1.72	45	0.848	5	5	0.840	.	3.2E	1.49E	0.769
5	4.56	43	36	5	3	0.830	.	2.7E	1.09E-2	0.742
6	4.50	42	0.788	6	2	0.783	.	2.0E	6.3 E-3	0.478
7	4.40	40	34	7	0.40	0.731	.	1.6E	4.0 E	0.342
8	1.93	39	0.685	7	0.39	0.682	.	1.4E	2.81E	0.275
9	4.37	38	79	7	37	77	.	1.2E	2.38E	0.268
0.10	4.29	0.36	0.639	9.98	0.36	0.637	10.0	1.0E-4	1.80E-3	0.227

$$r_{mean}(T) = 2.77; \quad var(r(T)) = 0.0061 \tag{15}$$

Comparison of the results of the three different schemes leads to the following remarks.

1. The optimal control functions in II and III are similar; they show the same number of discontinuities. In between two consecutive jumps,

the functions are about constant. The same feature was shown in [5], but there only one jump occured and in II and III there are more jumps.
2. In spite of the fact that in I on the one hand and in II and III on the other different optimization procedures were used, the variances of r(T) do not differ markedly. Of course the results in II and III are slightly worse. This was to be expected since in I a global minimization procedure was used.
3. Procedure II was repeated with different values for P(0), not shown here. It turned out that the optimal control function and var(r(T)) were very insensitive to these variations in P(0).

REFERENCES

[1] A.H. Jazwinsky, Stochastic processes and filtering theory. Academic Press, New York, 1970.
[2] G.C. Goodwin and K.S. Sin, Adaptive filtering, prediction and control. Prentice-Hall, Inc., Englewood Cliffs, New Jersey, 1984.
[3] W.H. Fleming and R.W. Rishel, Deterministic and stochastic optimal control. Springer Verlag, 1975.
[4] M. Hazewinkel and J.C. Willems (eds.), Stochastic systems: The mathematics of filtering and identifications and applications. Reidel Publishing Co, 1981.
[5] G.J. Olsder, On the optimal manoeuvering during bearings-only tracking. Proceedings 23rd IEEE Conference on Decision and Control, Las Vegas, 1984, pp. 935-940.

THE VALUE OF AN OPTION BASED ON AN AVERAGE SECURITY VALUE

A.G.Z. Kemna and A.C.F. Vorst
Erasmus University Rotterdam
P.O. Box 1738
3000 DR Rotterdam
The Netherlands

ABSTRACT. In this paper we shall discuss a financial option of which the payoff depends on the average value of the underlying security over some final time interval. After explaining what an option is about we will derive a partial differential equation for the option which is different from the partial differential equation of a simple European call option. From this we will get an expectation formula for the option value. We will give an economical as well as a mathematical argument for this expectation formula.

1. INTRODUCTION

The rapid development of option pricing theory and the application of that theory is caused by the path-breaking papers of Black and Scholes (Black and Scholes (1973)) and Merton (Merton (1973)). They derived a first explicit option pricing formula by using the theory of stochastic processes. This theory has also played a fundamental rôle in the further developments of option pricing and will also do so in this paper. A call option gives the owner the right to buy a specific share of stock at a specific future date (maturity date) for a fixed price (excercise price). In fact this is an European call option. One also has put options where the owner has the right to sell instead of to buy a share of stock and one has American options where the owner has the right to buy or sell at any time before the maturity date instead of just at the maturity date.

Now the Black-Scholes formula gives the price of such an European call option in the financial markets and Black and Scholes suggested that their solution to the option pricing problem could also be used for more complex options. This resulted in numerous extensions of the basic Black-Scholes model. In this paper we will also give an extension for a specific kind of option. To explain this kind of option we first remark that in practice the owner of the option doesn't actually buy the option at the maturity date but he simply gets the difference between the share price and the exercise price if this is positive and nothing if the difference is negative from the issuer of the option. Hence the issuer doesn't have to possess a share of stock.

S. Albeverio et al. (eds.), Stochastic Processes in Physics and Engineering, 193–203.

The option we want to discuss in this paper gives the owner the right of getting the difference between the average value of the underlying security (in our case a share of stock) over some final time interval preceeding the maturity date and the exercise price. This kind of option is often part of a commodity-linked bond contract. The value of a commodity-linked bond is determined by the value of a reference bundle of the commodity. A recent example is the guilder oil bond issued by the Dutch venture capital company Oranje Nassau of which the payoff is the maximum of the price of 10,5 barrels of North Sea oil and 1.000 guilders. By the price of 10,5 barrels of North Sea oil is meant the average price over the last year of the contract. A pricing model and more exact description can be found in Kemna (1986). The whole contract can be split in a normal bond of 1.000 guilders and an option of the kind we will consider here, where we take a share of stock as reference bundle.

The value of this option can be determined in two stages. In section 2 we derive a partial differential equation for the option price during the final time interval. The option price during the final time interval differs from a simple option price if the share price is high during the first part of the time interval. Then one can be sure of a positive pay off before the maturity date, while this is never true for a normal option. It will also be shown that we only need to solve the partial differential equation for the case that there is no positive payoff with certainty. The derivation will follow the lines of the Black-Scholes equilibrium approach. In section 3 we derive an expectation formula for this option. An economic as well as a mathematical argument is used. This approach is applied in section 4 for the option before the final time interval. Before the final time interval we have a simple option whose boundary value equals the starting value of the option during the final time interval. Due to this boundary condition there is no explicit formula for this simple option. In section 5 we explore under what conditions it would be possible to find an explicit formula for our complex option. It turns out that none of the conditions can be economically justified. This means that we have to solve the problem numerically or we have to use Monte-Carlo-simulation. Finally, section 6 summarizes the results and offers some concluding remarks.

2. A PARTIAL DIFFERENTIAL EQUATION FOR THE OPTION PRICE.

In this section we shall give a PDE for the value of an option of which the payoff is not only based on the value of the underlying security at the excercise time T but also on the average value of the security over some final time interval. If t_1 is the first moment from where on we take the average and hence $T-t_1$ is the length of the time interval we assume that the payoff of the option is equal to

$$\max(A_T-K,0) \tag{1}$$

with

$$A_T = \frac{1}{T-t_1} \int_{t_1}^{T} S_\tau \, d\tau \tag{2}$$

where K is the fixed excercise price and S_τ is the value of the underlying security at time τ.
In this section we will only give the PDE for $t_1 \le t \le T$ while in the next section we will give a PDE for $t < t_1$. Hence we will assume $t_1 \le t \le T$. If we put

$$A_t = \frac{1}{T-t_1} \int_{t_1}^{t} S_\tau \, d\tau \tag{3}$$

then it is clear that the value of the option at any time t will depend on t, S_t and A_t. As said in the preceeding section we will assume that our underlying security is a share of common stock and as usual we will assume that the stock price is governed by the following stochastic differential equation

$$dS_t = \alpha S_t dt + \sigma S_t dW_t \tag{4}$$

where W_t is a Wiener process and α and σ are constants. Since S_t is a stochastic process we have to interpret formulas (2) and (3) as stochastic integrals. If we put $X_t = (\ S_t\ A_t)'$ and $\beta = 1/(T-t_1)$ we can combine (3) and (4) to the following system of stochastic differential equations:

$$dX_t = \begin{pmatrix} \alpha & 0 \\ \beta & 0 \end{pmatrix} X_t dt + \begin{pmatrix} \sigma & 0 \\ 0 & 0 \end{pmatrix} X_t dW_t \tag{5}$$

If $C(S_t, A_t, t)$ is the value of an option at time t, where the underlying stock has a value S_t and the average up to t is given by A_t we have by Ito's formula that

$$dC_t = \left(\frac{\delta C}{\delta t} + \alpha S \frac{\delta C}{\delta S} + \beta S \frac{\delta C}{\delta A} + \tfrac{1}{2}\sigma^2 S^2 \frac{\delta^2 C}{(\delta S)^2}\right) dt + \sigma S \frac{\delta C}{\delta S} dW_t \tag{6}$$

Furthermore let r be the interest rate on riskless default free bonds. Hence if we invest an amount B_t in such a bond our investment is governed by the following differential equation

$$dB_t = r B_t dt \tag{7}$$

For the following argument we must assume that the stock markets are frictionless, that there are no transaction costs for buying or selling options, stocks or bonds and that the interest rates on lending and borrowing are equal. These are also the underlying assumptions for the Black-Scholes option pricing formula and are often used for deriving theoretical results.

Instead of buying an option we could also buy ($\delta C/\delta S$) shares of stock and borrow an amount of (($\delta C/\delta S)S$ - C). This last strategy has the same risk as holding the option. Or stated in another way : if we buy ($\delta C/\delta S$) shares of stock, borrow (($\delta C/\delta S)S$ - C) and sell to someone the option we would bring ourselves in a riskless position since the disturbance term would be

$$\left(\sigma S \frac{\delta C}{\delta S} - \sigma S \frac{\delta C}{\delta S} \right) dW_t = 0 \tag{8}$$

Furthermore we see that this would cost or bring us no money at this moment since

$$- \frac{\delta C}{\delta S} S + \left(\frac{\delta C}{\delta S} S - C \right) + C = 0 \tag{9}$$

Such a strategy of buying or selling shares of stock and lending or borrowing against a riskfree interest rate is called a hedging strategy if it comes with the same risk and initial investment. Bensoussan (Bensoussan (1984)) showed using martingales that a hedging strategy exists for more general claims as the option claim we discuss here. We also like to remark that in the hedging strategy one constantly has to adjust the amount of shares which one is holding and the amount one is borrowing. Hence the assumption that there are no transaction costs is essential. Since the hedging strategy has the same risk and the same investment costs as buying an option the expected profit of the option and the hedging strategy must be the same, otherwise one could make an arbitrage profit by buying the one with the highest expected profit and selling the other one. Hence in financial markets investment opportunities which require no investment and bring no risk must have a zero expected profit. And the combination of the option and the hedging strategy is such an opportunity. From this it follows that

$$\frac{\delta C}{\delta t} + \alpha S \frac{\delta C}{\delta S} + \beta S \frac{\delta C}{\delta A} + \tfrac{1}{2}\sigma^2 S^2 \frac{\delta^2 C}{(\delta S)^2} = \alpha S \frac{\delta C}{\delta S} + r\left(C - S \frac{\delta C}{\delta S}\right) \tag{10}$$

and we get the following PDE for the value of the option

$$\frac{\delta C}{\delta t} + \beta S \frac{\delta C}{\delta A} + \tfrac{1}{2}\sigma^2 S^2 \frac{\delta^2 C}{(\delta S)^2} + r\left(S \frac{\delta C}{\delta S} - C\right) = 0 \tag{11}$$

on the region $R_1 = \{(S,A,t) \mid S \geq 0,\ A \geq 0,\ t_1 \leq t \leq T\}$.
Of course we have some boundary conditions and these are

$$C(S,A,T) = \operatorname{Max}(A-K,0) \tag{12}$$

$$C(0,A,t) = \operatorname{Max}(e^{-r(T-t)}(A-K),0) \tag{13}$$

$$\frac{\delta C}{\delta S}(\infty,A,T) = \frac{T-t}{T-t_1}e^{-r(T-t)} \tag{14}$$

Before we can state the last boundary condition we first calculate $C(S,A,t)$ for $A \geq K$. If $A \geq K$ we know that we will get a positive payoff in the end. This payoff will be

$$(A-K) + \beta\int_t^T S_\tau d\tau \tag{15}$$

There is also another way to reach this payoff without using the option and that is the following : put $(A-K)e^{-r(T-t)}$ in riskless bonds and besides for every time interval $(\tau,\tau+\Delta\tau)$ convert $\beta e^{-r(T-\tau)}\Delta\tau$ number of shares of common stock into riskless bonds. If we do this for every time interval $(\tau,\tau+\Delta\tau)$ and let $\Delta\tau$ go to zero we will also end up with (15) as final amount (the factor $e^{-r(T-\tau)}$ reflects that we earn interest on our bonds). To do this we need the following amount of money, which must therefore be the option price :

$$C(S,A,t) = (A-K)e^{-r(T-t)} + S_t\int_t^T \beta e^{-r(T-\tau)}d\tau =$$

$$(A-K)e^{-r(T-t)} + \frac{\beta}{r}S_t(1-e^{-r(T-t)}), \tag{16}$$

where the first factor represents the riskless bond and the second factor the number of shares we converted in riskless bonds.
Hence (16) gives the value of $C(S,A,t)$ if $A \geq K$. Of course (16) fulfils the PDE (11). So we only have to solve (11) on the region $R_2 = \{(S,A,t) \mid S \geq 0,\ 0 \leq A \leq K,\ t_1 \leq t \leq T\}$ with boundary conditions (12), (13), (14) and

$$C(S,K,t) = \frac{\beta}{r}S_t(1-e^{-r(T-t)}) \tag{17}$$

3. THE OPTION PRICE AS AN EXPECTATION

There are two ways which lead to the same expectation formula for the option price $C(S,A,t)$. One uses an economic argument and the other is purely mathematically. We will start off with the economic argument which was first introduced by Cox and Ross (Cox and Ross (1976)). It is assumed that most investors are risk-averse, which means that if two financial objects have the same expected profit the investors prefer the object with the lowest risk i.e. with the smallest variance. Hence they will not buy the object with the higher risk and because of this the price of the object will decrease and hence its expected profit will increase. So in a risk-averse economy one assumes that objects with a higher risk must have a higher expected profit.

In a hypothetical risk-neutral economy investors only consider the expected profits and do not worry about the risk. Investors will prefer the objects with the highest expected profit. Because of this the object with lower expected profit will see their prices decrease and hence their expected profits increase until these are on the higher level. Hence in a risk-neutral economy all objects will have the same expected profit and the equilibrium rate of return must be the riskless interest rate. In the previous section we have seen that the share price and the bond price play a rôle in the forming of a hedging strategy and we have seen that such a riskless hedging strategy can be formed. But then the valuation of the objects should be independent of the investors attitude towards risk. Hence we might assume that all investors are risk neutral and we can find the option price in a risk-neutral economy. Hence we may assume that

$$\alpha = r \tag{18}$$

since in a risk neutral economy we must have

$$\exp(\alpha(t-t_0)) = E_{S_0}(S_t) = E_{B_0}(B_t) = \exp(r(t-t_0)) \tag{19}$$

and the price of the option must be the expected terminal payoff, discounted at the riskless interest rate r i.e. :

$$C(S,A,t) = e^{-r(T-t)} E_{S,A}\{\max(A_T - K, 0)\} \tag{20}$$

Now we will give a mathematical argument which leads to the same expectation formula. If we substitute

$$C(S,A,t) = e^{-r(T-t)}\, D(S,A,t) \tag{21}$$

then our PDE (11) becomes

$$\frac{\delta D}{\delta t} + \beta S \frac{\delta D}{\delta A} + \tfrac{1}{2}\sigma^2 S^2 \frac{\delta^2 D}{(\delta S)^2} + rS \frac{\delta D}{\delta S} = 0 \tag{22}$$

Now we see that this is the Kolmogorov backward equation of the following system of stochastic differential equations

$$dX_t = \begin{pmatrix} r & 0 \\ \beta & 0 \end{pmatrix} X_t dt + \begin{pmatrix} \sigma & 0 \\ 0 & 0 \end{pmatrix} X_t dW_t \tag{23}$$

with $X'_t = (S_t\ A_t)$.

But then it is well known that the solution of (20) is given by an expectation formula. Of course we have to use the boundary conditions for D instead of those for C but if we consider D on the region R_1 the first exit time will be T since we never hit the boundary before time T. Hence we find from (23) that

$$D(S,A,t) = E_{S,A}\,(D(S_T, A_T, T)) = E_{S,A}\,\{\max(A_T - K, 0)\} \tag{24}$$

and we see that the economical and mathematical argument give rise to the same formula. If $A \geq K$ (24) reduces of course to

$$D(S,A,t) = E_{S,A}\,(A_T) - K \tag{25}$$

and we will show in section 5 that in general

$$E_{S,A}\,(A_T) = A - \frac{\beta}{r} S(1 - e^{r(T-t)}) \tag{26}$$

And hence for $A \geq K$

$$D(S,A,t) = (A - K) + \frac{\beta}{r} S(e^{r(T-t)} - 1) \tag{27}$$

which is of course exactly the same as formula (14) for $A \geq K$. We can also give on expectation formula if we restrict ourselves to the region R_2 and this becomes

$$D(S,A,t) = E_{S,A}\,\{I_{\tau \leq T} \cdot \frac{\beta}{r} S(e^{(T-\tau)} - 1)\} \tag{28}$$

where τ is the first exit time from R_2 and I is the indicator function i.e.

$$I_{\tau \le T} = \begin{cases} 1 \text{ if } \tau \le T \\ 0 \text{ if } \tau > T \end{cases} \tag{29}$$

In practical applications one wants closed analytical formulas for the value of an option instead of (20), (24) or (28). If we had a simple option this would be possible and we would get the Black-Scholes formula. In the next section we will derive this formula. We do this for two reasons. The first is that for the Black-Scholes we need a PDE which also plays a rôle in our problem if $t < t_1$ and the second is of expositional nature.

4. THE BLACK-SCHOLES FORMULA

In the preceeding section we derived a PDE for the option price $C(S,A,t)$ if $t_1 \le t \le T$. In this section we are focussing on the option price if $t < t_1$ i.e. if we are not yet in the final time interval over which we take the average share price. It is clear that for $t < t_1$ A doesn't play a rôle and C only depends on S and t. Let us write $C^*(S,t)$ for the option price if $t < t_1$. By $C^*(S,t_1)$ we will of course mean $\lim_{t \to t_1} C^*(S,t)$ and we know that

$$C^*(S,t_1) = C(S,0,t_1) \tag{30}$$

We still assume that S_t follows the stochastic differential equation (4) and again we can apply Ito's lemma to $C^*(S,t)$. We can also form a riskless hedge and arguing as in section 2 we derive the following PDE for C^* if $t < t_1$:

$$\frac{\delta C^*}{\delta t} + \tfrac{1}{2}\sigma^2 S^2 \frac{\delta^2 C^*}{(\delta S)^2} + r(S\frac{\delta C^*}{\delta S} - C^*) = 0 \tag{31}$$

This is the PDE for a normal European call option which pays off $\max(S_T-K,0)$ instead of $\max(A_T-K,0)$ and has been derived by Black and Scholes. Hence for an European call option we have (31) with boundary condition

$$C^*(S,T) = \max(S_T-K,0) \tag{32}$$

The solution to (31) with our boundary condition (28) is by the same economical or mathematical argument as in the preceeding section given by

$$C^*(S,t) = e^{-r(t-t_1)}E_S\{C(S_{t_1},0,t_1)\} \tag{33}$$

with S_t given by the following stochastic differential equation

$$dS_t = rS_t dt + \sigma S_t dW_t \tag{34}$$

As we said before for a simple option we will also have (31) but now with boundary condition (32). Hence for a simple option the price $C^1(S,t)$ is given by

$$C^1(S,t) = e^{-r(T-t)} E\{\max(S_T-K,0)\} \tag{35}$$

but since S_t is again given by (34) we know that S_T is normally distributed and it easily follows that

$$C^1(S,t) = SN(x_1) - Ke^{-r(T-t)}N(x_2) \tag{36}$$

with $$x_1 = [\ln(S/K) + (r+\tfrac{1}{2}\sigma^2)(T-t)]/\sigma\sqrt{T-t} \tag{37}$$

$$x_2 = x_1 - \sigma\sqrt{T-t} \tag{38}$$

and where N is the normal distribution function. (36), (37) and (38) form the well-known Black-Scholes formula for the valuation of an European call option. For practical purposes it would be very convenient if we could find explicit formulas like (36), (37) and (38) for $C(S,A,t)$ and $C^*(S,t)$. We will comment on this in the next section.

5. PROBLEMS WITH AN EXPLICIT FORMULA

If we want to find explicit formula's for the option price if $t_1 \le t \le T$ then we have seen in section 3 formula (24) that we have to calculate expectations of a process given by the following system of stochastic differential equations

$$dX_t = (AX_t + a)dt + (BX_t + b)dW_t \tag{39}$$

with $a = b = 0$

$$A = \begin{pmatrix} r & 0 \\ \beta & 0 \end{pmatrix}, \quad B = \begin{pmatrix} \sigma & 0 \\ 0 & 0 \end{pmatrix} \tag{40}$$

Now it seems impossible to us to give explicit formulas for

$$D(S,A,t) = E_{S,A}\{\max(A_T-K,0)\} \tag{24}$$

since X_t is not a Gaussian process. In the case of a simple European call option we indeed have a Gaussian process as we have seen in the preceeding section and we could give the explicit formulas (36)-(38). Although X_t is not Gaussian in our case we can calculate the expectation and variance of X_t but this of course alone is not enough to give us an explicit formula for (24). To calculate the expectation and variance of X_t we know that (see Arnold (1974)) $m_t = EX_t$ follows the following differential equation

$$dm_t = Am_t dt \tag{41}$$

and $P_t = EX_tX_t'$ is the unique nonnegative definite symmetric solution of the system of differential equation

$$dP_t = AP_t + P_tA' + BP_tB' \tag{42}$$

where we have already used that $a = b = 0$.
Now (41) can be solved easily and gives

$$ES_t = S_{t_0}e^{r(t-t_0)} \tag{43}$$

$$EA_t = A_{t_0} + \frac{\beta}{r} S_{t_0}(e^{r(t-t_0)} - 1) \tag{44}$$

Since A and B have very special forms one can also find successively explicit formulas for ES_t^2, ES_tA_t and EA_t^2. The results are as follows

$$ES_t^2 = S_0^2\, e^{(2r+\sigma^2)(t-t_0)} \tag{45}$$

$$ES_tA_t = \frac{\beta S_0^2}{(r+\sigma^2)} e^{(2r+\sigma^2)(t-t_0)} + (A_0S_0 - \frac{\beta S_0^2}{(r+\sigma^2)})\, e^{r(t-t_0)} \tag{46}$$

$$EA_t^2 = \frac{2\beta^2 S_0^2}{(r+\sigma^2)(2r+\sigma^2)} e^{(2r+\sigma^2)(t-t_0)} + \frac{2\beta}{r}(A_0S_0 - \frac{\beta S_0^2}{(r+\sigma^2)})e^{r(t-t_0)} + C_0 \tag{47}$$

where C_0 is such that $EA_0^2 = A_0^2$.
As we said before this is not enough to find an explicit solution for (24) although (44) gives the solution of (24) if $A_{t_0} \geq K$ as we explained in section 3.

One might wonder whether it is possible to change the model such that we end up with a Gaussian process as for a simple European call option. This is indeed possible (see Arnold (1974) pg. 136) if we assume that S_t is governed by

$$dS_t = \alpha S_t dt + \sigma dW_t \tag{48}$$

instead of (4). In this case one easily sees that (37) gives rise to a Gaussian process and we could give explicit formulas for (24). However with a specification like (48) there is always a positive probability that S_t becomes negative and from an economic point of view this must be impossible since shares of common stock always have a positive value. An owner of a share of common stock is not responsible for the debt of the firm. Another way to get a Gaussian process would be to make a model with $a = b = 0$ and such that A and B commute (see Arnold (1974) pg. 144). This would happen for example if we assume

$$dA_t = \beta S_t dt + \sigma S_t dW_t \tag{49}$$

instead of (3). However this specification implies that there is a disturbance term in the measurement of the average share price. It will be clear that the owner of an option will never accept a downward disturbance, while the person who issued the option will never accept an upward disturbance. Hence from an economic point of view (49) doesn't make sense.

6. CONCLUSIONS

In this paper we have studied an option of which the payoff depends on the average value over some final time interval of the underlying security. We derived two PDE's for the option value, one for the case that we are already in the final time interval and one for the case where we are not yet in the interval. These PDE's can be seen as the Kolmogorov backward equations of two systems of stochastic differential equations. From this it follows that the option value can be written in an expectation formula. Unfortunately this formula is not as explicit as the Black-Scholes formula for a simple European call option. Hence if one wants to find the option value in a practical case one has to solve the PDE's numerically or one has to rely on simulation methods. Since the most important PDE has three variables it seems much cheaper to use simulation methods. All in all it seems interesting to find a reliable and fast method for computing the expectations and hence the option values.

REFERENCES

Arnold, L. (1974), Stochastic Differential Equations, Theory and Applications, John Wiley & Sons, New York.

Bensoussan, A. (1984), On the Theory of Option Pricing, Acto Applicandae Mathematicae 2, 139-158.

Black, F. and M. Scholes (1973), The Pricing of Options and Corporate Liabilities, Journal of Political Economics 81, 637-659.

Cox, J.C. and Ross, S.A. (1976), The Valuation of Options for Alternative Stochastic Processes, Journal of Financial Economics 3, 145-166.

Kemna, A.G.Z. (1986), The Value of Commodity-Linked Bonds : A Case Study, Centre For Research in Business Economics report 8609/F, Erasmus University Rotterdam.

Merton, R.C. (1973), Theory of Rational Option Pricing, Bell Journal of Economics and Management Sciences 4, 141-183.

ON ASYMPTOTIC LOCALIZATION BY PERTURBING OPERATORS FOR PARTIAL OBSERVATION OF A STOCHASTIC EVOLUTION EQUATION

T. Koski
Dept. of Mathematics
Åbo Akademi
SF-20500 Åbo
Finland

ABSTRACT. This paper considers some bounds for the Hellinger distance in spaces of probability measures induced by partial observations of a stochastic evolution equation using some results of F. Liese [18], [19] and a technique developed in [15]. The basic aim of the reported work is to discover those perturbations of the infinite dimensional system generator that will lead to calculations of Hellinger balls. The main result here is an asymptotic localization property, in terms of the variation distance (see [25]), derived under some fairly restrictive assumptions on a perturbing operator as the observation interval increases to infinity. A prerequisite for this is a study of the perturbation properties of the covariance operator of the filtering error under stability and stationarity assumptions.

1. A PARTIALLY OBSERVED STOCHASTIC EVOLUTION EQUATION AND SOME RESULTS ON THE CORRESPONDING LIKELIHOOD RATIOS

Let $w = \{w_t \mid t \geq 0\}$ be a Wiener process such that w_t is, for each $t > 0$, a random variable defined on a complete probability space $(\Omega, \mathcal{F}, \mathbf{P})$ and assuming values in a real and separable Hilbert space H. The scalar product in H is $\langle \cdot, \cdot \rangle$ and thus $\| \cdot \| = \langle \cdot, \cdot \rangle^{1/2}$. Then it holds that w is an H-valued Gaussian process such that $E\langle h, w_t - w_s \rangle^2 = |t - s| \langle h, Wh \rangle$ where W is known as the incremental covariance operator of w. (see e.g. [21] p. 141).

Assume that the linear, closed self adjoint operator A with domain $\mathcal{D}(A) \subseteq H$ is the generator of a strongly continuous semigroup of bounded operators $\{T_t \mid t \geq 0\}$ $(T_0 = I)$. Assume furthermore that the semigroup is stable in the sense that

$$\|T_t h\| \leq M e^{-\alpha t} \tag{1.1}$$

for every $h \in H$ and $t > 0$ with some $M \geq 1$, $\alpha > 0$. (see [6]).

We define on $(\Omega, \mathcal{F}, \mathbf{P})$ the random variables

$$x_t = T_t x_0 + \int_0^t T_{t-s} dw_s \tag{1.2}$$

where $x_0 \in H$ and may be allowed to be a random variable.

The process $x = \{x_t \mid t \geq 0\}$ is the a priori or mild sense solution of the stochastic evolution equation

$$dx_t = Ax_t dt + dw_t \tag{1.3}$$

S. Albeverio et al. (eds.), Stochastic Processes in Physics and Engineering, 205–222.

These concepts are considered in detail in e.g. [22] and [4] ch. 5. Suppose that C is a bounded linear functional (an observing operator) on H so that for every $h \in H$

$$Ch = \langle h_0, h \rangle \tag{1.4}$$

for some representer $h_0 \in H$. Assume also that $v = \{v_t \mid t \geq 0\}$ is a standard real valued Wiener process defined on $(\Omega, \mathcal{F}, \mathbf{P})$ and that it is independent of the process x.

Then we mean by partial observations of the stochastic evolution equation x in (1.2) or (1.3) the real valued Itô -process y defined by

$$y_t = \int_0^t Cx_s ds + v_t \tag{1.5}$$

Taken together the equations (1.2) and (1.5) comprise a standard mathematical model of a ***stochastic distributed parameter system.*** We refer to [24] and [26] for some extensive surveys of the concepts associated with these systems as well as of their industrial applications.

Let us now envisage a family of infinitesimal generators A_ε, $\varepsilon > 0$, of the corresponding semigroups $\{T_t(\varepsilon) \mid t \geq 0\}$ with the properties described above, in particular such that $\|T_t(\varepsilon)\| \leq Me^{-\alpha(\varepsilon)t}$. (It is assumed that M is independent of ε, which will be true in the sequel.) Furthermore we assume that $A_0 = A$, as is the case in the sections that follow. The particular functional dependence of A_ε on ε is, under these circumstances, of no interest whatsoever for the results to be presented in this section.

Then the process $y^s(\varepsilon) = \{y_t(\varepsilon) \mid 0 \leq t \leq s\}$ corresponds to the random variables

$$x_t(\varepsilon) = \int_0^t T_{t-u}(\varepsilon) dw_u \qquad (x_0 = 0) \tag{1.6}$$

and

$$y_t(\varepsilon) = \int_0^t Cx_u(\varepsilon) du + v_t \tag{1.7}$$

Set $\mu_\varepsilon^s = \mathbf{P}(y^s(\varepsilon)^{-1})$ i.e. μ_ε^s is the measure induced on$(C[0,s], \mathcal{B}_s)$ (= the Banach space of continuous functions on $[0,s]$ and the Borel sigma-algebra of sets in $C[0,s]$ w.r.t. the supremum norm) by the process $y^s(\varepsilon)$. Consequently, $\mu^s = \mu_0^s$ is induced by y^s from (1.5).

In addition we introduce $E^s = \{\mu_\varepsilon^s \mid \varepsilon > 0\}$, which is a statistical space and will in section 2 below be put to correspondence with perturbations of A by bounded operators.

Proposition 1.1.

The statistical space $E^s = \{\mu_\varepsilon^s \mid \varepsilon > 0\}$ is dominated by μ^s i.e. every μ_ε^s is absolutely continuous w.r.t. μ^s and the likelihood ratio $d\mu_\varepsilon^s / d\mu^s$ can be written as

$$\frac{d\mu_\varepsilon^s}{d\mu^s}(y^s) = \exp\left\{\int_0^s [C\hat{\bar{x}}_t(\varepsilon) - C\hat{x}_t] d\hat{v}_t - \frac{1}{2}\int_0^s [C\hat{\bar{x}}_t(\varepsilon) - C\hat{x}_t]^2 dt\right\} \tag{1.8}$$

where

$$\hat{v}_t = y_t - \int_0^t C\hat{x}_s ds \tag{1.9}$$

$$\hat{x}_t = \int_0^t T_{t-s} P C^* d\hat{v}_s \tag{1.10}$$

with C^* denoting the dual operator of C,

$$\hat{\bar{x}}_t(\varepsilon) = \int_0^t T_{t-s}(\varepsilon) P(\varepsilon) C^* d\hat{v}_s \tag{1.11}$$

where $P(\varepsilon)$ solves the inner product algebraic Riccati equation (ARE) in H for $\varepsilon \geq 0$ (i.e. $P(0) = P$), that

$$\langle P(\varepsilon)h, A_\varepsilon k\rangle + \langle A_\varepsilon h, P(\varepsilon)k\rangle + \langle Wh, k\rangle - \langle P(\varepsilon)C^*CP(\varepsilon)h, k\rangle = 0. \tag{1.12}$$

Proof: This result, in particular the representation (1.8), can be demonstrated as in [15]. The property of domination of E^s by μ^s is eventually derived from the results in [2]. ∎

The process $\hat{v}^s = \{\hat{v}_t \mid 0 \leq t \leq s\}$ constructed in (1.9) is the *innovation process* corresponding to the process y^s. As is well known, this process is a standard Wiener process w.r.t. the stochastic basis $(\Omega, \mathcal{F}, (\mathcal{F}_t^{y^s})_{0 \leq t \leq s}, \mathbf{P})$, where $(\mathcal{F}_t^{y^s})_t$ is the increasing family of subsigma-algebra of $\mathcal{F}$ generated by the process y^s i.e. $\mathcal{F}_t^{y^s} = \sigma(y_u^s \mid 0 \leq u \leq t)$. This property of $\hat{v}^s$ is the key for representing $d\mu_\varepsilon^s/d\mu^s$ in the form given in (1.8). In fact, invoking the process $\hat{v}$ we construct *a token process* $\overline{y}^s(\varepsilon) = \{\overline{y}_t^s(\varepsilon) \mid 0 \leq t \leq s\}$ by

$$\overline{y}_t(\varepsilon) = \int_0^t C\hat{\bar{x}}_s(\varepsilon) ds + \hat{v}_t \tag{1.13}$$

where the variables $\hat{\bar{x}}_s(\varepsilon)$ are given in (1.11). The process $\overline{y}^s(\varepsilon)$ is completely determined by the information in $(\mathcal{F}_t^{y^s})_{0 \leq t \leq s}$ corresponding to the assumed underlying null hypothesis, the parameter value A. But since $\hat{v}^s$ is a Wiener process, the covariance factorization results found e.g. in [13], ch. 9, show that the process $\overline{y}^s(\varepsilon)$ induces a measure that cannot be distinguished from μ_ε^s i.e. $\mu_\varepsilon^s = \mathbf{P}(\overline{y}^s(\varepsilon)^{-1})$. In other respects $\overline{y}^s(\varepsilon)$ and $y^s(\varepsilon)$ are, in general, two quite different processes. The advantages of the stochastic realization (1.13) for $y^s(\varepsilon)$ will become clear in the estimates for the Hellinger integrals in lemmas 2.2 and 2.3 below.

The purpose of this paper is to investigate *the asymptotic localization properties* of the statistical space $E^s \cup \{\mu^s\}$. Asymptotic localization is here understood in the following sense.

Definition 1.2.

Define the variation norm $\|\cdot\|_1$ as

$$\|\mu_\varepsilon^s - \mu^s\|_1 = \sup_{Q \in \mathcal{B}_s} |\mu_\varepsilon^s(Q) - \mu^s(Q)|$$

Suppose that ε decreases to zero as $s \to \infty$ i.e. $\varepsilon = \varepsilon(s)$. Then we say that E^s is *asymptotically localizable* at A, if

$$0 < \liminf_{s \to \infty} \|\mu_\varepsilon^s - \mu^s\|_1 \leq \limsup_{s \to \infty} \|\mu_\varepsilon^s - \mu^s\|_1 < 1 \tag{1.14}$$

(c.f. [25] p. 427).

The concept of inference behind this definition is that of testing hypotheses around a fixed parameter point A which is suggested by some physical consideration or by an initial estimate. Statistical inference for partial observations of a stochastic evolution equation aims here at dealing with deviations in terms of certain perturbations, of the true value from the generating operator assumed as null hypothesis. This is done by means of the asymptotically nondegenerate statistical testing problem inherent in (1.14). Since an operator plays here the role of the statistical parameter, the work might be regarded as abstract inference, in the sense of [10], for a space of unbounded operators. However, as soon as we have established the asymptotic localization property, the parameter under consideration will not be the unbounded operator A but a bounded perturbing operator. Furthermore, (1.14) will be verified in section 3 under the additional assumption that A has a compact resolvent, so that it is meaningful to construct a nuclear perturbation of the point spectrum of A.

The limit inequalities (1.14) will be verified in section 3 below by means of certain properties and inequalities connected with the *Hellinger distance* $d_2(\cdot,\cdot)$ between μ_ε^s and μ^s defined as

$$[d_2(\mu_\varepsilon^s,\mu)]^2 = 2\cdot(1-H_s(A_\varepsilon,A)) \tag{1.15}$$

where

$$H_s(A_\varepsilon,A) = \int_{C[0,s]} \sqrt{\frac{d\mu_\varepsilon^s}{dv}}\sqrt{\frac{d\mu^s}{dv}}dv \tag{1.16}$$

and v is some measure dominating both μ_ε^s and μ^s, (see lemma 2.3). An integral like $\int \sqrt{f_1}\sqrt{f_2}dv$ can obviously be considered for any pair of measures with v-densities f_1 and f_2 on an arbitrary measurable space. In this sense the defining integral in (1.16) is in different contexts known under varying names like the *Hellinger integral*, the *affinity* or the *Bhattacharyya coefficient*. Some nonoverlapping accounts of the significance of this integral in statistical theory are found e.g. in [12], [17], [23], [25]. In this work we shall be requiring some results by F. Liese about Hellinger integrals for diffusion type processes (c.f. [18], [19]).

The analysis of the behaviour of the Hellinger distance (1.15) for measures induced by partial observations of a stochastic evolution equation under perturbation of the generator A will be here performed with the following lemma as a technical tool. The lemma is a generalization of the result of K. Datta in [7] and gives an infinite series representation of the solution $P(\varepsilon)$ to (1.12). For simplicity of notation we shall formulate the result for $P(0) = P$.

Lemma 1.3.

Let $\|\cdot\|_{\mathcal{L}}$ be the operator norm on $\mathcal{L}(H)$. (In the sequel we set $\|\cdot\|_{\mathcal{L}} = \|\cdot\|$. The cases where operator norm is intended will be clear from the context.) Assume that

$$\frac{1}{2\alpha}\left\|\int_0^\infty T_t^* W T_t dt\right\|_{\mathcal{L}} M^2\|h_0\|^2 < \frac{1}{4} \tag{1.17}$$

and let

$$P_i = -\int_0^\infty T_t^* X_i T_t dt, \qquad i = 1,2,\ldots \tag{1.18}$$

$$P_0 = \int_0^\infty T_t W T_t dt$$

$$X_i = \sum_{k=0}^{i=1} P_k C^* C P_{n-k-1}, \qquad i = 1, 2, \ldots \tag{1.19}$$

Then the operator series $\sum_{i=0}^{\infty} P_i$ is convergent e.g. in the uniform topology of bounded operators in $\mathcal{L}(H)$ and if

$$P = \sum_{i=0}^{\infty} P_i \tag{1.20}$$

then P solves the equation (1.12) (= the algebraic Riccati-equation in Hilbert space).

Proof: The proof of these assertions is a straightforward generalization of the corresponding finite dimensional result in [7]. ∎

This lemma facilitates a fairly elementary analysis of the difference $P(\varepsilon) - P$. A restrictive feature in the technique to be presented is the assumption (1.17), which imposes a technical condition that does not seem to have any inherent statistical or system theoretic meaning.

2. PERTURBATION AND HELLINGER BALLS

Let us first recall the following perturbation result. A_ε will designate the operator defined by

$$A_\varepsilon := A + \varepsilon B \tag{2.1}$$

where B is a bounded and selfadjoint operator in H and ε is any positive number such that

$$\alpha > \varepsilon M \|B\| \tag{2.2}$$

Then A_ε generates a strongly continuous semigroup $T_t(\varepsilon)$, $t \geq 0$, and

$$(T_t(\varepsilon) - T_t)h = \varepsilon \int_0^t T_{t-s} B T_s(\varepsilon) h \, ds; \quad h \in H \tag{2.3}$$

see [4] pp. 38–39 and [9] p. 540. Furthermore, the stability bound (1.1) yields that

$$\|T_t(\varepsilon)\| \leq M e^{-(\alpha - \varepsilon M \|B\|)t} \tag{2.4}$$

Let now $P(\varepsilon)$ denote the solution of the ARE (1.12) corresponding to A_ε in (2.1). When ε is chosen small enough so that the representation (1.20) is valid for $P(\varepsilon)$, we obtain the following perturbation lemma for the covariance operator of the filtering error.

Lemma 2.1.

Suppose that $\alpha > \varepsilon M \|B\|$ and that

$$\left(\frac{M}{(\alpha - \varepsilon M \|B\|)} \right)^3 \|W\| \, \|C^* C\| < \frac{1}{4} \tag{2.5}$$

Let $P(\varepsilon)$ be the solution of (1.12) corresponding to A_ε and let $P = P(0)$. Then it holds that the limit

$$\lim_{\varepsilon\to 0} \frac{1}{\varepsilon}\|P(\varepsilon) - P\| \tag{2.6}$$

exists.

Proof: From (1.18) above it holds that

$$\begin{aligned} P_i(\varepsilon) - P_i = \int_0^\infty (T_t(\varepsilon) - T_t)\,(-X_i(\varepsilon))\,T_t(\varepsilon)dt + \\ + \int_0^\infty T_t\,(-X_i(\varepsilon))\,(T_t(\varepsilon) - T_t)\,dt + \\ + \int_0^\infty T_t\,(-(X_i(\varepsilon) - X_i))\,T_t dt \end{aligned} \tag{2.7}$$

Setting $\alpha' = \alpha - \varepsilon M\|B\|$ we have

$$\begin{aligned} \|P_i(\varepsilon) - P_i\| \le M\|X_i(\varepsilon)\| \int_0^\infty e^{-\alpha' t}\|T_t(\varepsilon) - T_t\|dt + \\ + M\|X_i(\varepsilon)\| \int_0^\infty e^{-\alpha t}\|T_t(\varepsilon) - T_t\|dt + \\ + M^2\|X_i(\varepsilon) - X_i\| \int_0^\infty e^{-2\alpha' t}dt \end{aligned} \tag{2.8}$$

by (1.1) and (2.4). Evoking (1.19) we obtain

$$\begin{aligned} &\| - \left(X_i(\varepsilon) - X_i\right)\| \\ &\le \| - C^*C\| \sum_{k=0}^{i-1} (\|P_{i-1-k}\| + \|P_{i-1-k}(\varepsilon)\|)\|P_k(\varepsilon) - P_k \end{aligned} \tag{2.9}$$

since $\sum_{k=0}^{i-1} \|P_{i-1-k}(\varepsilon) - P_{i-1-k}\|\|P_k\| = \sum_{k=0}^{i-1} \|P_{i-1-k}\|\|P_k(\varepsilon) - P_k\|$

Inserting (2.9) in the right hand side of (2.8) entails the inequality

$$\begin{aligned} \|P_i(\varepsilon) - P_i\| \le 2M\|X_i(\varepsilon)\| \int_0^\infty e^{-\alpha' t}\|T_t(\varepsilon) - T_t\|dt + \\ + \frac{M^2}{2\alpha}\|C^*C\| \sum_{k=0}^{i-1} (\|P_{i-1-k}(\varepsilon)\| + \|P_{i-1-k}\|)(\|P_k(\varepsilon) - P_k\|) \end{aligned} \tag{2.10}$$

where we have also used $\alpha > \alpha'$.

Now, by apparent identifications, (2.10) is recognized as a linear recurrent inequality of the Gronwall-type $x_i \le f_i + \sum_{k=0}^{i-1} z_k x_k$, which in view of the lemmata in [11] is solved as $x_i \le f_i + \sum_{k=0}^{i-1} \prod_{j=k+1}^{i-1}(1 + z_j)z_k f_k$. Applying these relationships on (2.10) we have

$$\|P_i(\varepsilon) - P_i\| \le 2M\|X_i(\varepsilon)\| \int_0^\infty e^{-\alpha' t}\|T_t(\varepsilon) - T_t\|dt +$$

$$+\|C^*C\|\frac{M^3}{2}\int_0^\infty e^{-\alpha' t}\|T_t(\varepsilon)-T_t\|dt$$

$$\sum_{k=0}^{i-1}\prod_{j=k+1}^{i-1}\left[1+\frac{M^2}{2\alpha}\|C^*C\|\{\|P_{i-1-j}(\varepsilon)\|+\|P_{i-1-j}\|\}\right]$$

$$\|x_k(\varepsilon)\|\{\|P_{i-1-k}(\varepsilon)\|+\|P_{i-1-k}\|\} \tag{2.11}$$

In view of (2.3) we obtain

$$\|P_i(\varepsilon)-P_i\|\le \varepsilon I(\varepsilon)\Big\{2M\|X_i(\varepsilon)\|\frac{M^3}{\alpha}\|C^*C\|\cdot R(\varepsilon)\sum_{k=0}^{i-1}\|x_k(\varepsilon)\|(\|P_{i-1-k}(\varepsilon)\|+$$

$$+\|P_{i-1-k}\|)\Big\} \tag{2.12}$$

where $R(\varepsilon)=\prod_{j=1}^{\infty}\left[1+\frac{M^2}{2\alpha}\{\|P_{j-1}(\varepsilon)\|+\|P_{j-1}\|\}\right]$ is a convergent infinite product since $\sum_{j=1}^{\infty}\|P_{j-1}(\varepsilon)\| < \infty$ as well as $\sum_{j=1}^{\infty}\|P_{j-1}\| < \infty$, and $I(\varepsilon) = \int_0^\infty e^{-\alpha' t}\|\int_0^t T_{t-s}BT_s(\varepsilon)ds\|dt$.

It can be seen, by working through the details underlying (1.17)–(1.19), c.f. [7], that

$$\|P_j(\varepsilon)\|\le t_j\|P_0(\varepsilon)\|(z(\varepsilon))^j \tag{2.13}$$

where $t_j=\sum_{k=0}^{j-1}t_kt_{j-1-k}$, $t_0=1$, and $z(\varepsilon)=\left(\frac{M}{2\alpha'}\right)^2\|P_0(\varepsilon)\|\|C^*C\|$. The same expressions hold, of course, for $\varepsilon=0$, too. In view of (1.18) and the stability bounds we have

$$\|P_0(\varepsilon)\|\le\left(\frac{M}{2\alpha'}\right)^2\|W\| \tag{2.14}$$

and from (1.19), (2.13) and 2.14)

$$\|X_i(\varepsilon)\|=\sum_{k=0}^{i-1}\|C^*C\|\|P_{i-1-k}(\varepsilon)\|$$

$$\le\left(\frac{M}{2\alpha'}\right)^4\|W\|^2\|C^*C\|(z(\varepsilon))^{i-1}\cdot t_i \tag{2.15}$$

Hence there exists a closed right-neighbourhood of zero $u=\{\varepsilon\mid 0\le\varepsilon\le a\}$, such that every ε in u satisfies (2.2) and (2.5), and that

$$\|P_i(\varepsilon)-P_i\|\le\varepsilon\cdot I(\varepsilon)S_i(a) \tag{2.16}$$

where $S_i(a)$ a bound for the expression inside the paranthesis in the right hand side of (2.12) obtained by (2.13)–(2.15) evaluated at $\varepsilon = a$. Also by (2.13)–(2.15) the sum $\sum_{i=1}^{\infty}S_i(a)$ is convergent.

It follows from (1.18) and (2.3) by a straightforward application of Lebesgue's theorem of dominated convergence that

$$\frac{1}{\varepsilon}\|P_0(\varepsilon)-P_0\|=\frac{1}{\varepsilon}\|\int_0^\infty(T_t(\varepsilon)-T_t)WT_tdt+$$

$$+(-1)\int_0^\infty T_t(\varepsilon)W(T_t(\varepsilon)-T_t)dt\|$$

converges to a limit as $\varepsilon \to 0$. But then the same is true for every expression $\frac{1}{\varepsilon}\|P_i(\varepsilon)-P_i(\varepsilon)\|$ in view of (1.18)–(1.19) and an induction argument. Thus (2.16) entails

$$\sum_{i=1}^{\infty}\frac{1}{\varepsilon}\|P_i(\varepsilon)-P_i\| \leq I(\varepsilon)\sum_{i=1}^{\infty}S_i(a) < \infty \tag{2.17}$$

The convergence of the infinite sum is thus uniform w.r.t. ε in u, and hence

$$\lim_{\varepsilon\to 0}\frac{1}{\varepsilon}\|P(\varepsilon)-P\| = \sum_{i=1}^{\infty}\lim_{\varepsilon\to 0}\frac{1}{\varepsilon}\|P_i(\varepsilon)-P_i\| \tag{2.18}$$

and the assertion (2.6) follows. ∎

A similar technique is used for proving the differentiability of the Riccati-operator w.r.t. a scalar parameter that appears linearly in A by [15], where works on differentiability of P are discussed in somewhat greater detail.

The following two lemmata give two-sided bounds for the Hellinger integrals $H_S(A_\varepsilon, A)$ for the measures $\mu_\varepsilon^S(\varepsilon)$ and μ^S induced on $(C[0,S],\mathcal{B})$ by the partial observations $y^S(\varepsilon)$ (in (1.7)) of the processes $dx_t(\varepsilon) = A_\varepsilon x_t(\varepsilon)dt + dw_t$ for $\varepsilon > 0$ and $\varepsilon = 0$, respectively, where A_ε is given in (2.1).

Lemma 2.2.

For A_ε as in (2.1)–(2.2) it holds that

$$H_S(A_\varepsilon, A) \geq \exp\Big\{-\frac{\varepsilon^2}{2}\int_0^S\int_0^t\Big[\langle u_{t-s}(\varepsilon)h_0, P(\varepsilon)h_0^*\rangle + \\ +\langle T_{t-s}h_0, \frac{1}{\varepsilon}\big(P(\varepsilon)-P\big)\rangle\Big]^2 ds\,dt\Big\} \tag{2.19}$$

where

$$u_{t-s}(\varepsilon)h_0 = \int_0^{t-s} T_{(t-s)-u}BT_u(\varepsilon)h\,du \tag{2.20}$$

and h_0, h_0^* are the representers in H of the observing operator C and its dual operator C^*, respectively.

Proof: By the definition of $H_S(A_\varepsilon, A)$ and the properties of natural logarithm we have

$$H_S(A_\varepsilon, A) \geq \exp(-\frac{1}{2}K_S(A_\varepsilon, A)) \tag{2.21}$$

where

$$K_S(A_\varepsilon, A) = E\left[-\ln\frac{d\mu_\varepsilon^S}{d\mu^S}(y^S(0))\right] \tag{2.22}$$

The ineqality (2.21) is noted e.g. in [5], and is obviously valid for any statistical space. The number $K_S(A_\varepsilon, A)$ is the well known Kullback-Leibler information number. (Some properties of the Kullback-Leibler-number for continuous time stochastic filtering systems are discussed in [16], where further references are available.)

It follows by (1.8), (1.10)–(1.11), (2.22) and well known properties of Itô 's stochastic integral that

$$K_S(A_\varepsilon, A) = \frac{1}{2}E\int_0^S \left(\int_0^t [CT_{t-s}(\varepsilon)P(\varepsilon)C^* - CT_{t-s}PC^*]d\hat{v}_s\right)^2 dt$$

$$= \frac{1}{2}\int_0^S \int_0^t [CT_{t-s}(\varepsilon)P(\varepsilon)C^* - CT_{t-s}PC^*]^2\, ds\, dt \tag{2.23}$$

where $\hat{v}$ is the innovations process corresponding to the null hypothesis i.e. the parameter value A.

Let us denote the squared, deterministic, expression in (2.23) by $J_{t-s}(\varepsilon)$. By introducing the representers h_0 and h_0^* of C and C^* respectively, see (1.4), and by making some identical rearrangements in view of the self adjointness of T_t and $T_t(\varepsilon)$, we obtain

$$J_{t-s}(\varepsilon) := \langle (T_{t-s}(\varepsilon) - T_{t-s})h_0, P(\varepsilon)h_0^*\rangle + \langle T_{t-s}h_0, P(\varepsilon) - P)h_0^*\rangle \tag{2.24}$$

Hence the assertion follows by (2.22) and (2.3). ∎

The following lemma is due to F. Liese (see [18], [19]), who derives bounds more refined than those required for the purposes of this paper.

Lemma 2.3.

For any $\rho \in]\frac{1}{2}, 1[$ and A_ε as in (2.1)–(2.2),

$$H_S(A_\varepsilon, A) \leq \left[E\left(D_\rho(y^S)\right)^{\frac{1}{2\rho-1}}\right]^{1-\frac{1}{2\rho}} \tag{2.25}$$

where $D_\rho(y^S) = \exp\{-\frac{1}{2}\rho(1-\rho)\int_0^S [C\hat{\bar{x}}_t(\varepsilon) - C\hat{x}_t]^2 dt\}$, $\hat{\bar{x}}_t(\varepsilon) = \int_0^t T_{t-s}(\varepsilon)P(\varepsilon)C^* d\hat{v}_s$ (as in (1.11)) and $\hat{x}_t \equiv \int_0^t T_{t-s}PC^* d\hat{v}_s$.

Proof: In view of the representation (1.13) and theorem 7.1 and its corollary (on pp. 237–239 of [20]) we obtain, since $\hat{\bar{x}}_t(\varepsilon)$'s are $\mathcal{F}_t^{y^s}$ $(= \mathcal{F}_t^{\bar{v}^s})$-measurable random variables,

$$l^S(\varepsilon) = \frac{d\mu_\varepsilon^S}{dv} = exp\left(\int_0^S C\hat{\bar{x}}_t(\varepsilon)d\hat{v}_t - \frac{1}{2}\int_0^S (C\hat{\bar{x}}_t(\varepsilon))^2 dt\right)$$

$$l^S = \frac{d\mu^S}{dv} = exp\left(\int_0^S C\hat{x}_t d\hat{v}_t - \frac{1}{2}\int_0^S (C\hat{x}_t)^2 dt\right)$$

where v denotes the Wiener measure on $(C[0,S], \mathcal{B})$, and $\hat{v}^S$ is the innovations process corresponding to y^S. Then $\hat{v}^S$ is a standard Wiener-process on $(\Omega, \mathcal{F}, (\mathcal{F}_t^{y^S})_{t\geq 0}^S, \mathbf{P})$. Therefore the computations of Liese for bounding the affinity $H_S(A_\varepsilon, A) = E(\sqrt{l^S(\varepsilon)\cdot l^S})$

The ineqality (2.21) is noted e.g. in [5], and is obviously valid for any statistical space. The number $K_S(A_\epsilon, A)$ is the well known Kullback-Leibler information number. (Some properties of the Kullback-Leibler-number for continuous time stochastic filtering systems are discussed in [16], where further references are available.)

It follows by (1.8), (1.10)–(1.11), (2.22) and well known properties of Itô 's stochastic integral that

$$K_S(A_\epsilon, A) = \frac{1}{2}E\int_0^S \left(\int_0^t [CT_{t-s}(\epsilon)P(\epsilon)C^* - CT_{t-s}PC^*]d\hat{v}_s\right)^2 dt$$

$$= \frac{1}{2}\int_0^S \int_0^t [CT_{t-s}(\epsilon)P(\epsilon)C^* - CT_{t-s}PC^*]^2 \, ds\, dt \tag{2.23}$$

where $\hat{v}$ is the innovations process corresponding to the null hypothesis i.e. the parameter value A.

Let us denote the squared, deterministic, expression in (2.23) by $J_{t-s}(\epsilon)$. By introducing the representers h_0 and h_0^* of C and C^* respectively, see (1.4), and by making some identical rearrangements in view of the self adjointness of T_t and $T_t(\epsilon)$, we obtain

$$J_{t-s}(\epsilon) := \langle (T_{t-s}(\epsilon) - T_{t-s})h_0, P(\epsilon)h_0^*\rangle + \langle T_{t-s}h_0, P(\epsilon) - P)h_0^*\rangle \tag{2.24}$$

Hence the assertion follows by (2.22) and (2.3). ∎

The following lemma is due to F. Liese (see [18], [19]), who derives bounds more refined than those required for the purposes of this paper.

Lemma 2.3.

For any $\rho \in]\frac{1}{2}, 1[$ and A_ϵ as in (2.1)–(2.2),

$$H_S(A_\epsilon, A) \leq \left[E\left(D_\rho(y^S)\right)^{\frac{1}{2\rho-1}}\right]^{1-\frac{1}{2\rho}} \tag{2.25}$$

where $D_\rho(y^S) = \exp\{-\frac{1}{2}\rho(1-\rho)\int_0^S [C\hat{\bar{x}}_t(\epsilon) - C\hat{x}_t]^2 dt\}$, $\hat{\bar{x}}_t(\epsilon) = \int_0^t T_{t-s}(\epsilon)P(\epsilon)C^* d\hat{v}_s$ (as in (1.11)) and $\hat{x}_t \equiv \int_0^t T_{t-s}PC^* d\hat{v}_s$.

Proof: In view of the representation (1.13) and theorem 7.1 and its corollary (on pp. 237–239 of [20]) we obtain, since $\hat{\bar{x}}_t(\epsilon)$'s are $\mathcal{F}_t^{y^\epsilon}$ $(= \mathcal{F}_t^{\bar{y}^\epsilon})$-measurable random variables,

$$l^S(\epsilon) = \frac{d\mu_\epsilon^S}{dv} = exp\left(\int_0^S C\hat{\bar{x}}_t(\epsilon)d\hat{v}_t - \frac{1}{2}\int_0^S (C\hat{\bar{x}}_t(\epsilon))^2 dt\right)$$

$$l^S = \frac{d\mu^S}{dv} = exp\left(\int_0^S C\hat{x}_t d\hat{v}_t - \frac{1}{2}\int_0^S (C\hat{x}_t)^2 dt\right)$$

where v denotes the Wiener measure on $(C[0,S], \mathcal{B})$, and $\hat{v}^S$ is the innovations process corresponding to y^S. Then $\hat{v}^S$ is a standard Wiener-process on $(\Omega, \mathcal{F}, (\mathcal{F}_t^{y^S})_{t\geq 0}^S, \mathbf{P})$. Therefore the computations of Liese for bounding the affinity $H_S(A_\epsilon, A) = E(\sqrt{l^S(\epsilon)\cdot l^S})$

upwards by means of Hölder's inequality and a martingale property go through without change providing us with (2.25) ∎

The process $\frac{1}{2}\int_0^S [C\hat{x}_t(\varepsilon) - C\hat{x}_t]^2 dt$ (letting S vary) is in fact the so called *Kullback-process* between μ_ε^S and μ^S. This manner of expression is justified by some recent general results and ideas about Kullback Leibler-information, c.f. [27].

When considering partial observations of a stochastic evolution equation one may encounter, due to the interference of a particular observing operator C, difficulties with the *identifiability of the induced measures.* This would make $H_S(A_\varepsilon, A) = 1$ and $K_S(A_\varepsilon; A) = 0$ in lemmas 2.2 and 2.3 above even though $A_\varepsilon \neq A$. This does not contradict the general properties of the Hellinger distance. A special study of the identifiability problem with some particular unbounded observing operators is found in [14]. The type of conclusions arrived at in [14] (concerning identifiability with regard to the point of measurement) are interpreted without difficulty for certain kinds of bounded observing operators, too.

Clearly, the results of this section convey some idea of the kind of perturbation in the "abstract" parameter A that give us Hellinger δ-balls in the sense of helping to determine sets that fulfill $d_2(A_\varepsilon, A) < \delta$.

3. ON ASYMPTOTIC LOCALIZATION OF THE STATISTICAL SPACE

We shall now derive the property of asymptotic localization by making some restrictions on the statistical space under consideration, as well as on the perturbing operator B.

First we assume that the operator A is self adjoint with compact resolvent (as well as that it generates an exponentially stable semigroup of operators). This is tantamount to the existence of a complete orthonormal system ϕ_i; $i = 1, 2, \ldots$ of eigenfunctions in H and a corresponding system of eigenvalues $\{-\lambda_i\}$ such that $\alpha \leq \lambda_1 \leq \lambda_2 \leq \cdots \lambda_i \uparrow +\infty$. The semigroup generated by A is for $t > 0$

$$T_t h = \sum_{i=1}^{\infty} e^{-\lambda_i t}\phi_i\langle\phi_i, h\rangle \tag{3.1}$$

This sort of representation is invoked in numerous studies of distributed parameter system c.f. [3] (see also [14] for references on identifiability in (3.1)), and has the advantages entailed by explicit computability.

Furthermore, we assume that the operator B in (2.1) is a positive trace-class operator with eigenfunctions ϕ_i and (positive) eigenvalues γ_i i.e. $B\phi_i = \gamma_i\phi_i$; $i = 1, 2, \ldots$

$$\sum_{i=1}^{\infty} \gamma_i < \infty \tag{3.2}$$

Then (2.1) is transformed into

$$-\lambda_i(\varepsilon) = -\lambda_i + \varepsilon\gamma_i; \qquad i = 1, 2, \ldots \tag{3.3}$$

i.e. we consider a perturbation of each of the eigenvalues λ_i with the amounts $\varepsilon\gamma_i$. For any h in the domain of definition of A we have $A_\varepsilon h = \sum_{i=1}^{\infty}(-\lambda_i(\varepsilon))\langle\phi_i, h\rangle\phi_i$. The perturbed semigroup is now ($h \in H$)

$$T_t(\varepsilon)h = \sum_{i=1}^{\infty} e^{-\lambda_i(\varepsilon)t}\langle\phi_i, h\rangle\phi_i \tag{3.4}$$

where we are assuming that $\{T_t(\varepsilon) \mid t \geq 0\}$ is an exponentially stable semigroup so that (2.2) holds for B.

Instead of (2.3) we may now represent the difference between the semigroups as

$$(T_t(\varepsilon) - T_t)h = \varepsilon t \sum_{i=0}^{\infty} \gamma_i e^{-(\lambda_i - \varepsilon\theta_i\gamma_i)t} \langle \phi_i, h \rangle \phi_i \tag{3.5}$$

by virtue of the mean value theorem of differential calculus, so that θ_i is for every i a number between 0 and 1. Obviously the results in section 2 above hold true for the perturbation thus defined. In particular, we have the inequality (2.19), where we now however are able to write

$$u_{t-s}(\varepsilon) = (t-s) \sum_{i=1}^{\infty} \gamma_i e^{-(\lambda_i - \theta_i \varepsilon \gamma_i)(t-s)} \tag{3.6}$$

The notion of asymptotic localization presupposes that ε decreases to zero at a particular rate as the length of the interval of observation, S, increases to infinity. In fact we set $\varepsilon = S^{-1/2}$ i.e.

$$A_S := A + (S)^{-1/2} B \tag{3.7}$$

with A and B as in (3.3). With this choice of ε as a function of S we have the following lemma

Lemma 3.1.

For A_S as in (3.7) it holds that

$$\lim_{S \to \infty} K_S(A_S; A) = \sigma_1 > 0 \tag{3.8}$$

Proof: The right hand side of (2.19) yields here

$$K_S(A_S; A) = \frac{1}{S} \int_0^S \int_0^t \Big[\langle u_{t-s}(S^{-1/2}) h_0^* \rangle +$$

$$+ \langle T_{t-s} h_0, \sqrt{S}(P(S^{-1/2}) - P) \rangle \Big]^2 ds\, dt$$

$$\leq \frac{2}{S} \int_0^S \int_0^t \langle u_{t-s}(S^{-1/2}) h_0 \rangle^2 ds\, dt +$$

$$+ \frac{2}{S} \int_0^S \int_0^t \langle T_{t-s} h_0, \sqrt{S}(P(S^{-1/2}) - P) h_0^* \rangle^2 ds\, dt \tag{3.9}$$

Let us consider the first term in the right hand side of the inequality (3.9), and denote

$$I_1(t, S) = \int_0^t \langle u_{t-s}(S^{-1/2}) h_0, P(S^{-1/2}) h_0^* \rangle^2 ds$$

By (3.6) we obtain

$$I_1(t,S) \leq \|P(S^{-1/2})\| \, \|h_0\|^4 \int_0^t (t-s)^2 \Big(\sum_{i=1}^{\infty} e^{-(\lambda_i - \frac{\theta_i}{\sqrt{S}}\gamma_i)(t-s)} \gamma_i\Big)^2 ds$$

$$\leq \|P(S^{-1/2})\|^2 \|h_0\|^4 \Big(\sum_{i=1}^{\infty} \gamma_i\Big)^2 \int_0^t (t-s)^2 e^{-2\beta(S^{-1/2})(t-s)} ds$$

where $\beta(S^{-1/2}) \leq \lambda_i - \frac{\theta_i}{\sqrt{S}}\gamma_i$ for all i. A straightforward integration shows that

$$\int_0^t (t-s)^2 e^{-2\beta(S^{-1/2})(t-s)} ds = -\frac{t^2}{2\beta(S^{-1/2})} e^{-\beta(S^{-1/2})t} -$$

$$-\frac{t}{(2\beta(S^{-1/2})} e^{-2\beta(S^{-1/2})t} + \frac{1}{(2\beta(S^{-1/2}))^3}(1 - e^{-2\beta(S^{-1/2})t}) \tag{3.10}$$

Hence

$$\frac{1}{S}\int_0^S I_1(t,S) ds \leq \Big[\frac{1}{S}\int_0^S \Big(-\frac{t^2}{2\beta} e^{-2\beta t}\Big) dt +$$

$$+\frac{1}{S}\int_0^S \frac{t}{2\beta} e^{-2\beta t} dt +$$

$$+\frac{1}{(2\beta)^3} - \frac{1}{S}\int_0^S \frac{e^{-2\beta t}}{2\beta^4} dt\Big] \|P(S^{-1/2})\| \, \|h_0\|^2 \tag{3.11}$$

where $\beta = \beta(S^{-1/2})$. But the same computations that produced (3.10) will also show that the upper bound in (3.11) will converge to a finite limit, as $S \to \infty$. (Obviously $\beta(S^{-1/2})$ can be chosen to converge to a positive limit as $S \to \infty$ and $\|P(S^{-1/2})\|$ converges to $\|P\|$ in view of the results in section 2. Quite analogously we see that

$$\frac{1}{S}\int_0^S \int_0^t \langle T_{t-s}h_0, \sqrt{S}(P(S^{-1/2}) - P)h_0^*\rangle^2 ds \, dt$$

$$\leq \|\sqrt{S}(P(S^{-1/2}) - P)^2\| \, \|h_0\|^4 \frac{1}{S}\int_0^S \int_0^t e^{-2\alpha(t-s)} ds \tag{3.12}$$

which has a finite limit as $S \to \infty$, by lemma 2.1. The fact that $\sigma_1 > 0$ is entailed by our next lemma. ∎

First we rewrite some of the preceding expressions by a few auxiliary notations. As in (2.24) we can write

$$C\hat{\bar{x}}_t(\varepsilon) - C\hat{x}_t = \int_0^t \Big[\langle (T_{t-s}(\varepsilon) - T_{t-s})h_0, P(\varepsilon)h_0^*\rangle +$$

$$+\langle T_{t-s}h_0, (P(\varepsilon) - P)h_0\rangle\Big] d\hat{v}_s \tag{3.13}$$

In view of (3.6) we obtain for $\varepsilon = S^{-1/2}$ that

$$C\hat{\bar{x}}_t(\varepsilon) - C\hat{x}_t = \frac{1}{\sqrt{S}}\int_0^t \Big[\langle u_{t-s}(S^{-1/2})h_0, P(S^{-1/2})h_0^*\rangle + \\ +\langle T_{t-s}h_0, \sqrt{S}(P(S^{-1/2}) - P)h_0^*\rangle\Big] d\hat{v}_s \tag{3.14}$$

Let us now introduce two real valued kernels

$$g_1(t-s;S) := \langle u_{t-s}(S^{-1/2})h_0, P(S^{-1/2})h_0^*\rangle \\ g_2(t-s;S) := \langle T_{t-s}h_0, \sqrt{S}(P(S^{-1/2}) - P)h_0^*\rangle \tag{3.15}$$

Then it follows that

$$\int_0^S ((\hat{\bar{x}}_t(\varepsilon) - (\hat{x}_t)^2 dt = \frac{1}{S}\int_0^S \Big(\int_0^t g_1(t-s;S)d\hat{v}_s + \\ + \int_0^t g_2(t-s;S)d\hat{v}_s\Big)^2 dt \tag{3.16}$$

To abbreviate the notation we write the squared integrand in the right hand side of (3.16) with $t \to J_t(S)$. Then we have

Lemma 3.2

$\frac{1}{S}\int_0^S J_t(S)dt \to \sigma_2$ in P-probability, as $S \to \infty$, where $\sigma_2 > 0$.

Proof: Let $g_1(u) = g_i(u;\infty) = \lim_{S\to\infty} g_i(u;S)$ for any $u \geq 0$; $i = 1,2$; and let

$$x_i(t;0) = \int_0^t g_i(t-s)d\hat{v}_s \qquad i = 1,2. \tag{3.17}$$

First we note that

$$\int_0^\infty (g_i(u))^2 du < \infty \qquad i = 1,2 \tag{3.18}$$

as is readily seen by the stability assumptions and by some straightforward computations (c.f. (3.10) above). Thus we can introduce the random variables

$$y_i(t) = \int_{-\infty}^t g_i(t-s)d\hat{v}_s, \quad z_i(t) = \int_{-\infty}^0 g_i(t-s)d\hat{v}_s \tag{3.19}$$

by defining $\hat{v}$ over the negative axis as a Gaussian zero mean process with the covariance $E\hat{v}_t\hat{v}_s = \frac{1}{2}(|s|+|t|-|t-s|)$, possibly by enlarging the original probability space. Evidently,

$$x_i(t;0) = y_i(t) - z_i(t) \quad \text{for all} \quad t \geq 0 \tag{3.20}$$

We shall now observe that as $t \to \infty$

$$x_i(t;0) \to y_i(t) \quad \text{almost surely;} \quad i = 1,2 \tag{3.21}$$

We consider the case of $x_1(t;0)$, the other case being treated similarly. By (3.19)

$$z_1(t) = \sum_{i=1}^{\infty} te^{-\lambda_i}\gamma_i\langle\phi_i, h\rangle\langle\phi_i, P(S^{-1/2})h_0^*\rangle \int_{-\infty}^{0} e^{\lambda_i s} d\hat{v}_s$$

where $z_1(t)$ converges obviously to zero as $t \to \infty$, almost surely. Thus (3.21) follows from (3.20). But $\{y_i(t) \mid t \geq 0\}$ is a strictly stationary and ergodic process. Hence, almost surely,

$$\frac{1}{S}\int_0^S (y_i^2(t))dt \to E(y_i(0))^2,$$

as $S \to \infty$, by [8] ch. IX. By this and (3.21) it can be seen that

$$\frac{1}{S}\int_0^S (x_i(t;0))^2 dt \to E(y_i(0))^2 \tag{3.22}$$

almost surely as $S \to \infty$.

Analogously to (3.17) we set $x_1(t;S) = \int_0^S g_1(t-s;S)d\hat{v}_s$ and demonstrate next that $H_S := \frac{1}{S}\int_0^S (x_1(t;S) - x_1(t,0))^2 dt \to 0$ in **P**-probability as $S \to \infty$.

Indeed, for any $\delta > 0$ we have

$$\mathbf{P}(\{H_S > \delta\} \leq \frac{1}{\delta}\frac{1}{S}\int_0^S [\int_0^t \{g_(t-s;S) - g_1(t-s)\}^2 ds]dt \tag{3.23}$$

by Markov's inequality and a property of the stochastic integral. Invoking the mean value theorem once more we obtain from (3.6), (3.15) and (3.17) that

$$\int_0^t \{g_1(t-s;S) - g_1(t-s)\}^2 ds$$

$$= \frac{1}{S}\int_0^t (t-s)^4 \Big(\sum_{i=1}^{\infty}(\gamma_i\theta_i)e^{-(\lambda_i-\xi_i)(t-s)}\Big)^2 ds \tag{3.24}$$

where ξ_i is a number such that $-\lambda_i < -\lambda_i + \xi_i < -\lambda_i + \varepsilon\theta_i\gamma_i$ for every i and $\xi_i \to 0$ as $\varepsilon = S^{-1/2}$ decreases to zero. Inserting the result of (3.24) in (3.23) gives an integral, which as a function of S is of the order

$$\frac{1}{S^2}\int_0^S\int_0^t (t-s)^4 e^{-C(S)(t-s)} ds\, dt \tag{3.25}$$

where $C(S) \to C > 0$ as $S \to \infty$. The integral in (3.25) can, however, be computed by partial integrations as in (3.10)–(3.11) with some additioned term of the type $S^k e^{-C(S)S}$ in the final result. (The "lowest" order integration i.e. $\frac{1}{S^2}\int_0^S (1 - e^{-C(S)s})ds$ provides also a term that converges to zero as $S \to \infty$. Hence $H_S \to 0$ as $S \to \infty$ in **P**-probability. A similar argument is valid for $x_2(t;s)$, too.

Now, let us consider (3.16) (through 3.17))

$$\frac{1}{S}\int_0^S J_t(S)dt \leq 4\Big[\frac{1}{S}\int_0^S (x_1(t;S) - x_2(t;0))^2 dt + \frac{1}{S}\int_0^S x_1^2(t;0)dt +$$

$$+\frac{1}{S}\int_0^S (x_2(t;S)-x_2(t;0))^2dt+\frac{1}{S}\int_0^S x_2^2(t;0)dt\Big] \tag{3.26}$$

By (3.25) the first term in the right-hand side of (3.26) converges in probability to zero as $S \to \infty$ and the same limit is obtained for the third term. The second and fourth terms converge to finite positive limits almost surely by the arguments summed up in (3.22). Hence the existence in probability of the limit $\sigma_2 = \lim_{S\to\infty} \frac{1}{S}\int_0^S J_t(S)dt$ is ascertained. The limit is a non-random number that actually represents the limiting variance of the asymptotically stationary and non degenerate random variable $x(t) = x_1(t;S) + x_2(t;S)$, and is thus non-zero. ∎

Hereafter we prove the chief result of this paper.

Proposition 3.3.

Let A be a selfadjoint generator of a stable semigroup and let A possess a compact resolvent, written as $A \in \mathcal{A}$. Let B be a positive nuclear operator, the eigenvectors of which coincide with those of A. Consider

$$A_S = A + S^{-1/2}B \tag{3.27}$$

Let μ_{A_S} be the measure induced on $(C[0,S], \mathcal{B}_S)$ by the process $y^S = \{y_t \mid t \in [0,S]\}$, where $y_s = \int_0^t Cx_t(\sqrt{S})ds + dv_t$ and μ_A be measure induced by the parameter A. Then

$$0 < \lim_{S\to\infty} \inf \|\mu_{A_S} - \mu_A\|_1 \leq \lim_{S\to\infty} \sup \|\mu_{A_S} - \mu_A\|_1 < 1 \tag{3.28}$$

i.e. the statistical space $\{C[0,S], \mathcal{B}_S, \{\mu_A^S \mid A \in \mathcal{A}\}\}$ is asymptotically localizable at A for deviations of the form (3.27).

Proof: There are a number of interesting relations between the Hellinger distance and the variation norm $\|\cdot\|_1$ as well as between $\|\cdot\|_1$ and the Kullback-Leibler number. Here we apply the following

$$2(1 - H_S(A_S, A)) \leq \|\mu_{A_S} - \mu_A\|_1 \leq \sqrt{8(1 - H_S(A_s, A))} \tag{3.29}$$

(see [5] and [18] for references). Then our lemmata 2.2 and 2.3 yield

$$2(1 - [ED_\rho(y^S)^{\frac{1}{2\rho-1}}]^{1-\frac{1}{2\rho}}) \leq \|\mu_{A_S} - \mu_A\|_1$$

$$\leq \sqrt{8[1 - \exp(-\frac{1}{2}K_S(A_S; A))]} \tag{3.30}$$

We might equally well apply the inequality $\|\mu_{A_A} - \mu_A\|_1 \leq \sqrt{2K_S(A_S; A)}$, a result proved by I. Csiszár and others.

By lemma 3.1 it holds that

$$\lim_{S\to\infty} \sqrt{8(1 - \exp(-\frac{1}{2}K_S(A_S; A))} = \sqrt{8(1 - \exp(-\frac{1}{2}\sigma_1))}$$

which verifies the right hand limit in the assertion (3.28). The lower bound is obtained by means of the lemma 3.2. We may write

$$D_\rho(y^S) = e^{-(\frac{\rho(1-\rho)}{2}\frac{1}{S}\int_0^S J_t(S)dt)}$$

in view of the notations used in the mentioned lemma. Since the expression $\frac{1}{S}\int_0^S J_t(S)ds$ converges in probability to σ_2 as $S \to \infty$ and $D_\rho(y^S) \leq 1$, we may verify uniform integrability of the r.v's $D_\rho(y^S)$, S large enough, so that (2.2) and (2.5) hold. Hence we have

$$\lim_{S\to\infty} 2\Big(1 - [ED_\rho(y^S)^{\frac{1}{2\rho-1}}\Big)\Big] \to 2(1 - \Big[e^{-(\frac{\rho(1-\rho)}{2})\sigma_2}\Big]^{\frac{1}{2\rho-1}})$$

which is clearly positive. Hence the complete assertion is proved. ∎

Acknowledgements:

The comments by participants of the BiBoS IV will be made use of in future work focusing, in particular, on a treatment less restrictive than that of section 3. The general influence of the author's previous discussions with Dr. L. Birgé from University of Paris are acknowledged with appreciation.

References

[1] M.G. Akritas: *Contiguity of probability measures associated with continuous time stochastic-processes.* Ph.D.-Thesis in Statistics. University of Wisconsin-Madison 1978.

[2] A. Bagchi & V. Borkar: 'Parameter Identification in Infinite-Dimensional Linear Systems'. Stochastics 1984 Vol. **12** pp. 201–213.

[3] R.J. Curtain: 'Spectral systems'. International Journal of Control 1984, Vol. **39** pp. 657–665.

[4] R.J. Curtain & A.J. Pritchard: *Infinite Dimensional Linear System Theory.* Springer New York 1978.

[5] D. Dacunha-Castelle: 'Vitesse de Convergence pour Certaines Problemes Statistique' in *Ecole d'Eté de Probabilites de Saint Flour VII–1977*: Lecture Notes in Mathematics Nr. 678 Springer Verlag Berlin 1970.

[6] R.M. Datko: 'Extending a Theorem of A.M. Lyapunov to Hilbert Space'. Journal of Mathematical Analysis and Applications 1970 Vol. **32**, pp. 610–616.

[7] K.B. Datta: 'Series solution of matrix Riccati equation'. International Journal of Control 1978 Vol. **27**, pp. 463–472.

[8] J. Doob: *Stochastic Processes.* John Wiley & Sons, New York 1973.

[9] J.S. Gibson: 'The Riccati Equations for Optimal Control Problems in Hilbert Space'. SIAM J. on Control and Optimization 1979, Vol. **17**, pp. 537–565.

[10] U. Grenander: *Abstract Inference.* John Wiley & Sons New York 1983.

[11] G.S. Jones: 'Fundamental inequalities for discrete and discontinuous functional equations'. Journal of SIAM 1964 Vol. **12**, pp. 43–57.

[12] T. Kailath: 'The Divergence and Bhattacharyya Distance Measures in Signal Selection'. IEEE Transactions on Communication Technology 1967, Vol. **COM-15**, pp. 52–60.

[13] G. Kallianpur: *Stochastic Filtering Theory.* Springer Verlag New York 1981.

[14] T. Koski: 'A note on the statistical identifiability of some stationary stochastic evolution equations'. Reports on computer science and mathematics Ser. A Nr. 35 Åbo Academy 1984.

[15] T. Koski: 'Local asymptotic normality for partial observations of a stochastic evolution equation'. TW-Memorandum Nr. 548 Twente University of Technology Enschede, The Netherlands 1986 (submitted for publication).

[16] T. Koski & W. Loges: 'On measures of information for continuous time stochastic processes'. BiBos Research Reports Nr. 157, Bielefeld 1986 (submitted for publication in an enlarged version).

[17] Yu.A. Kutoyants: *Parameter Estimation for Stochastic Processes.* Heldermann Verlag Berlin 1984.

[18] F. Liese: 'Hellinger integrals of Diffusion Processes'. Friedrich Schiller Universität Jena Preprint Nr. 83/39 1983.

[19] F. Liese: 'Chernoff Type Bounds of Errors in Hypothesis Testing of Diffusion Processes' in D. Rasch & M.L. Tiku: *Robustness of Statistical Methods and Nonparametric Statistics* VEB Berlin 1984.

[20] R.S. Liptzer & A.N. Shiryaev: *Statistics of Random Processes* Vol. I Springer Berlin 1977.

[21] M. Métivier: *Semimartingales.* W. de Gruyter Berlin 1985.

[22] M. Métivier & G. Pistone: 'Une formule d'isometrie pour l'intégrale stochastique hilbértienne et équations d'évolution linéaires stochastiques'. Z. Wahrscheinlichkeitstheorie verw. Geb. 1975 Vol. **33**, pp. 1-18.

[23] E.J.G. Pitman: *Some Basic Theory for Statistical Inference.* Chapman and Hall London 1979.

[24] W. Ray & D.G. Lainiotis (ed.): *Distributed Parameter Control Systems.* Marcel Dekker Inc, New York 1978.

[25] H. Strasser: *Mathematical Theory of Statistics.* W. de Gruyter Berlin 1985.

[26] S.G. Tzafestas (ed.): *Distributed Parameter Control Systems.* Pergamon Press London 1982.

[27] E. Valkeila: 'A survey of some recent developments in Kullback-Leibler information for a stochastic basis'. (A communication in Finnish, University of Helsinki, 1986).

ADAPTIVE GAMES

George P. Papavassilopoulos
Department of Electrical Engineering-Systems
University of Southern California
Los Angeles, California 90089-0781

1. INTRODUCTION

The area of adaptive control has received a lot of attention during recent years. Many different schemes have been proposed and studied and several interesting results have been obtained. In almost all the papers the single objective case is addressed: There is one decision maker with his own control objective or there are many controllers acting in a decentralized way who nonetheless have a common objective, i.e., they are a team. Nonetheless, there are cases where there exist many controllers, each one of which has his own objective. Such multiobjective control problems can arise after the decentralization of a large system or exist as such due to the inherent characteristics of the problem. Situations like these belong to the realm of game theory. It is only natural to try to extend the ideas of adaptive control to the area of game theory. As a matter of fact, ignorance of several parameters pertaining to an opponent for which parameters no apriori off line identification is feasible is quite natural in situations of conflict.

There are very few papers in the literature addressing such issues [1,2,4-6]. In the present paper we first introduce a simple example by which we demonstrate several ideas and subsequently we consider some more general situations and describe some results. Section 2 deals with the introductory example.

We consider a scalar linear deterministic evolution equation with two controllers each one being interested in minimizing a one step ahead quadratic cost. The matrix A (here is a scalar) as well as the weight with which each player (i.e., controller) penalizes his control effort are not known to the other player. At each instant of time, each player

This work was supported by ZWO, The Netherlands Organization for the Advancement of Pure Research, Contract Number B62-239, by the U.S. Air Force Office of Scientific Research under Grant AFOSR-85-0254 and by the National Science Foundation under Grant NSF-INT-85-04097.

S. Albeverio et al. (eds.), Stochastic Processes in Physics and Engineering, 223–236.

knows the previous states and his own previous control actions, but not those of his opponent (this is to be contrasted with [1] where the past actions of all the players are known to all of them, i.e., the players have the same information). Each player assumes such a model for the system evolution as if he were the sole decision maker and employs a least squares scheme to estimate the parameter of this model, based on which estimate he calculates his current control action. The question is how the system will behave if the control actions are thus calculated; one essentially has to compare the resulting behavior with that which would result if all the parameters were known to all the players and the Nash concept were employed. Our basic result is that if the closed loop matrix of the known parameter case is asymptotically stable, then the adaptive scheme for the unknown parameter case outlined above with produce an asymptotically stable system, and that the control gains will converge but not to those of the known parameter case. Thus, the closed loop system does not behave as time goes by, like in the known parameter case. This weakness is due to the fact that the standard least squares algorithm pertains really to time invariant systems, whereas the hypothetical system considered by each controller for estimation purposes should be thought of as time varying since the control gains of the ignored player are time varying and are being incorporated in the parameters of the hypothetical system. In Section 3 the more general ARMAX model is considered. The same rationale employed for the example of Section 2 is being utilized here and some results are briefly delineated.

Better or worse results are possible for schemes different than those proposed, but it should be stressed that the purpose of this paper is not as much to provide the best scheme and its complete analysis, but rather to introduce and explain some ideas and demonstrate that positive results are possible for dynamic adaptive games with different information available to the players. Finally, it should be stressed that here we are primarily interested in dynamic cases where the controllers have different information, whereas [1] assumes common information and [2], [3] deal with static cases. Related and more complete results are given in [4-6].

2. INTRODUCTORY EXAMPLE

Consider a system with evoluation equation

$$x_{k+1} = ax_k + u_{1k} + u_{2k}\ , \quad k = 0,1,2,\ldots,x_0 = \text{given} \tag{1}$$

and two costs

$$J_i = (x_{k+1})^2 + r_i u_{ik}^2\ , \quad r_i > 0, \quad i = 1,2 \tag{2}$$

All the quantities are scalars. At time k the players know $x_k, x_{k-1}, \ldots, x_0$ and thus the perfect (Nash) solution[1,2] is obtained by solving

[1]The perfect Nash solution for the known parameter case (5) does not change if player i has perfect recall of his past actions.

$$Q \begin{bmatrix} u_{1k} \\ u_{2k} \end{bmatrix} = -a \begin{bmatrix} 1 \\ 1 \end{bmatrix} x_k \tag{3}$$

$$Q = \begin{bmatrix} 1+r_1 & 1 \\ 1 & 1+r_2 \end{bmatrix} \tag{4}$$

Q has determinant $\Delta = r_1 r_2 + r_1 + r_2 \neq 0$ and thus (3) yields

$$\begin{bmatrix} u^*_{1k} \\ u^*_{2k} \end{bmatrix} = -\frac{1}{\Delta} \begin{bmatrix} r_2 \\ r_1 \end{bmatrix} a x_k \tag{5}$$

Substituting u^*_{1k}, u^*_{2k} from (5) into (1) results in the closed-loop system

$$x_{k+1} = a_c x_k \tag{6}$$

$$a_c = a \frac{r_1 r_2}{r_1 r_2 + r_1 + r_2} \tag{7}$$

The closed loop system is asymptotically stable if $|a_c| < 1$. (All the above generalize to the multivariable case with the only exception that the matrix corresponding to Q is not necessarily invertible and thus it has to be assumed to be.) The important thing to notice is that the solution (5) assumes that each player knows x_k, but that also has enough knowledge about a, r_1, r_2. Although it is reasonable for player 1 to know r_1, it is not reasonable to assume that he knows a and even more r_2. We are thus motivated to consider the following situation.

Player 1 knows r_1 and at time k has perfect recall of x_k, x_{k-1}, $\ldots, x_0$, $u_{1,k-1}, \ldots, u_{1,0}$. He assumes that the system obeys

$$x_{k+1} = a_i x_k + u_{ik} , \tag{8}$$

[2]The general definition of a Nash equilibrium is as follows: let J_i: $U_1 \times U_2 \to R$. $(u^*_1, u^*_2) \in U_1 \times U_2$ is a Nash equilibrium if $J_1(u^*_1, u^*_2) \leq (u_1, u^*_2)$, $\forall\, u^*_1 \in U_1$ and $J_2(u^*_1, u^*_2) \leq J_2(u^*_1, u_2)$, $\forall\, u_2 \in U_2$.

with $i=1$ in which a_1 is unknown to him. At time k he creates a least squares estimate of a_1, as if a_1 were an unknown constant, see (11), based on which estimate he minimizes J_1 and thus calculates u_{1k}, see (9). Of course, although it is easily seen by player 1 that a_1 is time varying (actually at time k, $a_1 = a - \frac{1}{1+r_2}\hat{a}_k^2$) it is treated for the least squares scheme as a constant in the hope that eventually a_1 converges to a constant. A similar scheme is employed by player 2, see (10), (12) and (8) with $i=2$. The thus calculated controls act in the real system (1) and therefore the resulting closed loop system is (13). The issue is if and where the sequences described in (9)-(13) converge and the relation of the limits to (5) and (6).

$$u_k^1 = -\frac{1}{1+r_1}\hat{a}_k^1 x_k \tag{9}$$

$$u_k^2 = -\frac{1}{1+r_2}\hat{a}_k^2 x_k \tag{10}$$

$$\hat{a}_{k+1}^1 = \hat{a}_k^1 + \frac{x_k}{\sum_{i=0}^{k} x_i^2}(x_{k+1} - \hat{a}_k^1 x_k - u_k^1), \quad \hat{a}_0^1 = \text{given} \tag{11}$$

$$\hat{a}_{k+1}^2 = \hat{a}_k^2 + \frac{x_k}{\sum_{i=0}^{k} x_i^2}(x_{k+1} - \hat{a}_k^2 x_k - u_k^2), \quad \hat{a}_0^2 = \text{given} \tag{12}$$

$$x_{k+1} = \left(a - \frac{1}{1+r_1}\hat{a}_k^1 - \frac{1}{1+r_2}\hat{a}_k^2\right)x_k, \quad x_0 = \text{given} \tag{13}$$

It should be pointed out that the difference in models assumed by the players — see (8), $i=1,2$, — is motivated by the information used, by which the players do not know their opponent's past actions; for the same reason the estimates $\hat{a}_k^1$, $\hat{a}_k^2$ are not identical, although both players use the same estimation scheme (contrast with [1]). In some sense, each player employs a "single objective" rationale, when he assumes (8) and minimizes J_i of (2). Let us introduce some notation.

$$\rho_1 = \frac{1}{1+r_1}, \quad \rho_2 = \frac{1}{1+r_2} \tag{14}$$

$$R = \begin{bmatrix} 1 & \rho_2 \\ \rho_1 & 1 \end{bmatrix}, \quad T = \begin{bmatrix} 1 & -\sqrt{\rho_2/\rho_1} \\ \sqrt{\rho_1/\rho_2} & 1 \end{bmatrix} \tag{15}$$

$$\lambda = 1 + \sqrt{\rho_1\rho_2}\ , \quad \lambda_2 = 1 - \sqrt{\rho_1\rho_2} \tag{16}$$

$$\hat{a}_k = \begin{bmatrix} \hat{a}_k^1 \\ \hat{a}_k^2 \end{bmatrix} \tag{17}$$

$$\beta_k = R\hat{a}_k - a\begin{bmatrix} 1 \\ 1 \end{bmatrix} \tag{18}$$

$$\gamma_k = T^{-1}\beta_k \tag{19}$$

$$\sigma_k^2 = \frac{x_k^2}{\sum_{i=0}^{k} x_i^2}\ , \quad \sigma_o^2 = 1 \tag{20}$$

It holds

$$T^{-1}RT = \begin{bmatrix} \lambda_1 & 0 \\ 0 & \lambda_2 \end{bmatrix}, \quad 2 > \lambda_1 > 1 > \lambda_2 > 0 \tag{21}$$

Substituting u_k^1, u_k^2 from (9), (10) into (11)-(13) and using (14)-(20) we obtain

$$\gamma_{k+1} = \left(I - \sigma_k^2 \begin{bmatrix} \lambda_1 & 0 \\ 0 & \lambda_2 \end{bmatrix} \right)\gamma_k, \quad \gamma_k = \begin{bmatrix} \gamma_{1k} \\ \gamma_{2k} \end{bmatrix} \tag{22}$$

$$x_{k+1} = \theta_k x_k \tag{23}$$

$$\theta_k = a_c - [\rho_1,\rho_2]\gamma_k \tag{24}$$

Our basic result is the following.

Proposition 1. Let $|a_c| < 1$ (i.e., the closed loop matrix of the known parameter perfect Nash solution is assumed to be asymptotically stable). Then the $\{x_k\}$ sequence generated by (23) converges to zero, i.e., the closed loop system resulting from the adaptive scheme (9)-(13) is also asymptotically stable for any initial conditions x_0, $\hat{a}_0^1$, $\hat{a}_0^2$. Also, the parameters $\hat{a}_k^1$, $\hat{a}_k^2$ converge.

Proof. $0 \le \sigma_k^2 \le 1$. If (i) $\liminf \sigma_k^2 = \sigma^* > 0$, then for $k \ge K(\varepsilon)$: $1 \ge \sigma_k^2 \ge \varepsilon > 0$ for some ε: $0 < \varepsilon < \sigma^*$. This yields: $1 - \lambda_i \le 1 - \lambda_i\sigma_k^2 < 1 - \lambda_i\varepsilon \Leftrightarrow \mp\sqrt{\rho_1\rho_2} \le 1 - (1\pm\sqrt{\rho_1\rho_2})\sigma_k^2 < 1 - \varepsilon(1\pm\sqrt{\rho_1\rho_2})$. It holds: $-1 < \mp\sqrt{\rho_1\rho_2}$ and $1 - \varepsilon(1\pm\sqrt{\rho_1\rho_2}) < 1$ and thus $\gamma_k \to 0$. Consequently, $\theta_k \to a_c$ and since $(\sigma_{k+1})^{-2} = 1 + \sigma_k^{-2}\theta_k^{-2} \cong 1 + \sigma_k^{-2}a_c^{-2}$ with $a_c^{-2} > 1$ we have $\sigma_k^{-2} \to +\infty$ and thus $\sigma_k^2 \to 0$, contradiction. If (ii) $\liminf \sigma_k^2 = 0$ but $\limsup \sigma_k^2 > 0$, then there is a subsequence $\sigma_{n_k}^2 \to \sigma^* > 0$ which implies $|1 - \lambda_i\sigma_{n_k}^{-2}| < 1 - \delta$ for some $\delta < 0$. Since the coefficients $1 - \lambda_i\sigma_{n_k}^2$ appear infinitely often in (22) (argument similar as in (i)), we have $\gamma_k \to 0$. Thus $\sigma_k^2 \to 0$; then $1 - \sigma_k^2\lambda_i \to 1$ so that eventually $\gamma_{i,k+1}$ and $\gamma_{i,k}$ have the same sign. It always holds: $|\lambda_{i,k+1}| < |\gamma_{i,k}|$ and thus $\gamma_{i,k} \to \gamma_i^*$ monotonically, for some $\gamma^* = (\gamma_1^*,\gamma_2^*)$. Consequently, $\theta_k \to a_c - [\rho_1,\rho_2]\gamma^* = \theta^*$. If $|\theta^*| > 1$, then

$$\sigma_k^2 \cong \frac{(\theta^*)^{2k}}{1 + (\theta^*)^2 + \ldots + (\theta^*)^{2k}} \to 1 - (\theta^*)^{-2}$$

and thus $\sigma_k^2 \to 1 - (\theta^*)^{-2} > 0$, contradiction. Thus $|\theta^*| \le 1$. If $|\theta^*| < 1$ then obviously $x_k \to 0$. If $\theta_k^2 \to 1^+$ then eventually $x_{k+1}^2 = \theta_k^2 x_k^2 \ge x_k^2$ and thus

$$\sigma_k^2 = \frac{1}{\left(\frac{x_o}{x_k}\right)^2 + \ldots + \left(\frac{x_{k-1}}{x_k}\right)^2 + 1} \ge \frac{1}{1 + \ldots + 1 + 1} = \frac{1}{k+1}$$

It then holds: $0 \le 1 - \lambda_i\sigma_k^2 \le 1 - \frac{\lambda_i}{k+1} \Rightarrow |\gamma_{i,k+1}| \le \left(1 - \frac{\lambda_i}{k+1}\right)|\gamma_{i,k}| \Rightarrow \gamma_{ik} \to 0 \Rightarrow \gamma_1^* = \gamma_2^* = 0 \Rightarrow \theta^* = a_c$. But $|a_c| < 1$ and this contradicts $|\theta^*| = 1$.

So θ_k^2 cannot converge to 1^+. If $\theta_k^2 \to 1^-$, then eventually, $x_{k+1}^2 \leq x_k^2$ and thus x_k^2 converges to some $(x^*)^2$. If $x^* \neq 0$ then $\sigma_k^2 = \frac{x_k^2}{x_o^2 + \ldots + x_k^2} \geq \frac{(x^*)^2}{x_o^2(k+1)}$ and an argument similar to the one above yields $\gamma^* = 0$, $\theta^* = a_c$ and thus $x^* = 0$, contradiction. Thus if $\theta_k^2 \to 1^-$ then $x_k \to 0$. Finally, consider the case $\theta_k^2 \to 1$ but not from above or below. In particular, let $\theta_k \to -1 = \theta^*$. It holds

$$\theta_{k+1} = \theta_k + \sigma_k^2[\rho_1\lambda_1, \rho_2\lambda_2]\gamma_k \tag{25}$$

Since θ_k does not converge to -1 from above or below it must be

$$[\rho_1\lambda_1, \rho_2\lambda_2]\gamma^* = 0 \tag{26}$$

$\theta^* = -1$ means

$$a_c - [\rho_1, \rho_2]\gamma^* = 0 \tag{27}$$

(26) and (27) can be solved for γ_1^*, γ_2^* to yield

$$\gamma^* = \frac{-1 + a_c}{2\rho_1\rho_2\sqrt{\rho_1\rho_2}} \begin{bmatrix} -\rho_2\lambda_2 \\ \rho_1\lambda_1 \end{bmatrix}, \quad \gamma_1^* > 0, \quad \gamma_2^* < 0 \tag{28}$$

From (22) we obtain

$$\frac{\gamma_{1,k+1} - \gamma_{1,k}}{\gamma_{2,k+1} - \gamma_{2,k}} = \frac{\lambda_1\gamma_{1k}}{\lambda_2\gamma_{2k}} \tag{29}$$

Since θ_k does not converge to -1 from above or below for infinitely many k's it holds

$$\theta_{k+2} < \theta_{k+1}, \quad \theta_{k+1} > \theta_k \tag{30}$$

which using (25) implies

$$\lambda_1\rho_1\gamma_{1,k+1} + \lambda_2\rho_2\gamma_{2,k+1} < 0 \tag{31}$$

$$\lambda_1\rho_1\gamma_{1k} + \lambda_2\rho_2\gamma_{2k} > 0 \tag{32}$$

Since $\gamma_1^* > 0$, $\gamma_2^* < 0$, for k sufficiently large, it holds $\gamma_{1k} > 0$, $\gamma_{2k} < 0$ and thus (31), (32) yields

$$0 > \frac{\gamma_{2k}}{\gamma_{1k}} > \frac{-\lambda_1\rho_1}{\lambda_2\rho_2} > \frac{\gamma_{2,k+1}}{\gamma_{1,k+1}} \tag{33}$$

(29) yields

$$\frac{\gamma_{2,k+1}}{\gamma_{1,k+1}} = \frac{\lambda_2}{\lambda_1}\frac{\gamma_{2k}}{\gamma_{1k}} - \frac{\lambda_2 - \lambda_1}{\lambda_1}\frac{\gamma_{1k}}{\gamma_{1,k+1}} \tag{34}$$

Inserting $\gamma_{2,k+1}/\gamma_{1,k+1}$ from (34) into (33) yields

$$0 > \frac{\gamma_{2k}}{\gamma_{1k}} > -\frac{\lambda_1\rho_1}{\lambda_2\rho_2} > \frac{\lambda_2}{\lambda_1}\frac{\gamma_{2k}}{\gamma_{1k}} + \frac{\lambda_1 - \lambda_2}{\lambda_1}\frac{\gamma_{1k}}{\gamma_{1,k+1}} \tag{35}$$

from which

$$\frac{\gamma_{2k}}{\gamma_{1k}}\frac{\lambda_1 - \lambda_2}{\lambda_2} > \frac{\lambda_1 - \lambda_2}{\lambda_2}\frac{\gamma_{1,k}}{\gamma_{1,k+1}}$$

or (since $\lambda_1 > \lambda_2$)

$$\frac{\gamma_{2k}}{\gamma_{1k}} > \frac{\gamma_{1k}}{\gamma_{1,k+1}}$$

which means that a negative number is greater than a positive one, contradiction. Thus if $\theta_k \to -1$ it must be $\theta_k^2 \to 1^-$. Similarly, we conclude for the case where $\theta_k \to +1$, that $\theta_k^2 \to 1^-$.

In conclusion, we have shown that $\sigma_k^2 \to 0$, $\gamma_k \to \gamma^*$, $x_k \to 0$, $\theta_k \to \theta^*$, $\theta^* \in [-1,1]$ and if $\theta^* = \pm 1$ then eventually $\theta_k \in (-1,1)$. Finally, let us notice that the convergence of γ_k implies that of $\hat{a}_k^1$, $\hat{a}_k^2$, see (17)-(19).

□

Remark 1. Ideally, we would like $\gamma_k \to \gamma^* = (0,0)'$, since then the control gains and the closed loop matrix of the adaptive scheme converge to those of the known parameter case. (This is easily verified by solving (18) with $\beta_k = (0,0)$, for $\hat{a}_k^1$, $\hat{a}_k^2$, using (9), (10) and comparing with (5).) This is not achieved by the scheme described here, as the following argument demonstrates (see also Examples in Remark 3): Let γ_{10}, γ_{20} be sufficiently small. Since $|\gamma_{1k}|$, $|\gamma_{2k}|$ are decreasing and converge, $|\theta^*| < 1$. Then, $\sigma_k^2 \cong (\theta^*)^{2k}/(1 + \ldots + \theta^{*2k}) \Rightarrow \sigma_k^2 \cong (1-\theta^{*2})\theta^{*2k}$ and thus $\gamma_{i,k+1} \cong \gamma_{i,k}[1 - \lambda_i(1-\theta^{*2})\theta^{*2k}]$. Let $\omega = \theta^{*2}$, $\beta = \lambda_i(1-\theta^{*2})$. We know that the sequence $y_{n+1} = (1-\beta\omega^n)y_n$, $\beta > 0$, $0 < \omega < 1$ does not go to zero except if $1 - \beta\omega^n = 0$ for some n. Thus, $\gamma_{i,k} \to \gamma_i \neq 0$ and thus $\theta^* \neq a_c$. Choosing γ_{io} small is obviously desirable, but this means β_k small, i.e., $\hat{a}_k$ close to $a\, R^{-1}\begin{bmatrix}1\\1\end{bmatrix}$, which is exactly what the players who choose $\hat{a}_0^1$, $\hat{a}_0^2$ do not know.

Remark 2. In our example we considered that the target trajectories are zero (see (2)). One can consider though $J_i = (x_{k+1} - x^*_{i,k+1})^2 + v_i u_{ik}^2$ instead of (2). Then (8) could be modified to $x_{k+1} = a_i x_k + u_{ik} + c_i$ where both a_i and c_i are to be estimated.

Remark 3, Examples. We conducted several examples which demonstrate the convergence properties of (20), (22), (23), (24).

i) For $(a_c, \gamma_{10}, \gamma_{20}, \rho_1, \rho_2) = (0.3, -1, 5, 0.5, 0.3)$, $(\sigma_k^2, \gamma_{1k}, \gamma_{2k}, \theta_k)$ converged to (0, 0.1891, 1.4762, -0.2374) in 6 iterations.

ii) For $(a_c, \gamma_{10}, \gamma_{20}, \rho_1, \rho_2) = (0.3, 3, 6, 0.5, 0.3)$, $(\sigma_k^2, \gamma_{1k}, \gamma_{2k}, \theta_k)$ converged to (0, 0.2767, 1.0231, -0.1453) in 5 ierations.

iii) For $(a_c, \gamma_{10}, \gamma_{20}, \rho_1, \rho_2) = (0.3, 3, 6, 0.5, 0.5)$, $(\sigma_k^2, \gamma_{1k}, \gamma_{2k}, \theta_k)$ converged to (0, 0.3733, 1.3430, -0.5582) in 12 iterations.

iv) For $(a_c, \gamma_{10}, \gamma_{20}, \rho_1, \rho_2) = (0.9, 3, 6, 0.5, 0.3)$, $(\sigma_k^2, \gamma_{1k}, \gamma_{2k}, \theta_k)$ converged to (0, 0.0954, 0.8223, 0.6056) in 12 iterations.

v) For $(a_c, \gamma_{10}, \gamma_{20}, \rho_1, \rho_2) = (-0.9, 3, 6, 0.5, 0.3)$ $(\sigma_k^2, \gamma_{1k}, \gamma_{2k}, \theta_k)$ converged to (0, 0.0011, 0.1176, -0.9358) in 76 iterations.

It should be noticed that θ_k does not converge to a_c (see Remark 1 also). It seems that a_c close to ±1 slows down the convergence as a comparison of (ii) to (iv) and (v) indicates.

3. THE ARMAX CASE

In this section we discuss the generalization of the introductory example to the ARMAX case. Consider the evolution equation

$$
\begin{aligned}
y_{t+1} &= a_o y_t + a_1 y_{t-1} + \ldots + a_n y_{t-n} + u_{1t} + u_{2t} + v_t \\
&= A(q^{-1}) y_t + u_1 + u_{2t} + v_t
\end{aligned}
\tag{36}
$$

and the two costs

$$
J_i = E[y_{t+1}^2 + r_i u_{it}^2], \quad i = 1,2 \tag{37}
$$

where $\{v_t\}$ is a sequence of independent gaussian random variables. If the parameter vector $\theta = (a_o, a_1, \ldots, a_n, r_1, r_2)'$ is known to both players and at each instant of time they know all the previous history $\{y_t, y_{t-1}, \ldots\}$, the Nash equilibrium is

$$
\begin{bmatrix} u_{1t} \\ u_{2t} \end{bmatrix} = - \begin{bmatrix} 1+r_1 & 1 \\ 1 & 1+r_2 \end{bmatrix}^{-1} \begin{bmatrix} 1 \\ 1 \end{bmatrix} A(q^{-1}) y_t \tag{38}
$$

and the resulting closed loop system is

$$
y_{t+1} = \left(1 - [1,1] \begin{bmatrix} 1+r_1 & 1 \\ 1 & 1+r_2 \end{bmatrix}^{-1} \begin{bmatrix} 1 \\ 1 \end{bmatrix} \right) A(q^{-1}) y_t + v_t \tag{39}
$$

(We assume that $r_1 r_2 + r_1 + r_2 > 0$ and that the closed loop system matrix of (39) is asymptotically stable.) Let us now consider that player 1 knows r_1, but not $a_o, \ldots, a_n, r_2$, so that he cannot find his control action u_{1t} at time t by using (38). We also assume that at time t he knows $y_t, y_{t-1}, \ldots$ and $u_{1,t-1}, u_{1,t-2}, \ldots$, but not $u_{2,t-1}, u_{2,t-2}, \ldots$.

We follow the same line of development as in the introductory example: player 1 assumes that the system obeys

$$y_{t+1} = a^1_o y_t + a^1_1 y_{t-1} + \ldots + a^1_n y_{t-n} + u_{1t} + v^1_t \tag{40}$$

This is a hypothetical system. If (40) were the true system and player's 1 cost were (37) with $i=1$, he would use at time t the control action:

$$-\frac{1}{1+r_1}(a^1_o y_t + \ldots + a^1_n y_{t-n}) \tag{41}$$

Using the true outputs of the system $y_t, y_{t-1}, \ldots$ and his own previous actions $u_{1,t-1}, \ldots$ he creates estimates — according to some scheme — of $a^1_o, \ldots, a^1_n$, call them $\hat{a}^{1t}_o, \ldots, \hat{a}^{1t}_n$ and he uses them in (41), so that he employs the control action:

$$u_{1t} = \bar{u}_{1t} = \det - \frac{1}{1+r_1}(\hat{a}^{1t}_o y_t + \ldots + \hat{a}^{1t}_n y_{t-n}) \tag{42}$$

Similarly, player 2 uses

$$u_{2t} = \bar{u}_{2t} = \det - \frac{1}{1+r_2}(\hat{a}^{2t}_o, y_t + \ldots + \hat{a}^{2t}_n y_{t-n}) \tag{43}$$

Both u_{1t}, u_{2t} of (42), (43) are applied to the real system, so that we have

$$y_{t+1} = a_o y_t + a_1 y_{t-1} + \ldots + a_n y_{t-n} + \bar{u}_{1t} + \bar{u}_{2t} + v_t \tag{44}$$

We also have

$$(\hat{a}^{1t}_o, \ldots, \hat{a}^{1t}_n) = F_1(\hat{a}^{1,t-1}_o, \hat{a}^{1,t-1}_1, \ldots, \hat{a}^{1,t-1}_n, u_{1,t-1}, u_{1,t-2}, \ldots, y_t, y_{t-1}, \ldots) \tag{45}$$

$$(\hat{a}^{2t}_o, \ldots, \hat{a}^{2t}_n) = F_2(\hat{a}^{2,t-1}_o, \ldots, \hat{a}^{2,t-1}_n, u_{2,t-1}, u_{2,t-2}, \ldots, y_t, y_{t-1}, \ldots) \tag{46}$$

where F_1, F_2 are determined by the estimation schemes employed by the two players. The question now is how the behavior of (42)-(46) is related to the behavior of (38), (39). The basic result, for the proof of which we refer to [5,6], is that if the closed loop system of the known parameter case (39) is asymptotically stable, then the control actions (42), (43) and the system (44) behave in the limit like (38),

(39), if the estimation schemes F_1, F_2 of (45), (46) pertain to time varying systems (such as weighted least squares, or stochastic approximation types). The important thing to notice is that the hypothetical system (40) should not be thought of as time invariant, since the $a_o^1,..,a_n^1$ incorporate in them not only the $a_o,..a_n$ but also the time varying gains of player 2, i.e., $a_k^1 \cong a_k - \frac{1}{1+r_2} a_k^{2,t}$.

A more general case is to consider

$$\begin{aligned} y_{t+1} &= a_o y_t + ..+ a_n y_{t-n} + b_{10} u_{1t} + ... + b_{1,k_1} u_{1,t-k_1} \\ &\quad + b_{20} u_{2t} + ... + b_{2,k_2} u_{2,t-k_2} + v_t \\ &= A(q^{-1}) y_t + B_1(q^{-1}) u_{1k} + B_2(q^{-1}) u_{2t} + v_t \end{aligned} \tag{47}$$

instead of (36), and the costs are still as in (37). Player 1 knows r_1, but ignores $(a_o,..,a_n, b_{10},..,b_{1,k_1}, b_{20},..,b_{2,k_2})$ and r_2, but he knows the basic structure of (47) and thus n, k_1, k_2. At time t, player 1 knows $y_t, y_{t-1},...$ and $u_{1,t-1}, u_{1,t-2},...$, but ignores the previous actions $u_{2,t-1}, u_{2,t-2},..$ of player 2. (Similarly for player 2.) Notice, that lack of knowledge of the other's previous actions is even more pertinent to the model (47) which has delays in the controls, than to (36). Nonetheless, the same rationale can be employed as before, but instead of the hypothetical system (40) which has the same number of delays in y_t as (36), player 1 considers the hypothetical system

$$\begin{aligned} y_{t+1} &= a_o^1 y_t + ..+ a_{\ell_1}^1 y_{t-\ell_1} + b_{10}^1 u_{1,t} + b_{1,1}^1 u_{1,t-1} + ... \\ &\quad + b_{1,m_1}^1 u_{1,t-m_1} + c_o v_t^1 + c_1 v_{t-1}^1 + ... + c_{1,t-p_1} v_{t-p_1}^1 \end{aligned} \tag{48}$$

where ℓ_1 does not necessarily equal n. The reason is the following — if all the parameters and previous control actions are known to both the players, the Nash equilibrium is given by solving for u_{1t}, u_{2t}:

$$(A(q^{-1}) y_t + B_1(q^{-1}) u_{1t} + B_2(q^{-1}) u_{2t}) b_{10} + r_1 u_{1t} = 0 \tag{49}$$

$$(A(q^{-1}) y_t + B_1(q^{-1}) u_{1t} + B_2(q^{-1}) u_{2t}) b_{20} + r_2 u_{2t} = 0 \tag{50}$$

from which we obtain:

$$u_{2t} = \frac{r_1 b_{20}}{r_1 r_2 + r_1 b_{20} B_2(q^{-1}) + r_2 b_{10} B_1(q^{-1})} A(q^{-1}) y_t$$

which, when substituted in (47), results to a system of the form (48). This substitution determines ℓ_1, m_1, p_1. Simulation studies based on the above rationale show that the scheme works well, under some assumptions such as asymptotic stability of the closed loop system of the known parameter case and some weak coupling conditions on the coefficients of B_1, B_2, but no rigorous theoretical analysis is currently available, see [6].

Remark. Cases where y_t is vector valued and the costs (37) are substituted by $E[(y_{t+1} - y^{*i}_{t+1})^T Q_i (y_{t+1} - y^{*i}_{t+1}) + r_i \|u_{it}\|^2]$, i.e., where the players have different target objectives — $\{y^{*i}_t\}$ — have been considered in [6].

4. CONCLUSIONS

The aim of this paper was to introduce some ideas pertaining to adaptive schemes for multiobjective control problems with information decentralization. Some basic models and partial results toward this end were presented. Further directions of research may involve the examination of different rationales for creating adaptive schemes, the completion of the theoretical analysis of some of the schemes presented, etc.

5. REFERENCES

[1] Y. M. Chan, "Self Tuning Methods for Multiple Controller Systems," Ph.D. Thesis, University of Illinois at Urbana-Champaign, Dept. of Electrical Engineering, 1981.

[2] T. L. Ting, J. B. Cruz, Jr. and R. A. Milito, "Adaptive Incentive Controls for Stackelberg Games with Unknown Cost Functionals," American Control Conference, San Diego, CA, June 1984.

[3] G. P. Papavassilopoulos, "Iterative Techniques for the Nash Solution in Quadratic Games with Unknown Parameters," accepted to appear in SIAM Journal on Optimization and Control, 1985; also, presented at the Istanbul Workshop on Large Scale Systems, Istanbul, Turkey, June 1984.

[4] G. P. Papavassilopoulos, "Adaptive Dynamic Nash Games: An Example," Seventh Annual Meeting of the Society for Economic Dynamics and Control, London, June 1985.

[5] W. Y. Yang and G. P. Papavassilopoulos, "Decentralized Adaptive Control in a Game Situation for Discrete Time Linear, Time In-

variant Systems," submitted for publication, 1986.

[6] W. Y. Yang, "Decentralized Adaptive Control in a Game Situation," Ph.D. Thesis, University of Southern California, Dept. of EE-Systems, Los Angeles, CA, August 1986.

NONLINEAR FILTERING AND SMOOTHING WITH HIGH SIGNAL-TO-NOISE RATIO

J. Picard
INRIA
Avenue Emile Hughes
Sophia Antipolis
F-06565 Valbonne (France)

ABSTRACT. This work deals with the problem of estimating a state process, the measurements of which are corrupted by an independent white noise of order ε. We derive a finite-dimensional filter which is asymptotically optimal as $\varepsilon \to 0$ and we estimate the rate of convergence with a new method. We also obtain an approximate solution of the smoothing problem.

1. INTRODUCTION

The problem of estimating a time-varying state which is only partially observed is involved in several applied problems and has been studied a great deal for several years; a mathematical framework has been developed and a lot of theoretical results are now available. In this work, we will assume that the process to be estimated (or signal process) is a Markovian diffusion process and that we observe a function of this process plus an independent white noise; we want to obtain a recursive estimate (or filter) of the signal. One generally tries to compute the best L^2 estimate of the signal at time t, given the observation up to time t, thus one is led to study a conditional expectation. It is now well-known that the problem of computing recursively this conditional expectation involves a stochastic partial differential equation (Zakaï's equation) and that this infinite-dimensional equation can be reduced to a finite-dimensional one only in particular cases, such as the linear case. More generally, if one disposes of the observation up to time t, one may want to estimate the signal process at a time $u \leq t$: this is the smoothing problem. Differential equations can also be derived for this problem ([6]), but their exact solutions are again difficult to reach.

Therefore, the practical problem consists of finding accurate and easily computable approximations of this optimal estimate. There are different means to achieve this purpose; one can try to discretize the Zakaï equation; another possibility, and this is the one we will be concerned with, consists of using some perturbation techniques: one assumes that the model is indexed by a parameter ε and that, as $\varepsilon \to 0$, the filtering problem tends to an easily solvable problem; then one looks for asymptotic filters which are efficient for ε small. Here we will assume that the measurement noise is of order ε and

S. Albeverio et al. (eds.), Stochastic Processes in Physics and Engineering, 237–251.

that for $\varepsilon = 0$, the signal process is exactly observed. To approach this problem, one can try to use some singular perturbation techniques on the Zakaï equation; this idea was worked out in [4]: assuming that the signal and the observation are one-dimensional, the authors formally derived an asymptotic expansion of the conditional density. In [7], we also studied the one-dimensional problem and obtained some suboptimal filters which are solutions of one- or two-dimensional equations; this was achieved by means of another method which enabled us to estimate the difference between these filters and the optimal one; we also studied some classical filters (extended Kalman filter, linearized filter, second-order filters) with the same technique; then it appeared ([1]) that at least a part of these estimates can be proved with more elementary tools. Finally, in [8], some of the results of [7] were extended to the multi-dimensional case.

The object of this work is to introduce a somewhat different method to prove these results; moreover, the assumptions are slightly weaker and we also study the smoothing problem. The argument is probabilistic and does not use the Zakaï equation. Whereas we used in [7] and [8] the theory of time reversal of diffusions, the proofs of this paper will rather rely on Malliavin's calculus, which is a stochastic calculus of variations. In §2, we make more precise our model, list the assumptions, introduce the reference probability method and state the results; the proofs are detailed in §3 and we conclude in §4 with further estimates which were proved in [7].

2. ASSUMPTIONS AND RESULTS

We suppose that we are given a probability space $(\Omega, \mathcal{F}, \mathbb{P})$, a right-continuous increasing family $(\mathcal{F}_t)_{t\geq 0}$ of sub-σ-algebras of $\mathcal{F}$, two independent $\mathcal{F}_t$ Brownian motions W_t and B_t with values in $\mathbb{R}^p$ and $\mathbb{R}^n$, an $\mathcal{F}_0$ measurable variable X_0 and three functions b, σ and h defined on $\mathbb{R}^n$, taking values respectively in $\mathbb{R}^n$, $\mathbb{R}^n \otimes \mathbb{R}^p$ and $\mathbb{R}^n$. We let ε be a positive parameter and we assume that the signal process X_t and the observation process Y_t satisfy the equation

$$\begin{cases} X_t = X_0 + \int_0^t b(X_u)du + \int_0^t \sigma(X_u)dW_u, \\ Y_t = \int_0^t h(X_u)du + \varepsilon B_t. \end{cases} \tag{2.1}$$

If f is a vectorial function of the variable $x \in \mathbb{R}^n$, f' will denote the Jacobian matrix of f; if f is scalar, f' will be a line vector; this notation is extended to functions which may also depend on other parameters. Except otherwise stated, the following assumptions will be supposed to hold throughout this work:
(i) The function b is C^2 and all its first and second derivatives are bounded.
(ii) The function σ is C^2, σ is bounded as well as its derivatives, and $a \equiv \sigma\sigma^*/2$ is uniformly elliptic.
(iii) The function h is C^3, all its derivatives are bounded and the two functions $h'b$ and $(h'\sigma\sigma^*h'^*)^{-1/2}h'b$ are Lipschitzian; moreover, there exists a constant number $\delta > 0$ such that

$$|h(x) - h(z)| \geq \delta|x - z|. \tag{2.2}$$

(iv) All the moments of the variable X_0 are finite.

Then let $\mathcal{Y}_t$ be the σ-algebra generated by Y_u, $0 \leq u \leq t$. The filtering problem consists of computing, for any measurable function g defined on $\mathbb{R}^n$, the conditional expectation

$$\widehat{g}_t = \mathbb{E}\left[g(X_t) \mid \mathcal{Y}_t\right]$$

and more generally, the smoothing problem consists of computing

$$\widehat{g}_u^t = \mathbb{E}\left[g(X_u) \mid \mathcal{Y}_t\right]$$

for $u \leq t$.

A useful tool in the study of these problems is the so-called reference probability method: we define a new probability $\mathring{\mathbb{P}}$, which is equivalent to $\mathbb{P}$ on each $\mathcal{F}_t$, by

$$\left.\frac{d\mathring{\mathbb{P}}}{d\mathbb{P}}\right|_{\mathcal{F}_t} = \exp\left\{-\frac{1}{\varepsilon}\int_0^t h^*(X_u)dB_u - \frac{1}{2\varepsilon^2}\int_0^t \left|h(X_u)\right|^2 du\right\}. \tag{2.3}$$

Then, under $\mathring{\mathbb{P}}$, the processes W and Y/ε are independent Brownian motions, the inverse of (2.3) is

$$L_t = \left.\frac{d\mathbb{P}}{d\mathring{\mathbb{P}}}\right|_{\mathcal{F}_t} = \exp\frac{1}{\varepsilon^2}\left\{\int_0^t h^*(X_u)dY_u - \frac{1}{2}\int_0^t \left|h(X_u)\right|^2 du\right\}$$

and it is easy to prove that if Γ is a $\mathcal{F}_t$ measurable $\mathbb{P}$ integrable variable, the relation between $\mathbb{P}$- and $\mathring{\mathbb{P}}$-conditional expectations is provided by the Kallianpur-Striebel formula

$$\mathbb{E}\left[\Gamma \mid \mathcal{Y}_t\right] = \frac{\mathring{\mathbb{E}}\left[\Gamma L_t \mid \mathcal{Y}_t\right]}{\mathring{\mathbb{E}}\left[L_t \mid \mathcal{Y}_t\right]}. \tag{2.4}$$

Application of this formula for $\Gamma = g(X_t)$ or $g(X_u)$ gives an expression for $\widehat{g}_t$ or $\widehat{g}_u^t$. Note that since X and Y are $\mathring{\mathbb{P}}$-independent, the conditional expectations of the right side are simply integrations with respect to the law of X. However, this result is not well adapted to the numerical computation of the filter: it needs the integration of a functional of the whole trajectory of X and, above all, it is not recursive so cannot be used in real-time applications. This is why one generally prefers Zakaï's equation: this is a linear partial differential equation, the solution of which is an unnormalized version of the density of X_t conditioned by $\mathcal{Y}_t$; it is also possible to obtain equations for the smoothing problem. But these equations are still numerically complicated, so one needs approximate easily computable solutions; it turns out that the Kallianpur-Striebel formula is a good tool for proving convergence results for the suboptimal filters. An example is provided by the results that we are now going to describe.

We will be concerned with the singular perturbation problem $\varepsilon \to 0$; in this case, from (2.2), the signal process X is nearly observed so one expects to derive good approximations of the filter. It is indeed not difficult to obtain $\mathcal{Y}_t$ adapted processes which are at a distance of order $\sqrt{\varepsilon}$ from X (see §3.1), but obtaining more precise results requires finer methods. We will here describe a filter which is ε-optimal. We now state the two results which will be proved in next section.

Main theorem. *Let m_0 be any vector of $\mathbb{R}^n$ and M_t the solution of*

$$dM_t = b(M_t)dt + \frac{1}{\varepsilon}h'^{-1}(h'\sigma\sigma^* h'^*)^{1/2}(M_t)(dY_t - h(M_t)dt) \tag{2.5}$$

with the initial condition $M_0 = m_0$. Then for every $k \geq 1$ and $t_0 > 0$,

$$\sup_{t \geq t_0} \left\| \hat{X}_t - M_t \right\|_k = O(\varepsilon). \tag{2.6}$$

Actually, the stronger following result holds: if in addition to the process M_t defined by (2.5), one considers for each t and $u \leq t$ the solution of the backward equation

$$\frac{dM_u^t}{du} = b(M_u^t) + \frac{1}{\varepsilon}h'^{-1}(h'\sigma\sigma^* h'^*)^{1/2}(M_u^t)(h(M_u^t) - h(M_u)) \tag{2.7}$$

with the final condition $M_t^t = M_t$, then

$$\sup_{t \geq u \geq t_0} \left\| \hat{X}_u^t - M_u^t \right\|_k = O(\varepsilon). \tag{2.8}$$

Note that even if the initial estimate M_0 is not close to the true value, it is quickly improved; this is a stability property of the filter: we say that the filter has short memory. One can prove that one can add to the drift part of (2.5) or to (2.7) a bounded process without modifying the order of convergence; therefore, if the function b is bouded, one can neglect it. One can also remark that the estimates are uniform as $t \to \infty$. With reference to [7] and [8], we have not assumed here that the initial law of X is absolutely continuous with respect to Lebesgue's measure, so we can consider the case of a deterministic $X_0 = x_0$; if, in this case, one chooses $m_0 = x_0$, one may wonder whether (2.6) and (2.8) hold with $t_0 = 0$; the answer is negative: the order of approximation becomes $\sqrt{\varepsilon}$ for small times and cannot be improved for the filter (2.5), as it can be seen for instance in the linear case.

3. PROOF OF THE MAIN THEOREM

In this section, we prove the main theorem in six steps: we first prove that the distance between X_u, M_u and M_u^t is of order $\sqrt{\varepsilon}$ and in §3.2, we apply a Girsanov transform which will be fundamental in our treatment; our next goal is to show (2.6) with an additional assumption about the law of X_0: this step is not necessary but the proof is more elementary in this case; the proof of this particular case, the proof of (2.6) and (2.8) in the general case are respectively achieved in §3.3, §3.4 and §3.5, except for the proof of a technical result which is rejected to §3.6. In [7] and [8], one of the main tools was the theory of time reversal of diffusion processes but the method that we are going to describe will not involve it. We first introduce some notations. Let

$$\gamma = (h'\sigma\sigma^* h'^*)^{1/2}.$$

From assumptions (ii) and (iii), this function is uniformly elliptic, and since the map $\alpha \mapsto \alpha^{1/2}$ is analytic on the space of symmetric definite positive matrices, γ is C_b^2. If $1 \le i \le n$, we will write γ_i for the ith column of γ, and this convention will also be used for other matrix-valued functions. Finally we define

$$\overline{W}_t = \int_0^t \gamma^{-1} h'\sigma(X_s) dW_s.$$

Then $\overline{W}_t$ is a $(\mathcal{F}_t, \mathbb{P})$ Brownian motion and

$$dX_t = b(X_t)dt + h'^{-1}\gamma(X_t)d\overline{W}_t.$$

3.1. Preliminary estimates

Lemma 1. *For any $k \ge 1$ and $t_0 > 0$,*

$$\sup_{t \ge t_0} \left\| X_t - M_t \right\|_k = O(\sqrt{\varepsilon}). \tag{3.1}$$

Sketch of the proof. If $\mathcal{L}$ denotes the infinitesimal generator of the diffusion X, we derive from Itô's formula

$$\begin{aligned} h(X_t) - h(M_t) =& h(X_0) - h(M_0) + \int_0^t (\mathcal{L}h(X_s) - \mathcal{L}h(M_s))ds \\ & - \frac{1}{\varepsilon}\int_0^t \gamma(M_s)(h(X_s) - h(M_s))ds + \int_0^t h'\sigma(X_s)dW_s - \int_0^t \gamma(M_s)dB_s. \end{aligned}$$

Since γ is uniformly elliptic, this is not difficult to prove (3.1), see [8]: one applies Itô's formula to compute $\left|h(X_t) - h(M_t)\right|^k$ for k even integer, one takes the expectation and one uses some estimates on ordinary inequations. □

Lemma 2. *For any $k \ge 1$ and $t_0 > 0$,*

$$\sup_{t \ge u \ge t_0} \left\| X_u - M_u^t \right\|_k = O(\sqrt{\varepsilon}). \tag{3.2}$$

Proof. One cannot apply directly the method described in the previous lemma to prove (3.2) because M_s^t is not $\mathcal{F}_s$ adapted, so it is difficult to deal with the stochastic integrals which appear in Itô's formula; nevertheless, one can use an auxiliary process which has absolutely continuous paths. So let us consider the process ξ_u solution of

$$\dot{\xi}_u = b(\xi_u) + \frac{1}{\varepsilon}(X_u - \xi_u); \qquad \xi_0 = X_0.$$

Then one can prove, as in lemma 1, that $\xi_u - X_u$ is of order $\sqrt{\varepsilon}$, so it is sufficient to estimate $M_u^t - \xi_u$; we have

$$\frac{\partial}{\partial u}\Big(h(M_u^t) - h(\xi_u)\Big) = \frac{1}{\varepsilon}\gamma(M_u^t)\big(h(M_u^t) - h(\xi_u)\big) + h'b(M_u^t) - h'b(\xi_u) + \frac{1}{\varepsilon}\gamma(M_u^t)\big(h(\xi_u) - h(M_u)\big) - \frac{1}{\varepsilon}h'(\xi_u)(X_u - \xi_u).$$

Considering this relation as an ordinary backward equation and using again the ellipticity of γ, we deduce the existence of constant numbers μ and $c > 0$ such that for ε sufficiently small,

$$\left|h(M_u^t) - h(\xi_u)\right| \leq \mu\left\{e^{-c\frac{t-u}{\varepsilon}}\left|M_t - \xi_t\right| + \frac{1}{\varepsilon}\int_u^t e^{-c\frac{s-u}{\varepsilon}}\left(\left|M_s - \xi_s\right| + \left|X_s - \xi_s\right|\right)ds\right\}.$$

From lemma 1, the moments of the right side are of order $\sqrt{\varepsilon}$. □

Remark 1. One can easily verify that when $X_0 = x_0 = m_0$, one can take $t_0 = 0$ in (3.1) and (3.2).

Remark 2. In the definitions of M_t and M_t^u, if one replaces $\left(h'\sigma\sigma^* h'^*\right)^{1/2}$ by any uniformly elliptic bounded function, (3.1) and (3.2) are still satisfied; moreover, in such a case, it is not necessary to suppose that a is uniformly elliptic.

3.2. Changes of probability measure

The method which is to yield the main theorem will essentially rely on a change of probability and an integration by parts formula. We actually use two changes of probability which affect respectively the laws of the observation noise and of the signal noise. The first one has been defined in (2.3) and leads to the reference probability $\mathring{\mathbb{P}}$. The second one is given by

$$\left.\frac{d\widetilde{\mathbb{P}}}{d\mathring{\mathbb{P}}}\right|_{\mathcal{F}_t} = \Lambda_t^{-1} = \exp\left\{\frac{1}{\varepsilon}\int_0^t \left(h(X_u) - h(M_u)\right)^* d\overline{W}_u - \frac{1}{2\varepsilon^2}\int_0^t \left|h(X_u) - h(M_u)\right|^2 du\right\}.$$

Under $\widetilde{\mathbb{P}}$, the processes

$$\widetilde{W}_t = \overline{W}_t - \frac{1}{\varepsilon}\int_0^t \left(h(X_u) - h(M_u)\right)du$$

and Y_t/ε are two independent Brownian motions and X_t is solution of

$$dX_t = \frac{1}{\varepsilon}h'^{-1}\gamma(X_t)\left(h(X_t) - h(M_t)\right)dt + b(X_t)dt + h'^{-1}\gamma(X_t)d\widetilde{W}_t. \tag{3.3}$$

On the other hand, for every $\mathcal{F}_t$ measurable $\mathbb{P}$-integrable variable Γ,

$$\mathbb{E}\left[\Gamma \mid \mathcal{Y}_t\right] = \frac{\widetilde{\mathbb{E}}\left[\Gamma L_t \Lambda_t \mid \mathcal{Y}_t\right]}{\widetilde{\mathbb{E}}\left[L_t \Lambda_t \mid \mathcal{Y}_t\right]} \tag{3.4}$$

and we now derive an expression for $L_t\Lambda_t$ which will be adequate for our further study. Note that

$$L_t\Lambda_t = \exp\Big\{-\frac{1}{\varepsilon}\int_0^t (h(X_u)-h(M_u))^* d\overline{W}_u + \frac{1}{\varepsilon^2}\int_0^t h^*(X_u)(dY_u - h(M_u)du) + \frac{1}{2\varepsilon^2}\int_0^t |h(M_u)|^2 du\Big\}.$$

Defining the function

$$F(x,m) = (h(x)-h(m))^*\gamma^{-1}(m)(h(x)-h(m))$$

and the differential operators

$$\mathcal{A}_x = \sum_{ij} a_{ij}(x)\frac{\partial^2}{\partial x_i \partial x_j} \qquad \text{and} \qquad \mathcal{A}_m = \sum_{ij} a_{ij}(m)\frac{\partial^2}{\partial m_i \partial m_j},$$

we obtain the Itô formula

$$\begin{aligned}
F(X_t,M_t) =& F(X_0,M_0) + 2\int_0^t (h(X_u)-h(M_u))^*\gamma^{-1}(M_u)h'(X_u)dX_u \\
&- 2\int_0^t (h(X_u)-h(M_u))^*\gamma^{-1}h'(M_u)dM_u \\
&+ \sum_i \int_0^t (h(X_u)-h(M_u))^*\frac{\partial\gamma^{-1}}{\partial m_i}(M_u)(h(X_u)-h(M_u))dM_u^i \\
&+ \int_0^t (\mathcal{A}_xF + \mathcal{A}_mF)(X_u,M_u)du \\
=& F(X_0,M_0) - 2\int_0^t (h(X_u)-h(M_u))^*(\gamma^{-1}(X_u)-\gamma^{-1}(M_u))h'(X_u)dX_u \\
&+ 2\int_0^t (h(X_u)-h(M_u))^* d\overline{W}_u - \frac{2}{\varepsilon}\int_0^t (h(X_u)-h(M_u))^*(dY_u - h(M_u)du) \\
&+ 2\int_0^t (h(X_u)-h(M_u))^*(\gamma^{-1}h'b(X_u) - \gamma^{-1}h'b(M_u))du \\
&+ \sum_i \int_0^t (h(X_u)-h(M_u))^*\frac{\partial\gamma^{-1}}{\partial m_i}(M_u)(h(X_u)-h(M_u))dM_u^i \\
&+ \int_0^t (\mathcal{A}_xF + \mathcal{A}_mF)(X_u,M_u)du
\end{aligned}$$

and deduce that

$$\begin{aligned}
L_t\Lambda_t = \exp\Big\{&-\frac{1}{2\varepsilon}(F(X_t,M_t)-F(X_0,M_0)) + \frac{1}{\varepsilon}\int_0^t \psi_1(X_u,M_u)du \\
&+ \frac{1}{\varepsilon}\int_0^t \psi_2^*(X_u,M_u)dM_u - \frac{1}{\varepsilon}\int_0^t \psi_3^*(X_u,M_u)dX_u \\
&+ \frac{1}{\varepsilon^2}\int_0^t h^*(M_u)dY_u - \frac{1}{2\varepsilon^2}\int_0^t |h(M_u)|^2du\Big\}
\end{aligned} \tag{3.5}$$

with the definitions

$$\psi_1(x,m) = \big(h(x) - h(m)\big)^*\big(\gamma^{-1}h'b(x) - \gamma^{-1}h'b(m)\big) + \frac{1}{2}\big(\mathcal{A}_xF + \mathcal{A}_mF\big)(x,m),$$

$$\psi_2^i(x,m) = \frac{1}{2}\big(h(x) - h(m)\big)^*\frac{\partial\gamma^{-1}}{\partial m_i}(m)\big(h(x) - h(m)\big)$$

and

$$\psi_3(x,m) = h'^*(x)\big(\gamma^{-1}(x) - \gamma^{-1}(m)\big)\big(h(x) - h(m)\big).$$

3.3. Proof of (2.6) with an additional assumption

In this subsection, we suppose that the law of X_0 has a density p_0 which is a C^1 function such that $p_0(x) > 0$ for every x and all the moments of $p_0'/p_0(X_0)$ are finite. Let $\mathcal{G}_t$ be the filtration generated by (X,Y); every $\mathcal{G}_t$ measurable variable Γ can be considered as a functional of $(X_0, \widetilde{W}, Y)$ so that one can consider $(x, \tilde{w}, y) \mapsto \Gamma(x, \tilde{w}, y)$; this map is actually defined almost everywhere and we will say that Γ is differentiable with respect to X_0 if there exists a version of this map which is partially differentiable with respect to x. Its Jacobian matrix taken at $(X_0, \widetilde{W}, Y)$ will be denoted $\partial\Gamma/\partial X_0$; it is a $\mathcal{G}_t$ measurable variable.

From our assumptions, if X_t^{sx} is, for $t \geq s$, the solution of (3.3) with initial condition $X_s^{sx} = x$, then $(s,t,x) \mapsto X_t^{sx}$ defines a stochastic flow of diffeomorphisms, so, in particular, the variable X_t is for every t differentiable with respect to X_0. Define

$$Z_t = \frac{\partial X_t}{\partial X_0} \qquad \text{and} \qquad Z_{st} = Z_t Z_s^{-1}.$$

Then, for $s \leq t$, Z_{st} can be viewed as the x-derivative of X_t^{sx} and Z_{ts} is the derivative of the reversed flow. Moreover, the jth column of Z_t is solution of the equation ([2])

$$dZ_t^j = \frac{1}{\varepsilon}h'^{-1}\gamma h'(X_t)Z_t^j dt + b'(X_t)Z_t^j dt + \sum_i Z_t^{ij}\frac{\partial(h'^{-1}\gamma)}{\partial x_i}\gamma^{-1}h'(X_t)\big(dX_t - b(X_t)dt\big) \quad (3.6)$$

with $Z_0 = I$. In particular, Z_t is adapted to the filtration of X_t.

On the other hand, it can be proved from (3.5) that $L_t\Lambda_t$ is also differentiable (the dX_u integral has to be developed using (3.3)) and that

$$\begin{aligned}\frac{\partial}{\partial X_0}\log(L_t\Lambda_t) = &-\frac{1}{2\varepsilon}F'(X_t,M_t)Z_t + \frac{1}{2\varepsilon}F'(X_0,M_0) + \frac{1}{\varepsilon}\int_0^t \psi_1'(X_u,M_u)Z_u du \\ &+ \frac{1}{\varepsilon}\int_0^t dM_u^*\psi_2'(X_u,M_u)Z_u - \frac{1}{\varepsilon}\int_0^t dX_u^*\psi_3'(X_u,M_u)Z_u \\ &- \frac{1}{\varepsilon}\int_0^t \psi_3^*(X_u,M_u)dZ_u\end{aligned}$$

where the dZ_u integral can be expressed by means of (3.6). Multiplication by $Z_t^{-1} = Z_{t0}$ yields

$$\frac{1}{2}F'(X_t, M_t) = \frac{1}{2}F'(X_0, M_0)Z_{t0} + \int_0^t \psi_1'(X_u, M_u)Z_{tu}du + \int_0^t dM_u^*\psi_2'(X_u, M_u)Z_{tu}$$
$$- \Big(\int_0^t dX_u^*\psi_3'(X_u, M_u)Z_{0u}\Big)Z_{t0} - \Big(\int_0^t \psi_3^*(X_u, M_u)dZ_{0u}\Big)Z_{t0}$$
$$- \varepsilon\frac{\partial}{\partial X_0}\log(L_t\Lambda_t)Z_{t0}. \tag{3.7}$$

Note that Z_t^{-1} can enter the dM_u integral because Z and M are $\widetilde{\mathbb{P}}$ independent, but it cannot enter the dZ_u integral because it depends on all the trajectory of X.

We now deduce from (3.7) that the $(\mathcal{Y}_t, \mathbb{P})$ conditional expectations of the two sides are equal. For the left side, we obtain

$$\mathbb{E}\Big[(h(X_t) - h(M_t))^*\gamma^{-1}(M_t)h'(X_t) \;\Big|\; \mathcal{Y}_t\Big].$$

By developing h and h' around M_t and using lemma 1, this expression is shown to be equal to $(\widehat{X}_t - M_t)^*(h'^*\gamma^{-1}h')(M_t)$ with an error of order ε, so in order to prove (2.6), it is sufficient to prove that the $\mathbb{P}$-moments of the conditional expectation of the right side of (3.7) are of order ε. To this end, we first write the stochastic differential equation satisfied by Z_{tu} for u fixed and $t \geq u$:

$$dZ_{tu} = -\frac{1}{\varepsilon}Z_{tu}(h'^{-1}\gamma h')(X_t)dt + Z_{tu}(\beta - b')(X_t)dt - \sum_i Z_{tu}(h'^{-1}\gamma_i)'(X_t)d\overline{W}_t^i$$

where $\beta = \sum_i (h'^{-1}\gamma_i)'^2$ (this term appears because we use the Itô formalism and not the Stratonovich one). By looking at the equation of $Z_{tu}h'^{-1}(X_t)$, since γ is uniformly elliptic, one can prove that for every $k \geq 1$, there exist μ and $c > 0$ such that

$$\|Z_{tu}\|_k \leq \mu e^{-c(t-u)/\varepsilon} \tag{3.8}$$

(the notation $\|.\|_k$ will always denote the norm in $L^k(\Omega, \mathcal{F}, \mathbb{P})$). From this inequality, the term $F'(X_0, M_0)Z_{t0}$ taken from the right side of (3.7) is very small; since the moments of $\psi_1'(X_u, M_u)$ are bounded, we also check that the Lebesgue integral is of order ε; the same property holds for the dM_u integral: we decompose this integral into a Lebesgue one and a dB_u one, the latter one is estimated by means of its quadratic variation and we use the property $\psi_2'(X_u, M_u) = O(\sqrt{\varepsilon})$ which is valid except for u small. The dX_u and dZ_u integrals can also be decomposed into Lebesgue and dW_u parts; after some calculation, the Lebesgue integrals can be estimated like previously, so the only difficult part concerns the dW_u integrals; it turns out that this problem can be reduced to the following question: if V_t is a $\mathbb{R}^p \otimes \mathbb{R}^n$ valued $\mathcal{F}_t$ adapted process with uniformly bounded moments such that for every $k \geq 1$ and $t_0 > 0$,

$$\sup_{t \geq t_0}\|V_t\|_k = O(\sqrt{\varepsilon}),$$

does the estimate

$$\sup_{t\geq t_0}\left\|\mathbb{E}\left[\left(\int_0^t dW_u^* V_u Z_u\right) Z_{t0}\;\middle|\;\mathcal{Y}_t\right]\right\|_k = O(\varepsilon) \tag{3.9}$$

hold? This technical result will be proved in §3.6.

To complete the study of (3.7), we have to prove that the moments of

$$\Phi_t = \mathbb{E}\left[\frac{\partial}{\partial X_0}\log(L_t\Lambda_t)Z_{t0}\;\middle|\;\mathcal{Y}_t\right] = \frac{\widetilde{\mathbb{E}}\left[\frac{\partial}{\partial X_0}(L_t\Lambda_t)Z_{t0}\;\middle|\;\mathcal{Y}_t\right]}{\widetilde{\mathbb{E}}\left[L_t\Lambda_t\;\middle|\;\mathcal{Y}_t\right]}$$

are bounded. From the equation satisfied by Z_{t0}, one can see that it is differentiable with respect to X_0, so that one can write the ith component of Φ_t as

$$\Phi_t^i = -\mathbb{E}\left[\mathrm{Trace}\left(\frac{\partial Z_{t0}^i}{\partial X_0}\right)\;\middle|\;\mathcal{Y}_t\right] + \mathrm{Trace}\frac{\widetilde{\mathbb{E}}\left[\frac{\partial}{\partial X_0}(L_t\Lambda_t Z_{t0}^i)\;\middle|\;\mathcal{Y}_t\right]}{\widetilde{\mathbb{E}}\left[L_t\Lambda_t\;\middle|\;\mathcal{Y}_t\right]}. \tag{3.10}$$

We now study the first term of the right side; if ς_t^i is the trace of $\partial Z_{t0}^i \;/\; \partial X_0$, it can be seen by differentiating the equation of Z_{t0} that the line vector ς_t is solution of

$$\begin{aligned} d\varsigma_t = &-\frac{1}{\varepsilon}\varsigma_t(h'^{-1}\gamma h')(X_t)dt + \varsigma_t(\beta - b')(X_t)dt - \sum_j \varsigma_t(h'^{-1}\gamma_j)'d\overline{W}_t^j \\ &+ \frac{1}{\varepsilon}\phi_0(X_t)dt + \bar{\phi}_0(X_t)dt + \sum_j \phi_j(X_t)d\overline{W}_t^j \end{aligned}$$

with the notations

$$\phi_0^k = -\mathrm{div}\left(\frac{\partial(h'^{-1}\gamma h)}{\partial x_k}\right), \quad \bar{\phi}_0^k = \mathrm{div}\left(\beta_k - \frac{\partial b}{\partial x_k}\right), \quad \phi_j^k = -\mathrm{div}\left(\frac{\partial(h'^{-1}\gamma_j)}{\partial x_k}\right).$$

It follows from this equation by an application of Itô's formula (using again the ellipticity of γ) that the $\mathbb{P}$ moments of ς_t are bounded. Let us turn to the second part of (3.10); since X_0 and Y are $\widetilde{\mathbb{P}}$ independent, we can write the numerator as

$$\begin{aligned} \int \widetilde{\mathbb{E}}\left[\frac{\partial}{\partial X_0}(L_t\Lambda_t Z_{t0}^i)\;\middle|\;\mathcal{Y}_t, X_0 = x\right]p_0(x)dx &= \int \frac{\partial}{\partial x}\widetilde{\mathbb{E}}\left[L_t\Lambda_t Z_{t0}^i\;\middle|\;\mathcal{Y}, X_0 = x\right]p_0(x)dx \\ &\quad \text{(exchanging an integral and a derivative with respect to a parameter)} \\ &= -\int \widetilde{\mathbb{E}}\left[L_t\Lambda_t Z_{t0}^i\;\middle|\;\mathcal{Y}_t, X_0 = x\right]p_0'(x)dx \\ &\quad \text{(using an integration by parts formula)} \\ &= -\widetilde{\mathbb{E}}\left[L_t\Lambda_t Z_{t0}^i\frac{p_0'}{p_0}(X_0)\;\middle|\;\mathcal{Y}_t\right]. \end{aligned}$$

All these steps can be made rigorous by a study of the moments of the involved variables. Then we normalize the last line to get a $\mathbb{P}$ conditional expectation, and by using the assumption about p_0'/p_0 and the estimate (3.8) on Z_{t0}, we can complete our estimation.

3.4. Proof of (2.6) in the general case

We now want to extend our result to the case of an initial state X_0 which has not necessarily a density. In last subsection, we have used an integration by parts formula for derivatives with respect to the initial value; here, we use derivatives with respect to perturbations on $\widetilde{W}$, namely the Malliavin calculus, and we also apply an integration by parts formula. Actually, we need a partial Malliavin's calculus, since we want to differentiate with respect to $\widetilde{W}$ functionals of $(X_0, \widetilde{W}, Y)$. If Γ is a real function defined on the canonical space $\mathbf{C}(0,t;\mathbb{R}^n)$ endowed with the Wiener measure $\mathcal{W}$, we can define the notion of weak H-differentiability as follows: we say that Γ is a smooth cylinder function if it is a smooth function of a finite number of $(l_i, .)$, where $l_i \in \mathbf{C}^*$; in this case, we let $D_u\Gamma$ be the function such that, for any deterministic path ρ_u of $L^2(0,t,\mathbb{R}^n)$, $\int_0^t D_u\Gamma\rho_u du$ is the derivative of Γ with respect to perturbations in the direction $\int_0^{\cdot} \rho_u du$; then for each $q > 1$, we consider the closure of the set of smooth cylinder functions with respect to the norm

$$\Big(\int_{\mathbf{C}} |\Gamma|^q d\mathcal{W}\Big)^{1/q} + \Big(\int_{\mathbf{C}}\int_0^t |D_u\Gamma|^q du\, d\mathcal{W}\Big)^{1/q}.$$

A variable in this closure will be said to be differentiable (see [5], except for a change of notation: $D_u\Gamma$ is $[D\Gamma]'(u)$ in [5]). Then we can transport this notion on our probability space Ω endowed with the probability $\widetilde{\mathbb{P}}$ by considering derivatives of functionals of $\widetilde{W}$. Finally, since (X_0, Y) is independent from $\widetilde{W}$, we can consider partial derivatives of $\mathcal{G}_t$ measurable variables (one has to prove that these derivatives are $\mathcal{G}_t$ measurable). If Γ is a functional of $(X_u, u \geq s)$, we can also define $\partial\Gamma \,/\, \partial X_s$ as the partial derivative with respect to X_s with $\widetilde{W}$ fixed (when it exists).

With these definitions, one can prove from (3.3) and classical results about differentiability of solutions of stochastic differential equations that X_u is differentiable, as well as $L_t\Lambda_t$. One has for $s \leq u$

$$D_sX_u = Z_{su}h'^{-1}\gamma(X_s)$$

and from (3.5),

$$\begin{aligned} D_s \log(L_t\Lambda_t) = &-\frac{1}{2\varepsilon}F'(X_t, M_t)D_sX_t + \frac{1}{\varepsilon}\int_s^t \psi_1'(X_u, M_u)D_sX_u du \\ &+ \frac{1}{\varepsilon}\int_s^t dM_u^*\psi_2'(X_u, M_u)D_sX_u - \frac{1}{\varepsilon}\int_s^t dX_u^*\psi_3'(X_u, M_u)D_sX_u \\ &- \frac{1}{\varepsilon}\int_s^t \psi_3^*(X_u, M_u)d(D_sX_u). \end{aligned}$$

If we multiply on the right by $\gamma^{-1}h'(X_s)Z_{ts}$, we obtain an expression for $\frac{1}{2}F'(X_t, M_t)$ which looks like (3.7): the term

$$\frac{\partial}{\partial X_0}\log(L_t\Lambda_t)Z_{t0} \quad \text{has been replaced by} \quad D_s\log(L_t\Lambda_t)\gamma^{-1}h'(X_s)Z_{ts},$$

the term containing $F'(X_0, M_0)$ has disappeared and in the other terms, the integrals on $[0,t]$ are replaced by integrals on $[s,t]$. If we integrate this expression with respect to s on

$[0,t]$, we obtain

$$\frac{1}{2}F'(X_t,M_t) = -\frac{\varepsilon}{t}\int_0^t D_s \log(L_t\Lambda_t)\gamma^{-1}h'(X_s)Z_{ts}ds + \frac{1}{t}\int_0^t (\dots)ds$$

where we have put in the dots all the integrals on $[s,t]$; now the $(\mathcal{Y}_t, \mathbb{P})$ conditional expectations of these integrals are estimated as in §3.3, so that we only have to prove that the $\mathbb{P}$ moments of

$$\begin{aligned}\overline{\Phi}_t =& \frac{1}{t}\mathbb{E}\Big[\int_0^t D_s \log(L_t\Lambda_t)\gamma^{-1}h'(X_s)Z_{ts}ds \;\Big|\; \mathcal{Y}_t\Big] \\ =& \frac{1}{t}\tilde{\mathbb{E}}\Big[\int_0^t D_s(L_t\Lambda_t)\gamma^{-1}h'(X_s)Z_{ts}ds \;\Big|\; \mathcal{Y}_t\Big] \Big/ \tilde{\mathbb{E}}\big[L_t\Lambda_t \mid \mathcal{Y}_t\big]\end{aligned}$$

are bounded.

Thus let us study the ith component $\overline{\Phi}_t^i$ of $\overline{\Phi}_t$; remark that

$$\begin{aligned}D_s(L_t\Lambda_t)\gamma^{-1}h'(X_s)Z_{ts}^i &= D_s(L_t\Lambda_t)\gamma^{-1}h'(X_s)Z_sZ_{t0}^i \\ &= \operatorname{Trace}\Big(D_s(L_t\Lambda_t)Z_{t0}^i\gamma^{-1}h'(X_s)Z_{0s}\Big) \\ &= \operatorname{Trace}\Big(D_s\big(L_t\Lambda_tZ_{t0}^i\big)\gamma^{-1}h'(X_s)Z_{0s}\Big) \\ &\quad - L_t\Lambda_t\operatorname{Trace}\Big(\big(D_sZ_{t0}^i\big)\gamma^{-1}h'(X_s)Z_{0s}\Big)\end{aligned}$$

But the integration by parts formula of Malliavin's calculus yields (actually, the formula of [5] is not directly applicable since $L_t\Lambda_tZ_{t0}$ is not necessarily square integrable but one can use a truncation argument)

$$\begin{aligned}&\tilde{\mathbb{E}}\Big[\int_0^t D_s\big(L_t\Lambda_tZ_{t0}^i\big)\gamma^{-1}h'(X_s)Z_{0s}ds \;\Big|\; \mathcal{Y}_t \vee X_0\Big] \\ &\qquad = \tilde{\mathbb{E}}\Big[L_t\Lambda_tZ_{t0}^i\int_0^t d\widetilde{W}_s^*\gamma^{-1}h'(X_s)Z_{0s} \;\Big|\; \mathcal{Y}_t \vee X_0\Big]\end{aligned}$$

and

$$D_sZ_{t0}^i = Z_{s0}\frac{\partial Z_{ts}^i}{\partial X_s}h'^{-1}\gamma(X_s)$$

so

$$\overline{\Phi}_t^i = \frac{1}{t}\mathbb{E}\Big[\Big(\int_0^t d\widetilde{W}_s^*\gamma^{-1}h'(X_s)Z_{0s}\Big)Z_{t0}^i \;\Big|\; \mathcal{Y}_t\Big] - \frac{1}{t}\int_0^t \mathbb{E}\Big[\operatorname{Trace}\frac{\partial Z_{ts}^i}{\partial X_s} \;\Big|\; \mathcal{Y}_t\Big]ds.$$

The first part is decomposed into a Lebesgue integral which is order 1 and a dW_s integral which is shown to be of order $\sqrt{\varepsilon}$ from (3.9); finally, the study of the partial derivative of the second part is similar to the case $s = 0$ which was dealt with in §3.3, so we can conclude.

3.5. The smoothing problem

We now explain briefly how the previous method can be extended to the study of the smoothing problem. Our goal is to prove that the $(\mathcal{Y}_t, \mathbb{P})$ conditional expectation of $F'(X_u, M_u^t)$ is of order ε: from lemma 2, this will imply (2.8). For each $u \leq t$ define the probability $\widetilde{\mathbb{P}}_t^u$ by

$$\begin{aligned}\frac{d\widetilde{\mathbb{P}}_t^u}{d\mathring{\mathbb{P}}} = (\Lambda_t^u)^{-1} \equiv \exp\Big\{\frac{1}{\varepsilon}\int_0^u (h(X_s) - h(M_s))^* d\overline{W}_s - \frac{1}{2\varepsilon^2}\int_0^u |h(X_s) - h(M_s)|^2 ds \\ - \frac{1}{\varepsilon}\int_u^t (h(X_s) - 2h(M_s^t) + h(M_s))^* d\overline{W}_s \\ - \frac{1}{2\varepsilon^2}\int_u^t |h(X_s) - 2h(M_s^t) + h(M_s)|^2 ds\Big\}.\end{aligned}$$

Note that we have put a negative sign in front of the stochastic integral on $[u,t]$ and not on $[0,u]$; this will cause time u to be the most important one: roughly speaking, the derivative Z_{us}^u of X_s with respect to value at time u with be small in both cases $s \ll u$ and $s \gg u$. Now, the previous arguments work with adequate modifications; one writes Itô's formula to obtain an expression for $F(X_t, M_t) - F(X_u, M_u^t)$; this implies that

$$\begin{aligned}L_t\Lambda_t^u =& L_t\Lambda_t \exp\Big\{\frac{2}{\varepsilon}\int_u^t (h(X_s) - h(M_s^t))^* d\overline{W}_s \\ &+ \frac{2}{\varepsilon^2}\int_u^t (h(X_s) - h(M_s^t))^*(h(M_s) - h(M_s^t))ds\Big\} \\ =& L_t\Lambda_t \exp\Big\{\frac{1}{\varepsilon}(F(X_t, M_t) - F(X_u, M_u^t)) - \frac{2}{\varepsilon}\int_u^t \bar{\psi}_1(X_s, M_s^t)ds \\ &- \frac{2}{\varepsilon}\int_u^t \psi_2^*(X_s, M_s^t)\dot{M}_s^t ds + \frac{2}{\varepsilon}\int_u^t \psi_3^*(X_s, M_s^t)dX_s\Big\}\end{aligned}$$

where $\bar{\psi}_1 = \psi_1 - \frac{1}{2}\mathcal{A}_m F$ and $L_t\Lambda_t$ is given by (3.5).

Then we differentiate; in the Malliavin framework, we compute

$$\frac{1}{u}\int_0^u D_s \log(L_t\Lambda_t^u)\gamma^{-1}h'(X_s)Z_{us}^u ds.$$

By proceeding like previously, we obtain, after estimating some integrals and using an integration by parts formula, that

$$\mathbb{E}\big[F'(X_u, M_u^t) - \frac{1}{2}F'(X_t, M_t)Z_{ut}^u \mid \mathcal{Y}_t\big] = O(\varepsilon)$$

uniformly for $t \geq u \geq t_0$. To conclude, we still have to prove that the conditional mean of $F'(X_t, M_t)Z_{ut}^u$ is of order ε; this is easy when $t - u$ is not too small because in this case, Z_{ut}^u is very small; however, one has to be finer if one wants to obtain a uniform bound when $u \to t$. So let η_s be the matrix-valued observable process solution of the forward equation for $s \geq u$

$$\dot{\eta}_s = -\frac{1}{\varepsilon}\gamma(M_s)\eta_s; \qquad \eta_u = h'(M_u).$$

By writing the equation satisfied by $h'(X_s)Z^u_{us} - \eta_s$, one can prove that it is of order $\sqrt{\varepsilon}$ and

$$\begin{aligned}\mathbb{E}\big[F'(X_t, M_t)Z^u_{ut} \mid \mathcal{Y}_t\big] =& \mathbb{E}\big[F'(X_t, M_t)\big(Z^u_{ut} - h'^{-1}(X_t)\eta_t\big) \mid \mathcal{Y}_t\big] \\ &+ \mathbb{E}\big[F'(X_t, M_t)h'^{-1}(X_t) \mid \mathcal{Y}_t\big]\eta_t.\end{aligned}$$

The first term is the conditional expectation of the product of two terms of order $\sqrt{\varepsilon}$ and the second term is of order ε from results on the filtering problem, so the proof is complete.

3.6. Proof of (3.9)

We now verify the technical result which was not proved in previous subsections, namely the proof of (3.9) with a process V_s of order $\sqrt{\varepsilon}$. The difficulty is that Z_{t0} cannot enter the stochastic integral; to solve this problem, one could think of using some results about time reversal of diffusion processes. We here introduce another method which is again derived from Malliavin's calculus, but under the probability $\mathring{\mathbb{P}}$; X_0, W and Y are independent and we note $\mathring{D}$ the derivative with respect to perturbations of W. The integrations by parts formula can then be written for $(\mathring{\mathbb{P}}, \mathcal{Y}_t)$ conditional expectations to yield

$$\mathring{\mathbb{E}}\Big[\Big(\int_0^t dW_s^* V_s Z_s\Big) Z^i_{t0} L_t \Bigm| \mathcal{Y}_t\Big] = \mathring{\mathbb{E}}\Big[\int_0^t \mathrm{Trace}\big(V_s Z_s \mathring{D}_s(L_t Z^i_{t0})\big)\, ds \Bigm| \mathcal{Y}_t\Big]$$

so that

$$\mathbb{E}\Big[\Big(\int_0^t dW_s^* V_s Z_s\Big) Z^i_{t0} \Bigm| \mathcal{Y}_t\Big] = \mathbb{E}\Big[\int_0^t \mathrm{Trace}\Big(V_s\big(Z_s \mathring{D}_s Z^i_{t0} + Z^i_{ts}\mathring{D}_s \log L_t\big)\Big)\, ds \Bigm| \mathcal{Y}_t\Big]. \quad (3.11)$$

The variable Z^i_{ts} can be viewed as a functional of $(X_s; W_u - W_s, u \geq s)$, and the term $Z_s \mathring{D}_s Z^i_{t0}$ is its partial derivative with respect to X_s multiplied by $\sigma(X_s)$; thus, by differentiating the equation of Z_{ts}, $t \geq s$, one proves that it is of order $\frac{t-s}{\varepsilon}\exp(-c(t-s)/\varepsilon) + \sqrt{\varepsilon}$. The other derivative is

$$\mathring{D}_s \log L_t = \frac{1}{\varepsilon^2}\int_s^t \big(dY_u^* - h^*(X_u)du\big)h'(X_u)\mathring{D}_s X_u = \frac{1}{\varepsilon}\int_s^t dB_u^* h'(X_u)\mathring{D}_s X_u.$$

After multiplication by Z_{ts}, the moments of this expression are of order $\frac{\sqrt{t-s}}{\varepsilon}\exp\big(-c(t-s)/\varepsilon\big)$ so multiplication by V_s and integration on $[0,t]$ yields a term of order ε. This completes the proof.

4. CONCLUSION

In this work, we have obtained asymptotic solutions for filtering and smoothing problems in the case of a high signal-to-noise ratio. The general case was dealt with by means of Malliavin's calculus, but in the case of absolutely continuous initial conditions, this calculus is only involved in the proof of the technical result (3.9); note however that in the one-dimensional case, if one replaces the function F by

$$\overline{F}(x,m) = 2\int_0^x \frac{h(v)-h(m)}{\sigma(v)}dv,$$

one can get rid of the dW_u integrals in (3.7), so one does not need (3.9); actually, it has been pointed out to the author by A. Bensoussan that this situation can also be studied with another elementary method ([1]).

Then one may look for generalizations of these results. For instance, the case where both signal and observation noises are of order ε behaves quite differently and is studied in [3]. One may also wonder if it is possible to obtain more accurate solutions. We have proved in [7] that in the one-dimensional case $n=1$, if (M_t, R_t) is solution of

$$\begin{cases} dM_t = b(M_t)dt + \dfrac{R_t}{\varepsilon}\sigma^2 h'(M_t)\big(dY_t - h(M_t)dt - \dfrac{\varepsilon}{2}\dfrac{\sigma h''}{h'}(M_t)dt\big) \\ \dfrac{dR_t}{dt} = -\sigma^2 h'^2(M_t)\dfrac{R_t^2}{\varepsilon} + 2\big(\dfrac{b}{\sigma}\big)'\sigma(M_t)R_t + \dfrac{1}{\varepsilon} \end{cases}$$

with $M_0 = m_0$ and $R_0 = r_0 > 0$, then $\widehat{X}_t - M_t$ is of order $\varepsilon^{3/2}$. Three classical filters (extended Kalman filter, linearized filter, modified second-order filter) were also considered. The extension of these results to the multi-dimensional case is still in progress.

REFERENCES

[1] A. Bensoussan, On some singular perturbation problems arising in nonlinear filtering, to appear in *Proc. Stoch. Control Conf.* (Minneapolis 1986).

[2] J.M. Bismut, *Mécanique Aléatoire*, Lect. N. in Math. **866**, Springer, 1981.

[3] O. Hijab, Asymptotic Bayesian estimation of a first order equation with small diffusion, *Annals of Proba.* **12** (1984), 890–902.

[4] R. Katzur, B.Z. Bobrovsky and Z. Schuss, Asymptotic analysis of the optimal filtering problem for one-dimensional diffusions measured in a low noise channel, *SIAM J. Applied Math.* **44** (1984), Part I: 591–604, Part II: 1176–1191.

[5] D. Ocone, Malliavin's calculus and stochastic integral representations of functionals of diffusion processes, *Stochastics* **12** (1984), 161–185.

[6] E. Pardoux, Equations du filtrage non linéaire de la prédiction et du lissage, *Stochastics* **6** (1982), 193–231.

[7] J. Picard, Nonlinear filtering of one-dimensional diffusions in the case of a high signal-to-noise ratio, to appear in *SIAM J. Applied Math.*

[8] J. Picard, Filtrage de diffusions vectorielles faiblement bruitées, *Proc. 7th Int. Conf. on Analysis and Optimization of Systems* (Antibes 1986), Lect. N. in Control and Inf. Sci. **83**, Springer, 1986.

FRACTAL AND MULTIFRACTAL STRUCTURES IN KINETIC CRITICAL PHENOMENA

L. Pietronero, C. Evertsz and A.P. Siebesma
Solid State Physics Laboratory, University of Groningen
Melkweg 1
9718 EP Groningen
The Netherlands

ABSTRACT. Kinetic Critical Phenomena consist of irreversible growth models that generate fractal structures. In the last few years a large amount of activity has been devoted to these problems that represent one of the most challenging fields in today's theoretical physics. The idea is to understand the physical origin for the development of the many fractal structures one can observe in nature. In this respect one of the most interesting models is the one formulated to describe the patterns of dielectric breakdown. It is based on a combination of Laplace equation and a probability field that give rise to random fractals. An essential element for the understanding of these results is the proper characterization of the self-similar properties of the growth probability field. To this purpose the more general concept of multifractal is necessary. This is discussed in some detail and applied to the above model showing that the growth probability should be described by a continuous distribution of singularities. The origin of multifractality is then elucidated by showing that, despite its apparent complexity, it arises in a natural way even in simple random multiplicative processes. This allows to consider apparently unrelated problems like the properties of self-similarity of Anderson localized wavefunctions for which a complete characterization is given. The main open problems as well as the perspectives for future development are briefly discussed for each topic.

1. INTRODUCTION.

In the last few years the word "fractal" has been one of the most common expressions in physics literature [1-3]. Fractal identifies systems in which increasing detail is revealed by increasing magnification, and the newly revealed structure looks the same as what one can observe at lower magnification [4]. This implies invariance under changes of the length-scale or self-similarity.

It is clear that systems which possess this property cannot be described by analytical methods. The requirement of analiticity for a shape implies the possibility to define derivatives at each point.

S. Albeverio et al. (eds.), Stochastic Processes in Physics and Engineering, 253–278.

This means that in the limit of small lenght scales the shape can be approximated by a line. This is not possible for self-similar shapes because at all scales they manifest the same complexity and a single tangent line cannot be defined. This may be the reason why, despite the many examples of self-similar structures that can be found in nature, this field has substantially developed only recently.

A notable exception to the above situation is represented by the large amount of activity that has led to a rather complete understanding of the critical properties of second order phase by means of renormalization group methods [5]. In these problems self-similarity develops exactly at the transition temperature T_c because of the loss of the characteristic length-scale for the fluctuations. For a while scale-invariance in statistical physics was associate only to these peculiar properties of second order transitions around T_c. Recently the study of the origin of fractal structures has led to a variety of growth models that generate patterns with self-similar properties. These models can be called *Kinetic Critical Phenomena* [1] because, in analogy with second order transitions at T_c, they give rise to critical exponents (non analytic power law behavior) that can be interpreted as fractal dimensions. However, in contrast with phase transitions, these growth models are kinetic or fully irreversible and in general they cannot be reduced to the study of the quilibrium properties of a Hamiltonian. The formulation of a theory for these processes is still lacking, however some nontrivial progress has been made recently by realizing that the growth probability field cannot be described simply by a fractal dimension (a single exponent) but the more general concept of multifractal is needed. This concept describes systems in which the local fractal dimension is different from point to point and a continuum distribution of exponents is necessary for their characterization.

This paper consists essentially in a report on these recent progresses. In particular:

In Sect. 2. we give a brief summary of the impact of the concept of fractal dimension in physics.

In Sect. 3. we introduce the most studied growth model that gives rise to stochastic fractal structures. This model is based on a suitable combination of Laplace equation and a stochastic field and it is believed to contain the essential ingredients for the development of fractal patterns in a variety of physical phenomena like dielectric breakdown, dendritic growth, propagation of fractals and viscous fingers [1].

In Sect. 4. the concept of multifractal is introduced and discussed in some examples.

In Sect. 5. the multifractal method of analysis is applied to the growth probability of Laplacian fractals. The relevance of these results with respect to the formulation of a theory for these processes is briefly discussed.

In Sect. 6. we finally show how multifractal properties arise naturally in random multiplicative processes and how these properties characterise the fluctuations of wavefunctions localized by disorder (Anderson localization).

2. FRACTALS IN PHYSICS

It is easy to realize that many features in nature appear rather fragmented and manifest, at least embrionically, features of self-similarity and scale invariance. The most notable examples are trees, coastlines, rivers, lightnings, clouds, etc. [4]. Despite these evident examples a scientific approach to the study of these patterns was undertaken only in the last few years following the compelling arguments presented by Mandelbrot [4]. He discusses many mathematical algoritms that give rise to deterministic and random fractals and presents convincing arguments for a close relationship between these structures and many natural phenomena. This work has been indeed very inspiring, however it does not contain a theory of fractal structures in the standard sense. The formulation of such a theory is the objective of the present activity and represents one of the most challenging problems in today's theoretical physics [1].

Let us make clear what the problem is with an example. Consider the well known Sierpinski gasket [2,4]. This can be constructed by a simple iterative method and gives rise to a structure in which the number of occupied bonds (volume or mass M) varies with the length L one consider according to a power law

$$M \sim L^D \qquad (2,1)$$

where $D = \ln 3/\ln 2 = 1.58496\ldots$. In the sense of the volume-lenght relation we can interpret D as a non-integer fractal dimension that describes our systems. It should be noted that this is a rather restrictive definition of dimension and in order to describe the properties of this system one cannot simply take the known physical laws and replace the standard, euclidean dimension d with the fractal dimension D.

Now let us consider for example the phenomenon of dielectric breakdown. If we analyze the experimental patterns obtained one can also conclude that they manifest features of self-similarity and that a fractal dimension can be associated to the pattern. However, the relation with the Sierpinski gasket stops here because we cannot learn from this analogy which are the features of the physical laws that govern dielectric breakdown that give rise to the fractal features. The Sierpinski gasket analogy does not help because its construction does not have any relation with physical laws. This basic missing link between the laws of physics and the generation of fractal structures is what we call the theory of fractals and in these notes we are going to discuss the state of the art for this problem.

An additional line of research is also present in the field of fractals. Since fractal structures exist in nature, one can neglect the questions about their origin and study their physical properties. This consists in assuming a simple fractal structure and study problems like diffusion, elasticity or phase transitions on this structure [2]. We are not goint to discuss this point of view here.

The simplest non-trivial models that give rise to non-trivial fractal structures are random walks with memory [1,6]. These models represent an interesting bridge between the more general class of kinetic growth models and polymer statistics. They represent a playground on which theoretical approaches can be tested and have led to some interesting developments that we have recently reviewed [6]. We do not discuss them here and we move directly to the more complex case of branched Laplacian fractals that represent at the moment the most challenging model systems.

3. LAPLACIAN FRACTALS

In this section we describe a well defined physical model that gives rise to random fractal structures. It is based on a suitable combination of Laplace equation and a probability field. This model was originally introduced to describe the patterns generated by dielectric breakdown [7,8] but it was soon realized that it bears close relation with the Diffusion Limited Aggregation (DLA) model [9,10] that describes dendritic growth and others growth models [11,12].

Dielectric breakdown gives rise to complex ramified structures like the one shown in Fig. 1 corresponding to a planar discharge in a gas. For a more detailed discussion see Refs. [7,8,12]. What happens is that a dielectric medium (gas, liquid, solid) breakes down under the effect of an electric field and gives rise to channels of ionized plasma. The propagation of these channels proceeds by steps of more or less unitary length. The process of addition of a step can be described in a simplified way as follows. An initiator (say an electron) should be available in the vicinity of a tip where the electric field is high. This initiator is then attracted towards the tip and accelerated until it can generate by ionization another free electron. Under suitable conditions this process generates an avalanche that gives rise to the addition of a new plasma element to the previous structure. Note that there are *two* essential elements: the electric field and the presence of the initiator.

Let us consider now Fig. 2 in which we give a schematic lattice description of the process. The black pattern represents the plasma lines at a given time while the dashed bonds connecting a black point to a while one represent the possible candidates for the propagation of the pattern.

Considering that the pattern dischange is made of plasma channels with high conductivity we can consider the whole black pattern equipotential with the same potential as the central electrode, say, $\phi = 0$. If we want to know the local field around this pattern we have to solve Laplace equation

$$\nabla^2\phi = 0 \qquad (3,1)$$

on the whole lattice. We consider here a two dimensional (planar) geometry, with the boundary conditions $\phi = 0$ for all the points of the black pattern, and $\phi = 1$ on a far away circle that represents the other electrode. This problem can be solved by discretizing Laplace equation on the lattice and using iterative numerical methods. Convergency to-

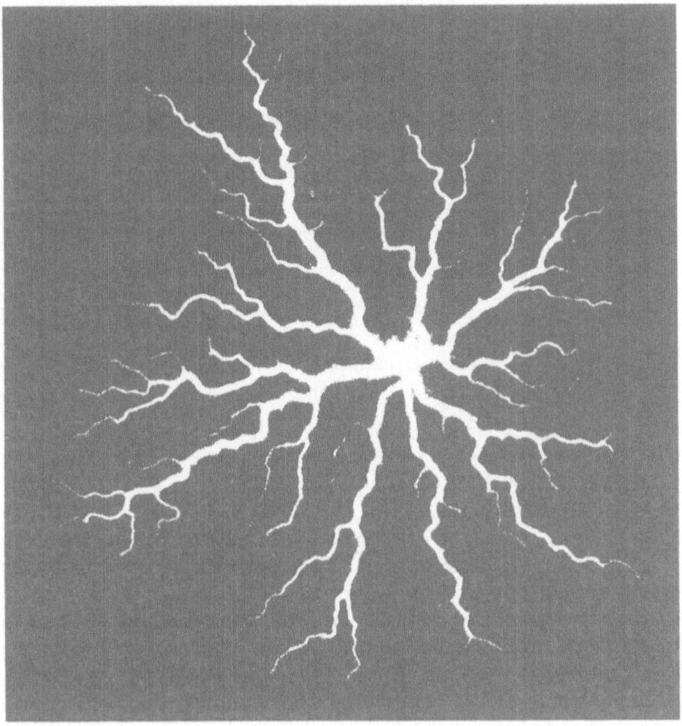

Figure 1. Time-integrated photograph of a surface leader discharge (Lichtenberg figure). For a discussion see Refs. [7,13]. This experiment corresponds to an equipotential channel system growing in a plane with radial electrode.

wards the solution can be accelerated substantially by the use of techniques of over-relaxation. Once we have the potential at all points we can compute the modulus of the local electric field around the pattern

$$E_j = |\nabla\phi|_j = |\phi_j(\text{black}) - \phi_j(\text{white})| \; ; \qquad (3,2)$$

where j labels all the dashed bonds that connect a black to a white point.

Given the local field we have to assign a stepping rule for the growth of the pattern. The simplest would be just to say that the growth occurs where the field is the largest. This does not take into account the randomness for the distribution of the initiator and would predict that lightnings propagate along straight lines that is not a very appealing result. In view of the randomness of the initiator the relation between growth and local field should be *probabilistic* rather than

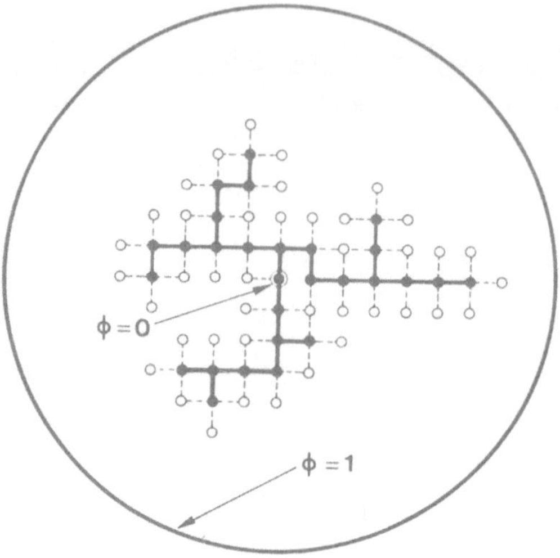

Figure 2. Illustration of the stochastic model we introduce to simulate dielectric breakdown on a lattice. The central point represents one of the electrodes while the other electrode is modeled as a circle at large enough distance. The discharge pattern is indicated by the black dots connected with thick lines and it is considered equipotential ($\phi = 0$). The dashed bonds indicate all the possible growth processes. The probability for each of these processes is proportional to the local electric field (see text).

deterministic. The simplest assumption is then to define a normalized growth probability per dashed bond p_j that depends on E_j as follows

$$p_j = \frac{(E_j)^\eta}{\sum_j (E_j)^\eta} \quad . \tag{3,3}$$

This growth probability gives rise to a class of models characterized by the exponent η. The case of simple proportionality ($p_j \propto E_j$) appears to be a reasonable choice for discharges in gases on the basis of microscopic considerations [7]. For other systems a different value of η may be more appropriate. Of course the microscopic link between E_j and p_j may be also of a different structure then Eq. (3,3). We intend to discuss these complications elsewhere [13].

Now the stochastic model is completely defined. It consist of the following steps:

(a) For a given pattern (black dots) solve Laplace equations with appropriate boundary conditions. This gives the local field corresponding to the bonds that are candidates for growth.

(b) Construct the probability field as given by Eq. (3,3) and select by a Monte Carlo procedure based on these probabilities one

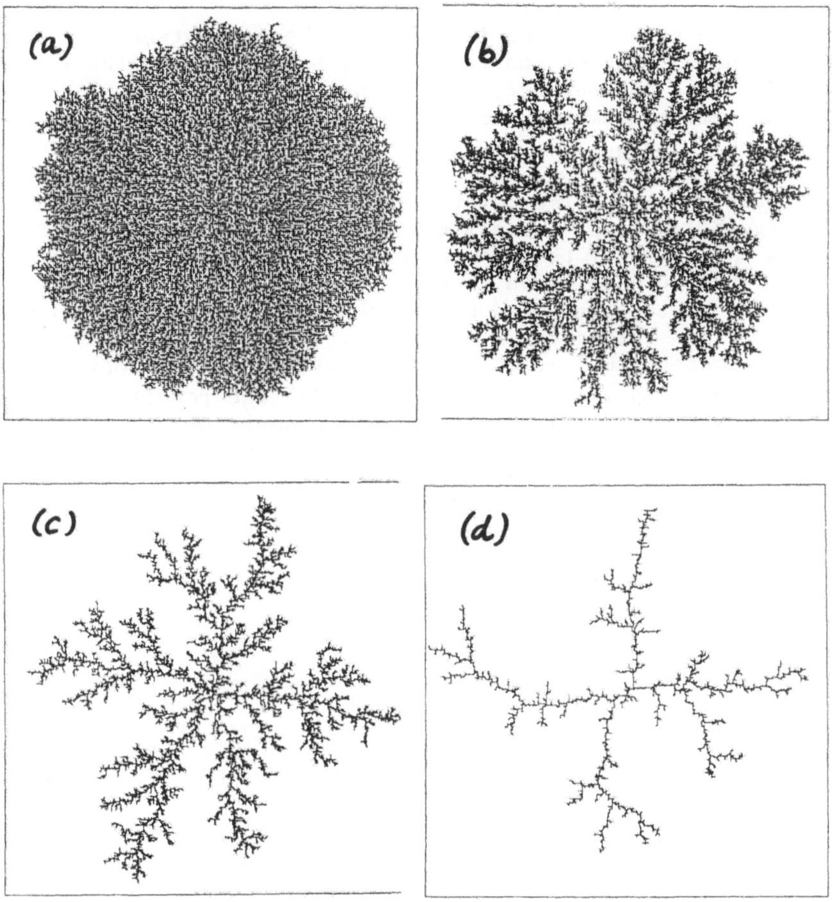

Figure 3. Examples of computer generated discharge patterns whose fractal dimensions are given in Table I [d = 2]. The size of each frame is of 300 x 300 lattice steps. The case (a) refers to $\eta = 0$ for which we obtain a version of the Eden model that gives rise to a space filling structure. The cases b, c and d correspond respectively to η = 0.5, 1 and 2, and give rise to non trivial fractal structures. [It is a pleasure to thank H.J. Wiesmann for permission to reproduce these figures].

single bond.

(c) The selected bond is added to the equipotential pattern (set of black dots) and one starts again with (a) solving the Laplace equation with this new boundary condition.

The time consuming parts of this program is the iterative solution of the Laplace equation. This limits the size of the system one can consider.

For $\eta = 0$ the effect of the local field is irrelevant and the probability is identical for all the bonds that are candidates for growth. This is a version of the Eden model [8,9] and gives rise to a space filling structure as shown in Fig. (3a). The fractal dimension D is then equal to the euclidean dimension: D = d = 2. More interesting is the case $\eta = 1$ that gives rise to Fig. (3c). This corresponds to a fractal dimension $D \simeq 1.7$. For other values of η one obtains a smooth variation of D as shown by Fig. 3b and 3d. The results shown in Fig. 3 refer to planar (d = 2) geometry, analogous results can be obtained for d = 3. A summary of these results is presented in Table I. The fractal dimension of a pattern was here computed by considering the total number of black dots inside a radius R from the center. Alternative definitions have been shown to give very similar results.

TABLE I

	η	0	0.5	1	2
[d = 2]	D	2	1.92	1.70	1.43
[d = 3]	D	3	2.78	2.65	2.26

Dielectric Breakdown Model. Results for the fractal dimensions (D) of the stochastic pattern generated as a function of the variable η. The results for growth in two dimensions [d = 2] (see Fig. 3) refer also to a two dimensional field equation. For three dimensions [d = 3] the growth takes place within a sphere. For more details on these results see Ref. [15].

Before closing this discussion one should mention that also different Laplacian fractals than those discussed here have been studied. An interesting one is the Laplacian Random Walk in which growth can only occur at the tip of the last added bond. This gives rise to a topologically linear structure without loops that is interesting also for polymer statistics. We will not discuss this model here [14].

From the study of the Dielectric Breakdown Model we have learned that: (a) A suitable combination of *Laplace equation* and a *stochastic field* is able to generate stochastic fractal structures. This shows how *fractal structures* can be generated by a *well defined physical model*. The model was discussed here in relation to dielectric breakdown but, since Laplace equation appears in many fields of physics, it can easily be applied to other problems like dendritic growth (DLA model), viscous fingers in liquids and fracture propagation [12]. (b) From a mathematical point of view this model corresponds to a stochastic process with infinite memory and long range interaction. In addition the dependence of D on η shows a low degree of universality. Despite the simplicity of the formulation we have therefore a problem of formidable mathematical complexity that is not treatable with the available mathematical methods.

The task of a theory is to define the universality classes of the model (if any) and be able to compute the fractal dimensions of Table I without resorting to computer simulations. An essential point of this program is the understanding of the nature of the growth probability. It is useful therefore to look at a potential chart around a given structure. In Fig. 4 we show a perspective view of the potential (vertical probability is proportional to the drop of the potential in the vicinity of the equipotential pattern. This gives rise to a distribution of possible values for this probability at each point as contrasted to the pattern itself that is characterized simply by a set. A bond is either occupied (part of pattern) or not. As we are going to see in the next section the analysis of self-similar properties of distributions requires the more general concept of multifractals.

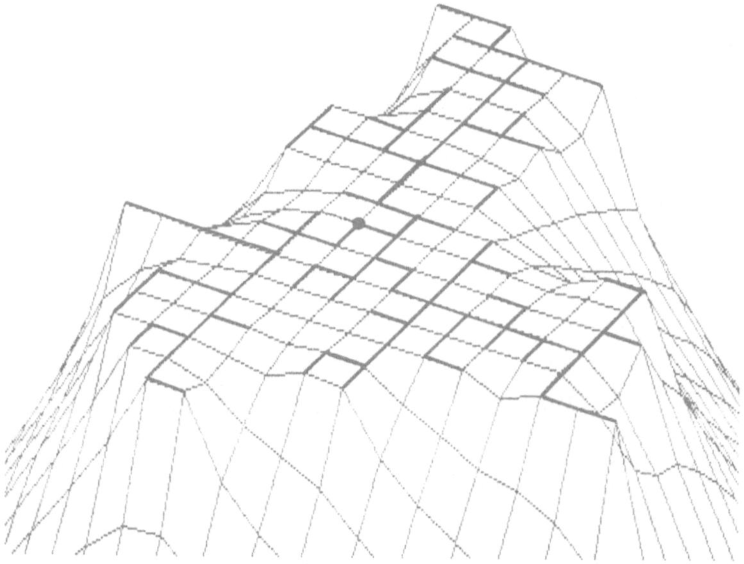

Figure 4. Behavior of the potential (vertical axis) around a given equipotential pattern (thick lines). The growth probability is related to the drop of the potential in the vicinity of the equipotential pattern.

4. MULTIFRACTALS: GENERAL DISCUSSION

In this section we introduce and describe in some detail the concept of multifractal. This concept is useful to characterize systems with different local properties of self-similarity and it is a generalization of the usual fractal dimension. It was introduced by Mandelbrot [16] and Renyi [17] in different fields. It has been then developed by various other authors [18-23] and we use it here in the sense of Ref. [23]. As we are going to see it is an essential concept for the study of self-similar properties of distributions.

It is useful to start the discussion with a simple example. Let us consider a deterministic construction of self-similar probability distributions that preserve normalization. This can be obtained as a generalization of the Cantor set construction that gives different weights to the points of the Cantor set [4,24]. We start with a Cantor set $C_k(\lambda,\eta)$ where λ is the number of intervals, η their lenght and k is the order of the iteration. Define $U_{k=1}(x)$ by giving a weight to each of the λ intervals of $C_1(\lambda,\eta)$:

$$U_{k=1}(x) = \begin{cases} w_1 & \text{if } x \ [0,\eta] \\ w_2 & \text{if } x \ \left[\frac{1-\eta}{\lambda-1}\ ;\ \frac{(\lambda-1)\eta+1-\eta}{\lambda-1}\right] \\ w_\lambda & \text{if } x \ [1-\eta;1] \\ 0 & \text{otherwise} \end{cases} \qquad (4,1)$$

and, because of normalization

$$\sum_{i=1}^{\lambda} w_i\eta = 1 \quad . \qquad (4,2)$$

Within the next iteration we procede in the same way with the same weight distribution. For simplicity we consider the specific case $\lambda=3$, $\eta=1/4$ and $(w_1, w_2, w_3) = (w, 2w, w)$ as shown in Fig. 5. Because of normalization the relation between η and w is simply $4w\eta = 1$. We have then the scaling property

$$U_{k+1}(\eta x) = \frac{1}{4\eta} U_k(x) \quad . \qquad (4,3)$$

For small x (note that the origin of our coordinates is at the edge of Fig. 5) U(x) can be described by the smooth function

$$\bar{U}(x) \sim x^{(D-1)} \quad (x\to 0) \qquad (4,4)$$

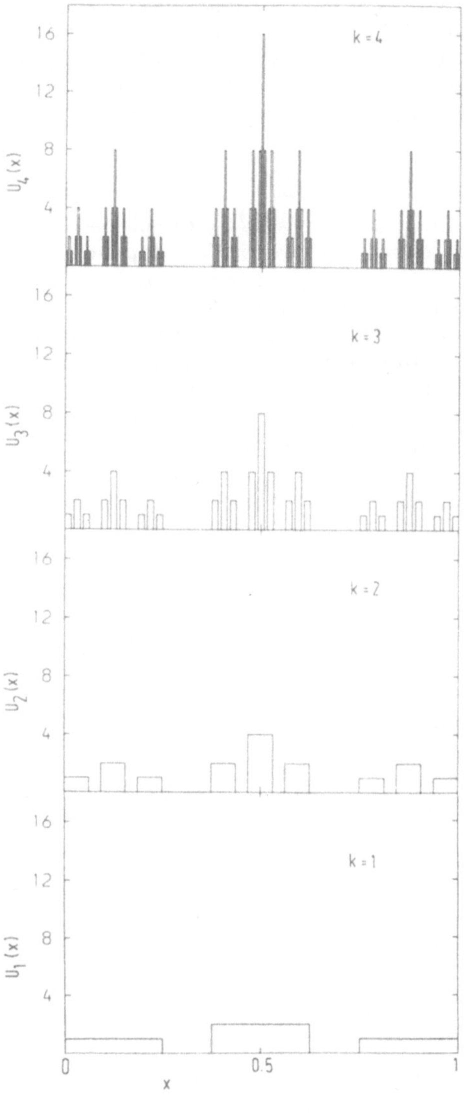

Figure 5. Example of a self-similar probability distribution obtained as a generalization of the Cantor set construction. This distribution shows a simple example of multifractal.

where $D = -\ln 4/\ln \eta = 1$ for $\eta = 1/4$. These well defined scaling properties have been derived for a particular origin. We now show that by changing the origin one obtains different scaling properties. Consider the translation $x \to x - \frac{1}{2}$ so that the origin lies in the center of the distribution. The scaling relations are now given by

$$U_{k+1}(\eta x) = \frac{1}{2\eta} U_k(x) \quad , \qquad (4,5)$$

so that $D = -\ln 2/\ln \eta = 0.5$ (for $\eta = 1/4$) which is a factor of two smaller then the previous case. This origin dependence of the fractal dimension is due to the introduction of different weights and shows that a single fractal dimension is not enough to describe self-similar distributions like the one we have considered. We are going to see in the following that this distribution is actually a simple case of multifractal and that it is characterized by a continuous spectrum of singularities (fractal dimensions) that can be computed exactly.

The basic idea is to develop a mathematical tool that is able to detect singularities of different nature. Let us start by considering the simple example of a probability distribution given by

$$p(x) = \alpha_o\, x^{(\alpha_o - 1)} \qquad (4,6)$$

and defined on the interval $x \in [0,1]$, as shown in Fig. 6. In order to be integrable we require $\alpha_o > 0$. For $0 < \alpha < 1$ we have a singularity at $x \to 0$. For $\alpha > 1$ $p(x) \to 0$ when $x \to 0$ and for $\alpha = 1$ the function is regular and analytic in the whole interval. We say then that the function $p(x)$ has a singularity of type α_o at the point $x = 0$ generalizing this concept also to the cases $\alpha_o > 1$. Let us now devide our unit interval in ℓ^{-1} segments each of size ℓ as indicated schematically in Fig. 6.

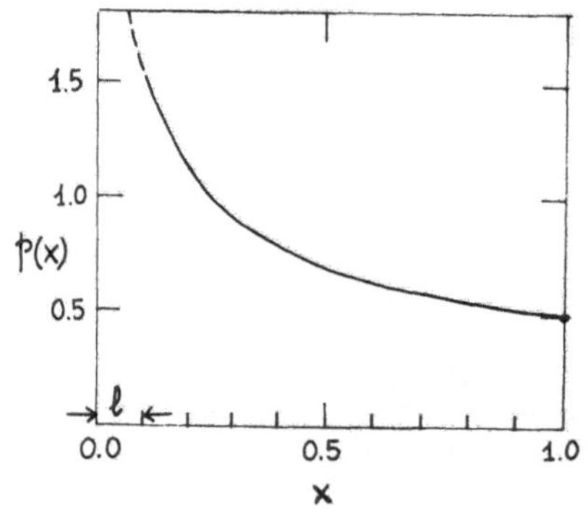

Figure 6. Schematic picture of a probability distribution $p(x)$ on the interval $[0,1]$ with an integrable singularity at $x=0$.

We then integrate p(x) over each segment to obtain ℓ^{-1} values defined as

$$p_i(\ell) = \int_{(i-1)\ell}^{i\cdot\ell} p(x)dx \qquad (4,7)$$

where $i = 1, 2, \ldots, \ell^{-1}$ is an index that characterizes a particular segment. The first segment contains the singularity and gives

$$p_{i=1}(\ell) = p_s(\ell) = \int_0^{\ell} p(x)dx = \ell^{\alpha_o} . \qquad (4,8)$$

The remaining $\ell^{-1}-1$ segments give each a contribution of order

$$p_{i\neq 1}(\ell) \simeq \ell \quad . \qquad (4,9)$$

We then construct the function

$$\chi(q;\ell) = \sum_{i=1}^{\ell^{-1}} [p_i(\ell)]^q \quad . \qquad (4,10)$$

This function is usually called partition function in the literature but one could name it "hunting function" because, as we are going to see, it acts like a hunter of singularities. It is easy to see that in the limit of small ℓ this functions behaves as a power of ℓ

$$\chi(q;\ell) \simeq \ell^{\tau(q)} \quad . \qquad (4,11)$$

In fact the contribution of the segment that contains the singularity gives

$$\chi_s(q;\ell) = \ell^{q\alpha_o} \qquad (4,12)$$

while the normal contribution of all the other $\ell^{-1} - 1 \simeq \ell^{-1}$ segments gives

$$\chi_N(q;\ell) \simeq \ell^{(q-1)} \quad . \qquad (4,13)$$

In the limit of small ℓ the smallest of these two exponents will provide the leading contribution to Eq. (4,10) and therefore

$$\tau(q) = \min\,[(q-1);q\alpha_o] \quad . \qquad (4,14)$$

We can interpret this result by saying that our original distribution $p(x)$ consists of two types of singularities: one with $\alpha=\alpha_o$ defined on a fractal set of dimension $f = 0$ (the point $x = 0$) and one of type $\alpha = 1$ defined on a set of fractal dimension $f = 1$ (the whole interval). By varying q from $-\infty$ to $+\infty$ in Eq. (4,10) one of these two contributions dominates at the beginning but at the value of q defined by

$$q - 1 = q\alpha_o \tag{4,15}$$

the contribution due the other type of singularity will take over. In this sense the variation of q in Eq. (4,10) is able to evidentiate all the different types of singularities of our distribution.

It is now easy to generalize the simple example we have discussed to the more complex case of a distribution characterized by a continuum of singularities whose nature is defined by the variable α. Also we assume that the singularities of type α are defined on a set with fractal dimension $f(\alpha)$. Our system is therefore defined by all the values (in principle a continuum) that the pair

$$\{\alpha \; ; \; f(\alpha)\} \tag{4,16}$$

can take. Let us now divide the volume on which our distribution $p(\vec{r})$ is defined in boxes (in principle of any dimension) of linear size ℓ. We label the boxes by the index i and construct for each box the function

$$p_i(\ell) = \int_{\substack{i\text{-th}\\ \text{box}}} p(\vec{r})d\vec{r} \quad . \tag{4,17}$$

In analogy with Eq. (4,10) we then define

$$\chi(q;\ell) = \sum_{\substack{i \\ [\text{all boxes}]}} [p_i(\ell)]^q \simeq \ell^{\tau(q)} \tag{4,18}$$

and assume that for small ℓ it behaves as a power of ℓ with exponent $\tau(q)$. If the box i contains a singularity of type α (in the sense of Eq. (4,6)) we have

$$[p_i(\ell)]^q \simeq \ell^{q\alpha} \quad . \tag{4,19}$$

Since the singularities of type α are defined on a set with fractal dimension $f(\alpha)$ the number of boxes that contain this type of singularities will vary with ℓ like $\ell^{-f(\alpha)}$. Therefore defining $n(\alpha,\ell)$ as the density of singularities of type α for boxes of size ℓ we have

$$n(\alpha,\ell)d\alpha = \rho(\alpha)\ell^{-f(\alpha)}d\alpha \quad , \tag{4,20}$$

where $\rho(\alpha)$ is a regular function. We can now rewrite Eq. (4,18) as

$$\chi(q) = \int d\alpha\ \rho(\alpha)\ \ell^{[q\alpha - f(\alpha)]} \ . \tag{4,21}$$

For ℓ very small this integral is dominated by the minimum of the possible exponents. This means that for a given q there will be a single pair of values $\alpha(q)$ and $f[\alpha(q)] = f(q)$ that give the main contribution. The saddle point condition gives

$$\frac{d}{d\alpha}\left[q\alpha - f(\alpha)\right]\Big|_{\alpha = \alpha(q)} = 0 \ , \tag{4,22}$$

that implies

$$\frac{df}{d\alpha} = q \ . \tag{4,23}$$

We obtain finally

$$\tau(q) = q\cdot\alpha(q) - f(q) \tag{4,24}$$

$$\frac{d\tau(q)}{dq} = \alpha(q) \tag{4,25}$$

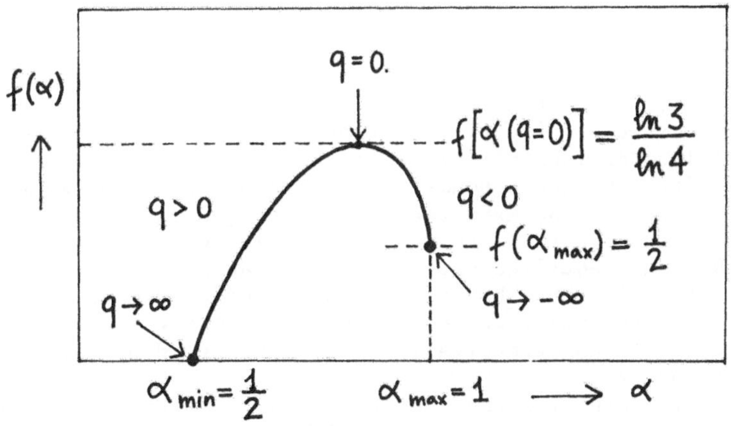

Figure 7. Multifractal spectrum (f,α) obtained by the analysis of the self-similar distribution shown in Fig. 5.

that corresponds to a Legendre transformation of the same type as used in thermodynamics to define the free energy. Interesting relations have been recently derived between the multifractal analysis of a probability distribution described here and the properties of generalized Lyapunov exponents of dynamical systems [25].

We can now go back to the self-similar distribution of Fig. 5 and analyze it as a multifractal. Since the iteration scheme it strictly self-similar we can consider a single iteration. This splits a constant distribution over a given interval into three peaks with probabilities $p_1 = 1/4$, $p_2 = 1/2$ and $p_3 = 1/4$, each occupying an interval of length 1/4 with respect to the initial segment. We obtain then for Eq. (4,18)

$$\sum_{i=1}^{3} [p_i]^q = 2\left(\frac{1}{4}\right)^q + \left(\frac{1}{2}\right)^q = \left(\frac{1}{4}\right)^{\tau(q)} \quad . \qquad (4,26)$$

This gives us directly $\tau(q)$ from which, using Eqs. (4,24) and (4,25) we obtain the (f,α) spectrum that is shown in Fig. 7. This makes clear that a continuum of singularities (α) and corresponding fractal dimensions ($f(\alpha)$) is necessary to characterize even the very simple self-similar distribution of Fig. 5.

5. MULTIFRACTALS: APPLICATION TO THE GROWTH PROBABILITY.

We can now go back to the problem discussed at the end of Sect. 3 and reconsider the distribution of growth probabilities (Fig. 4) for the dielectric breakdown model from the point of view of multifractals. To this purpose it is more convenient to use a cylinder geometry instead of the usual circle geometry in two dimensions. Consider two parallel electrodes, each of lenght ℓ_c on a square lattice, having separation distance h_c. The top electrode is kept at potential value 1 while the bottom one is kept at potential 0. The left and right-hand side of the so formed rectangle are identified, resulting in a cylinder with height h_c. As initial conditions all sites on the lower electrode are taken to be occupied. The subsequent bonds are then added with the same rules as discussed in Sect. 3. The main reason for using a cylinder geometry is that it allows the study of the multifractality of the growth probability distribution in equilibrium, i.e. not entangled with the time development of the cluster. This is due to the existence of an asymptotic steady state growth phase, characterized by a constant average thickness. This property has been extensively studied in the Eden model with the same geometry [26,27].

In order to determine the average (f,α) spectrum in the steady state we compute $\chi(q)$ as given by Eq. (4,18). For the case q = 1 this expression is not well defined and one has to use [19]

$$\chi(q = 1, \ell) = \sum_{\substack{i \\ [\text{all boxes}]}} [p_i(\ell)]\cdot[\ln p_i(\ell)] \quad . \qquad (5,1)$$

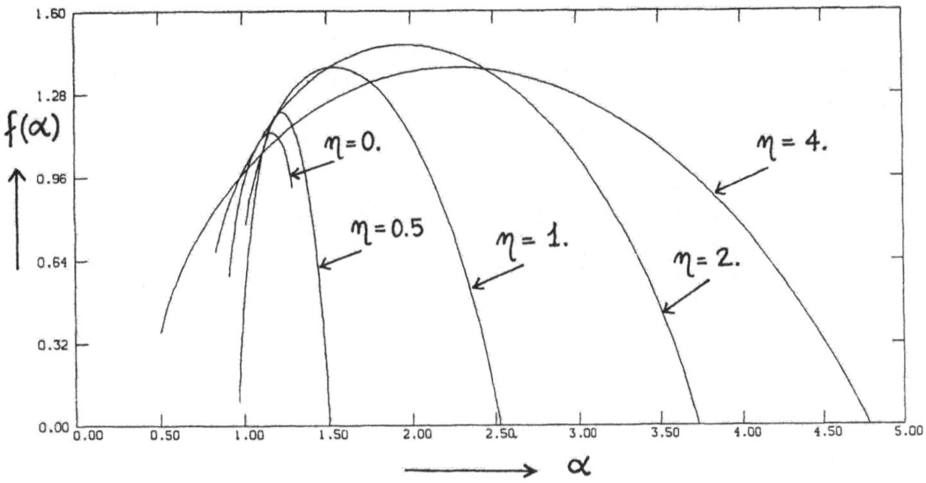

Figure 8. Spectrum of singularities (f,α) obtained by a multifractal analysis of the growth probability for the dielectric breakdown model for different values of the parameter η.

The behavior of $\chi(q)$ at small ℓ allows to define the exponents $\tau(q)$. All clusters were grown in a cylinder with total height $h_c = 48$ and electrode length $l_c = 16$. For the determination of $\chi(q)$ we have used boxes of sizes 1, 2, 4, 8 and 16. The results where averaged over 100 clusters in steady state regime (constant interface thickness). The entire calculation was repeated for different values of η.

The final result of these calculations is the determination of the series of exponents $\tau(q)$ for different values of η. Note that for $\eta\neq 1$ the growth probability is not given just by the local field but by the local field to the power η as given by Eq. (3,3). The corresponding (f,α) spectra as shown in Fig. 8. For $\eta=0$ the spectrum consists essentially of a single point $f = \alpha = 1$ confirming the relatively trivial properties of the Eden model. For $\eta \neq 0$ we observe the develpment of a continuum (f,α) spectrum that is broader for larger values of η. These calculations show the multifractal nature of the growth probability as opposed to the fractal nature of the pattern generated. These results are in contrast with a preliminary analysis of the growth probability (based on the calculation of $\tau(q)$ only for $q > 2$) from which a single (f,α) point was obtained [28]. They agree however with more recent improved analysis by various authors [29,30]. It is important to note that the results shown in Fig. 8 correspond to clusters of relatively small size. In view of the expected asymptotic invariance of the (f,α) spectrum it is important to check how these results depend on the size of the clusters considered. Preliminary results show an appreciable dependence of these spectra on the size and therefore the results of Fig. 8 cannot be considered asymptotic.

The open problems for the formulation of a theory of these processes are now:

(a) The relation between the multifractal properties of the growth probability and the fractal dimension of the generated pattern.

(b) The origin and the stabilization of the multifractal spectrum for the growth probability.

Various preliminary arguments have been formulated with respect to the problem (a) [31]. The problem (b) is at the moment rather understood but the concept we are going to develop in the next section will show how easily one can obtain a multifractal spectrum from a multiplicative process. In the present case the multipricative nature of the process could be identified in the phenomenon of tip splitting during growth but, at the moment, these analogies are rather speculative.

A last question one may consider before closing this section is what happens if one applies the multifractal analysis used for the growth probability to the pattern itself as shown for example in Fig.3c. For $\eta = 0$ we obtain from such an analysis a single point for the (f,α) spectrum:

$$f = \alpha = 1.7 \quad . \tag{5,2}$$

This means that, at least as far as the *density of occupation* , the generated pattern is a homogeneous fractal.

6. MULTIFRACTALS: MULTIPLICATIVE PROCESSES AND LOCALIZED WAVEFUNCTIONS.

In this section we show how a multifractal spectrum can be easily derived for certain fluctuations of random multiplicative processes. These processes appear in many problems related to the physics of disordered systems. Classical examples are provided by the Schrödinger or the elastic equation on a lattice with disorder [32] but also in turbulence and in other fields, like the tip splitting mentioned before, fragmentation phenomena appear to be relevant [16,33,34] The study of these processes can be often reduced to the products of matrixes with random elements. These studies up to now have concerned mainly the growth of some average quantity characterized by Lyapunov exponents [32]. Here we focus our attention on the fluctuations of these processes and, in particular, on the possibility of a self-similar structure for these fluctuations. We consider now the simplest of these processes: *the mul-*

tiplicative random walk.

Let us consider the random multiplicative process for the scalar function $\psi(x)$ of the variable x (that we call space variable)

$$\psi(x+1) = e^{\Delta}\ \psi(x) \tag{6,1}$$

where Δ is a random variable that for simplicity is assumed to be described by the probability distribution

$$P(\Delta) = \tfrac{1}{2}\delta(\Delta-\Delta_o) + \tfrac{1}{2}\delta(\Delta+\Delta_o). \tag{6,2}$$

Generalizations to more complex cases are possible. Such a process is a simplification of the much studied problem of the products of random matrix [32]. Up to now the main objective in this field has been the analysis of the Lyapunov exponents that govern the average growth of the various moments of ψ. Here we characterize the fluctuations of these processes in terms of their properties of self-similarity. These properties will be shown to be quite remarkable already for the very simple case described by Eqs. (6,1) and (6,2).

Starting from $\psi(x=0)=1$ the probability destribution P_N for the values of $\psi(x=N)$ is obviously given by the log-binomial distribution

$$P_N(\psi_{N,k}) = (\tfrac{1}{2})^N \cdot \binom{N}{k} \tag{6,3}$$

where $\psi_{N,k}$ can take the values

$$\psi_{N,k} = e^{(-N+2k)\Delta}\ ;\ k = 0,\ 1,\ 2,\ \ldots\ldots\ N\ . \tag{6,4}$$

From this distribution the various moments of $\psi(N)$ and the corresponding Lyapunov exponents can be easily computed by standard methods [34].

The new problem we intend to consider here is whether the fluctuations of such a process posess properties of self-similarity. First of all one has to distinguish between two types of fluctuations: The first one that we call *disorder fluctuations* refers to the different values of ψ (x) at a fixed point x=N for different sequences of the random process (realizations of the disorder). The second type, that we call *space fluctuations* refers to a sequence of values of $\psi(x)$ for different values of the space variable x but for a single realization of the disorder. In order to consider the possibility of a self-similar structure one has to define a metric with respect to which these properties may be analyzed. It is clear that this information is not provided by the log-binomial (or log-normal when it is allowed to use it) distribution of Eq. (6,3). These smooth distributions correspond in fact to global averages for the possible values of ψ at a given point x=N. They do not provide information on the internal correlations of the process that are averaged out. The nature of these internal correlations will become clear in the following.

We show now how an appropriate metric structure can be defined by mapping all the possible trajectories of $\psi(x)$ (realizations of the dis-

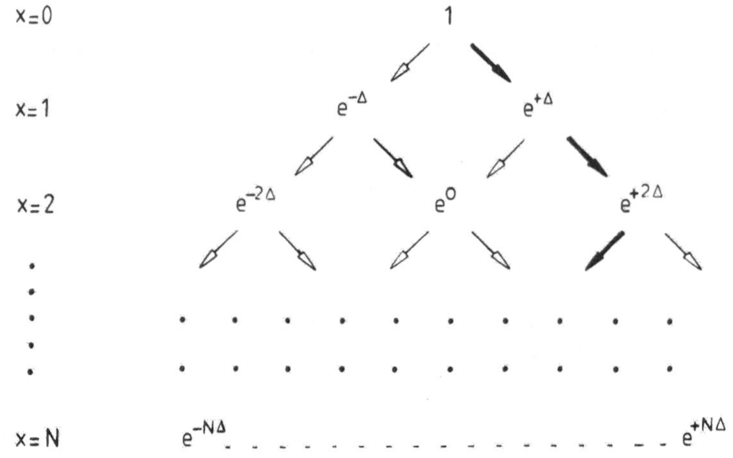

Figure 9. Schematic view of the tree of all possible trajectories (realizations of the disorder) for the multiplicative process defined by Eqs. (6,1) and (6,2). A trajectory is defined by the complete set of arrows that link the point x = 0 to x = N. An example of a trajectory is given by the thick arrows.

order) on the interval $[0,1]$. In this way the process described by Eq. (6,1) will be shown to correspond to the iteration of a distribution on this interval. It is useful to look at the tree of possible trajectories as shown in Fig. 9. Our objective is to map an *entire trajectory* from x = 0 to x = N (and in principle $N \to \infty$) into a point of the $[0,1]$ interval. A trajectory corresponds in Fig. 9 to a complete sequence of arrows like for example the series of thick arrows.

Consider the point x = 0. At this point all trajectories correspond to $\psi(0) = 1$. In Fig. 10a we plot the value of $\psi(x=0;T)$ as a function of the variable T on $[0,1]$ that as we are going to see corresponds to an entire trajectory from x = 0 to x = N. Since for any trajectory $\psi(x=0) = 1$,we obtain simply a constant. At the next point (x = 1 of Fig. 9) the set of all trajectories splits into two groups. One group corresponds to $\psi(x=1) = e^{-\Delta_o}$ and the other one to $\psi(x=1) = e^{\Delta_o}$. So we define our T's in such a way that all the trajectories of the first group have $0 < T < \frac{1}{2}$ while for the second group $\frac{1}{2} < T < 1$. Therefore $\psi(x=1;T)$ as a function of T will consist of a distribution with two values. (Note that Fig. 10 shows the case $\Delta_o = \ln 2$ that implies dividing or multiplying by two at each step). At the next point (x = 2) each of these two groups of trajectories is on its turn split into two subgroups and so on as illustrated in Fig. 10. The precise value of T for each trajectory can be obtained with the following construction. For a given trajectory (a) form x = 1 to x = N we consider the set of values $\Delta^a(x)$ that define this trajectory. We then introduce the variable

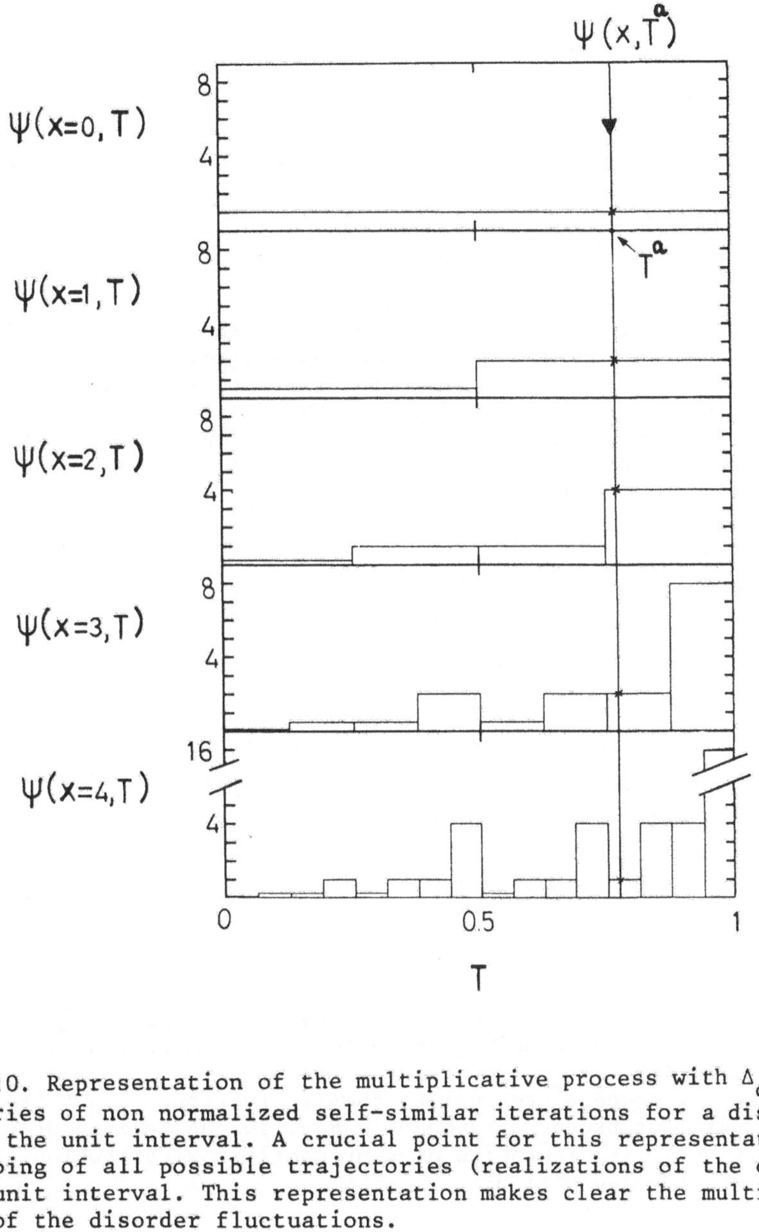

Figure 10. Representation of the multiplicative process with $\Delta_o = \ln 2$ as a series of non normalized self-similar iterations for a distribution on the unit interval. A crucial point for this representation is the mapping of all possible trajectories (realizations of the disorder) on the unit interval. This representation makes clear the multifractal nature of the disorder fluctuations.

$C^a(x)$ given by

$$C^a(x) = \begin{cases} 1 \text{ if } \Delta(x) = \Delta_o \\ 0 \text{ if } \Delta(x) = -\Delta_o \end{cases} . \tag{6,5}$$

The variable T^a corresponding to this trajectory is then defined as

$$T^a = \sum_{x=1}^{N} C^a(x) \, (\tfrac{1}{2})^x \quad . \tag{6,6}$$

For $N \to \infty$ this definition leads to a point in the interval $[0,1]$. For a finite N one has to consider the segment defined by $T^a < T < T^a + (\frac{1}{2})^N$ that includes all the trajectories from $x = 0$ to $x = \infty$ that have the same subtrajectory between $x = 0$ and $x = N$.

We have therefore mapped the whole evolution of our multiplicative process into the iteration of a distribution on the interval (Fig. 10). This distribution contains much more information then the simple log-binominal distribution for the end point (Eq. (6,3)). Equation (6,3) can be of course recovered by integrating the new distribution over the variable T in the following sense

$$P_N(\psi) = \int_0^1 dT \, \delta[\psi - \psi \, (x = N; \, T)] \quad . \tag{6,7}$$

The average with respect to the possible realizations of the disorder $\langle .... \rangle$ corresponds now simply to an integral over the variable T. One obtains then immediately $\langle\psi(x=1;T)\rangle = (5/4)\langle\psi(x=0;T)\rangle$ and therefore

$$\langle\psi(x=N;T)\rangle = \int_0^1 dT \; \psi(x=N;T) = (5/4)^N \langle\psi(x=0;T)\rangle =$$

$$= e^{L(1).N} \langle\psi(x=0;T)\rangle \tag{6,8}$$

where $L(1) = \ln(5/4)$ is the Generalized Lyapunov exponent [21] for the first moment of ψ. The generalization to other moments is straightforward.

We can now proceed to analyze the fluctuations. Apart from the average growth our process as shown in Fig. 10 can be interpreted as a strictly self-similar iteration of a probability distribution. If we consider the rescaled variable

$$\tilde{\psi}(x) = e^{-L(1)x} \; \psi(x) \tag{6,9}$$

the corresponding distribution is also normalized at each step.

As we have shown in Sect. 4 self-similar probability distributions can be described as multifractals. It is a simple exercise to perform such an analysis for our problem. If we divide our interval into seg-

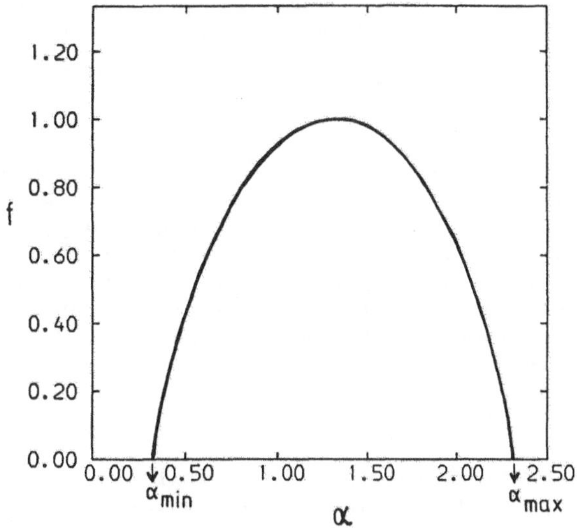

Figure 11. Multifractal spectrum of the disorder fluctuations derived from the analysis of the iterated distribution of Figure 10. The variable α characterizes the type of singularity and $f(\alpha)$ is the corresponding fractal dimension.

ments of length ℓ we can construct

$$\chi(q) = \sum_{n=0}^{1/\ell-1} \left[\int_{n\ell}^{(n+1)\ell} dT\ \tilde{\psi}(N \to \infty; T) \right]^q = \ell^{\tau(q)} \qquad (6,10)$$

This defines the exponents $\tau(q)$. In view of the strict self-similarity of our process $\tau(q)$ can be obtained by considering a single iteration. This gives

$$\tau(q) \frac{1}{\ln 2} \left[q\ln\ (e^{-\Delta_o} + e^{\Delta_o}) - \ln(e^{q\Delta_o} + e^{-q\Delta_o}) \right] . \qquad (6,11)$$

Using then the relations given by Eqs. (4,24) and (4,25) we obtain the multifractal spectrum characterized by a continuum of singularities defined by α and the fractal dimensions $f(\alpha)$ of the sets on which they are defined. A plot of $f(\alpha)$ is shown in Fig. 11 and extends between $\alpha_{min} = (\ln 2)^{-1} \cdot \ln[(e^{-\Delta_o} + e^{\Delta_o})/e^{\Delta_o}]$ and $\alpha_{max} = (\ln 2)^{-1} \cdot \ln\ [(e^{-\Delta_o} + e^{\Delta_o})/e^{-\Delta_o}]$.

We have therefore identified a metric (Fig. 10) with respect to which the fluctuations of our process are described by a multifractal spectrum. These are the disorder fluctuations, namely those defined at fixed x for different configurations of the disorder. We can now consider the properties of the space fluctuations. These correspond to a single trajectory (T^a) for different values of x. In Fig. 10 they are identified by the series of points (indicated by crosses) in which a vertical line difined by T^a encounters the distributions corresponding to different values of x. Since these values grow on average exponentially (Eq. (6,8)) the study of their fluctuations should be done for the rescaled variable $\tilde{\psi}$ given by Eq. (6,9). We have performed a multifractal analysis of these fluctuations numerically by interpreting the values of $\tilde{\psi}(x)$ on a given interval as a probability distribution. The surprising result is

$$f = \alpha = 1 \quad . \tag{6,12}$$

This implies that,contrary to the multifractal structure of the disorder fluctuations,the space fluctuations do not possess properties of self-similarity.

We finally discuss briefly the extension of these results to the properties of fluctuations of wavefunctions localized by disorder. Recently various interesting ideas about fractal properties of these wavefunctions have been put forwards [24,35-37]. The task of analyzing this issue is however complicated by the obvious difficulty of dealing with finite size numerical realizations. In this respect the most accurate results can be obtained in one dimension where the range of length scales between the lattice unit and the localization length can be maximized. From what we have discussed until here it is clear that a multifractal analysis is necessary to resolve this question while up to now only the simple fractal concept has been used [35,36].

Considering the iterative solutions of Schrödinger equation in a random (diagonal) potential V(x) described by a uniform distribution of width W we have

$$\psi_A(x+1) = \left[\frac{E-V(x)}{J}\right] \psi_A(x) - \psi_A(x-1) \tag{6,13}$$

where E is the energy and J is the off diagonal matrix element. This iteration process corresponds to the multiplication of 2 x 2 matrixes with random elements. Guided by the previous results for random scalar multiplications we have performed numerically a multifractal analysis for both space and disorder fluctuations of these functions. We do not report here about the details of the analysis [38] but we only quote the results. These are:

(a) Contrary to previous claims [35] the *space fluctuations* of Anderson localized wavefunctions in one dimension are *not* self-similar (fractal).

(b) In full analogy with the simple multiplicative process discussed before the fluctuations with respect to *disorder configurations* are characterized by a multifractal spectrum. A result that is in agreement with similar calculations performed in the meantime by other authors [25].

Acknowledgements: It is a pleasure to thank E. Hafkenscheid and R. Harmsma for their contribution to the multifractal analysis of the growth probability.

REFERENCES

(1) L. Pietronero and E. Tosatti Eds. Fractals in Physics, North-Holland, Amsterdam, New York (1986).

(2) A. Aharony, Europhys. News, **17**, 41 (1986).

(3) L.P. Kadanoff, Physics Today, Feb. 1986, p. 6.

(4) B.B. Mandelbrot, The Fractal Geometry of Nature, W.H. Freeman, New York (1983).

(5) D.J. Amit, Field Theory, the Renormalization Group and Critical Phenomena, Mc Graw Hill Int., (1978).

(6) L. Peliti and L. Pietronero, 'Random Walks with Memory', La Rivista del Nuovo Cim., to appear.

(7) L. Niemeyer, L. Pietronero and H.J. Wiesmann, Phys. Rev. Lett., **52** 1033 (1984).

(8) L. Pietronero and H.J. Wiesmann, J. Stat. Phys. **36**, 909 (1984).

(9) T.A. Witten and L.M. Sander, Phys. Rev. Lett. **47**, 1400 (1981).

(10) P. Meakin, Phys. Rev. A **27**, 1495 (1983).

(11) J. Nittmann and H.E. Stanley, Nature **321**, 663 (1986).

(12) For an updated discussion see Parts V(A) and V(B) of Ref. (1).

(13) L. Niemeyer, L. Pietronero and H.J. Wiesmann, Phys. Rev. Lett. (Response to a Comment), 4-th Aug. 1986, to appear.

(14) Lyklema, C. Evertsz and L. Pietronero, Europhys. Lett **2**, 77 (1986).

(15) H.J. Wiesmann and L. Pietronero, Ref. (1) p. 151.

(16) B.B. Mandelbrot, J. Fluid. Mech. **62**, 331 (1974).

(17) A. Renyi, Probability Theory, North-Holland, Amsterdam (1970).

(18) H.G.E. Hentschel and I. Procaccia, Physica 8D, 435 (1983).

(19) P. Grassberger and I. Procaccia, Physica 13D, 34 (1984).

(20) V. Frisch and G. Parisi in Turbulence and Predictability of Geophysical Flows and Climate Dynamics, M. Ghil, R. Benzi and G. Parisi Eds. North-Holland, New York (1985), p. 84.

(21) R. Benzi, G. Paladin, G. Parisi and A. Vulpiani, J. Phys. A17, 3521 (1984); J. Phys. A18, 2157 (1985).

(22) L. de Arcangelis, S. Redner and A. Coniglio, Phys. Rev. B31, 4725 (1985) and preprint.

(23) T.C. Halsey, M.H. Jensen, L.P. Kadanoff, I. Procaccia and B.I. Shraiman, Phys. Rev. A33, 1141 (1986).

(24) A.P. Siebesma and L. Pietronero, submitted to Phys. Rev. Lett.

(25) G. Paladin and A. Vulpiani, preprint.

(26) M. Plischke and Z. Racz, Phys. Rev. Lett. 53, 415 (1984). R. Jullien and R. Botet, Phys. Rev. Lett. 54, 2055 (1985) and in Ref. (1), p. 251.

(27) P. Meakin, R. Jullien and R. Botet, Europhys. Lett. 1, 609 (1986). M. Kardar, G. Parisi and Y.C. Zhang, Phys. Rev. Lett. 56, 889, (1986).

(28) T.C. Halsey, P. Meakin and I. Procaccia, Phys. Rev. Lett. 56, 854, (1986).

(29) C. Amitrano, A. Coniglio and F. di Liberto, preprint.

(30) P. Meakin, preprint.

(31) C. Evertsz, E. Hafkenscheid, R. Harmsma and L. Pietronero, preprint.

(32) K. Ishii, Progr. Theor. Phys. Suppl. 53, 77 (1973). J.P. Bouchaud and P. Le Doussal, J. Phys. A19, 797 (1986). J.L. Richard, J. Phys. C19, 1519 (1986).

(33) D. Schertzer and S. Lovehoy, in Ref. [1], p. 457.

(34) G. Paladin and A. Vulpiani, in Ref. [1], p. 447.

(35) C.M. Soukoulis and E.N. Economou, Phys. Rev. Lett. 52, 565 (1984).

(36) H.E. Roman, J. Phys. C19, L285 (1986).

(37) C. Castellani and L. Peliti, J. Phys. A19, L. 429 (1986).

(38) L. Pietronero, A.P. Siebesma, E. Tosatti and M. Zanetti, preprint.

GENERALIZED LEVY DISTRIBUTIONS AS LIMIT LAWS

W.R. Schneider
Brown Boveri Research Center
5405 Baden
Switzerland

ABSTRACT. For each $m \in N$ and each $\alpha \in (0,1)$ there exists a unique probability distribution $F_{m,\alpha}$ with support R_+ whose Laplace transform $\tilde{f}_{m,\alpha}$ satisfies the differential equation

$$(-\frac{d}{dp})^m \tilde{f}(p) = \alpha p^{\alpha-1} \tilde{f}(p) \quad .$$

These $F_{m,\alpha}$ generalize the one-sided Lévy distributions F_α obtained for $m = 1$. Existence is proved by constructing a solution, uniqueness by appealing to general theorems of function theory. The generalized Lévy distributions are limit laws in the following sense: Let $(X_n)_{n\geq 0}$ be a sequence of i.i.d. random variables in the domain of attraction of F_α. For each $m \geq 2$ and $t > 0$ there exists a random variable Y_t^m depending nonlinearly on (X_n) which tends in distribution to $F_{m,\alpha}$ as t tends to zero. The random variable Y_t^m is obtained as limit of a recursion. Its distribution satisfies a homogeneous integral equation whose Laplace transformed version tends to the above differential equation as t tends to zero.

1. INTRODUCTION

Stable distributions [1-3] were introduced in 1924 by Paul Lévy. They occur as limit laws of properly normalized sums of independent random variables. Their characteristic functions are known explicitly [2] but only recently a direct characterization in terms of Fox

S. Albeverio et al. (eds.), Stochastic Processes in Physics and Engineering, 279–304.

functions [4-8] has been found [4,5].

A Lévy distribution F is called one-sided if its support is R_+. It is characterized by a parameter α, $0 < \alpha < 1$ and its Laplace transform is given by [2]

$$\tilde{f}_\alpha(p) = \int_0^\infty e^{-px}\, dF_\alpha(x) = e^{-p^\alpha} \quad . \tag{1.1}$$

It follows [2] that F_α has a density f_α whose series expansion is given by [2]

$$f_\alpha(x) = -\frac{1}{\pi x} \sum_{n=1}^\infty \frac{\Gamma(n\alpha+1)}{n!} (-x^{-\alpha})^n \sin\pi n\alpha \quad . \tag{1.2}$$

The domain of attraction D_α of F_α consists of all distributions G on R_+ for which there exist norming constants c_n, $n \in N$, such that

$$\lim_{n\to\infty} G^{n*}(c_n x) = F_\alpha(x) \quad . \tag{1.3}$$

Here, G^{n*} is the n-fold convolution of G with itself. It can be shown [2] that G belongs to D_α iff asymptotically

$$1-F(x) \sim \frac{x^{-\alpha}\, L(x)}{\Gamma(1-\alpha)} \quad , \quad x \to \infty \quad , \tag{1.4}$$

where L is a slowly varying function [2], i.e.

$$L(x) > 0 \quad , \quad \lim_{t\to\infty} \frac{L(tx)}{L(t)} = 1, \quad x \in R_+ \quad . \tag{1.5}$$

It can also been shown [2] that apart from rescaling ($x \to \lambda x$) the only possible limit in (1.3) (not concentrated at a single point) is

F_α for some α, $0 < \alpha < 1$. The Laplace transform $\tilde{f}_\alpha$ of f_α satisfies the first order ordinary differential equation

$$- \frac{d}{dp} \tilde{f}(p) = \alpha\, p^{\alpha-1} \tilde{f}(p) \tag{1.6}$$

as is seen from (1.1). The question then arises whether the following generalization

$$(- \frac{d}{dp})^m \tilde{f}(p) = \alpha\, p^{\alpha-1} \tilde{f}(p) \tag{1.7}$$

with $m = 2,3,4,\ldots$ has solutions which are Laplace transforms of probability densities. By the theorem of Bernstein [2] this is equivalent with

$$\tilde{f}(0) = 1 \quad , \quad (- \frac{d}{dp})^k \tilde{f}(p) \geq 0 \quad , \quad p > 0 \tag{1.8}$$

(i.e. normalization and complete monotonicity).

In Section 2 it is shown that there exists exactly one solution $\tilde{f}_{m,\alpha}$ with these properties. The associated densities and distributions will be denoted by $f_{m,\alpha}$ and $F_{m,\alpha}$, respectively.

A further question is whether these generalized one-sided Lévy distributions [4,5] are also limit laws in some sense. It is clear from the outset that a more complicated expression than G^{n*} will be needed to attain such a goal. Such an expression (for each $m \geq 2$) is exhibited in Section 3. It attributes in a nonlinear way a random variable Y_t^m, $t > 0$, to a sequence of i.i.d. random variables with distribution G. If G belongs to the domain of attraction D_α of the stable law F_α then Y_t^m tends to $F_{m,\alpha}$ in distribution as t tends to zero. A "linear" version for $m = 1$ is also exhibited where the limit is the ordinary Lévy distribution F_α. Some technicalities are deferred to Section 4.

2. GENERALIZED LEVY DISTRIBUTIONS

We are looking for solutions of the differential equation

$$(-\frac{d}{dp})^m \tilde{f}(p) = \alpha\, p^{\alpha-1} \tilde{f}(p) \quad , \qquad 0 < \alpha < 1 \quad , \tag{2.1}$$

which are completely monotone, i.e.

$$(-\frac{d}{dp})^k \tilde{f}(p) \geq 0 \quad , \qquad p > 0 \quad , \qquad k \in Z_+ \quad , \tag{2.2}$$

and satisfy the normalization condition

$$\tilde{f}(0) = 1 \quad . \tag{2.3}$$

By the theorem of Bernstein [2] this is equivalent to the representability of $\tilde{f}$ as Laplace transform of a probability distribution F.

The differential equation (2.1) has a fundamental system of solutions ϕ_ν, $\nu = 0,1,2,\ldots, m-1$, with the asymptotic behaviour

$$\phi_\nu(p) \sim \exp(-\xi_\nu\, Bp^{a/m}) \tag{2.4}$$

where

$$a = m+\alpha-1 \quad , \qquad B = \frac{m}{a}\, \alpha^{1/m} \tag{2.5}$$

and

$$\xi_\nu = \exp(i\, \frac{2\pi}{m}\, \nu) \quad , \qquad \nu = 0,1,\ldots,m-1 \quad . \tag{2.6}$$

The general solution is a linear combination of $\phi_0, \phi_1, \ldots, \phi_{m-1}$. The requirement (2.2), however, excludes the presence in this combination of those ϕ_ν where Re $\xi_\nu \geq 0$.

As $a/m < 1$ the function $\tilde{f}$ is rapidly decreasing (faster than $|p|^{-\lambda}$ for all $\lambda > 0$) in Re $p > 0$ as $|p| \to \infty$. Hence, it is the Laplace transform of an ordinary function f [9] (i.e. F has a density), which is rapidly vanishing near the origin [9] ($x^{-\lambda} f(x)$ remains bounded as $x \downarrow 0$, for all $\lambda > 0$). From the differential equation (2.1) for $\tilde{f}$ one may easily derive the integral equation

$$x^m f(x) = \frac{\alpha}{\Gamma(1-\alpha)} \int_0^x dy(x-y)^{-\alpha} f(y) \tag{2.7}$$

for f. It is more convenient to consider the integral equation

$$g(x) = \frac{\alpha}{\Gamma(1-\alpha)} x^{m+\alpha-2} \int_0^\infty dy\ y^\alpha (y-x)^{-\alpha} g(y) \tag{2.8}$$

for

$$g(x) = x^{-2} f(x^{-1}) \quad , \qquad x > 0 \quad , \tag{2.9}$$

which is rapidly decreasing for $x \to \infty$. Together with f also g is a probability density, satisfying the normalization

$$\int_0^\infty g(x)dx = 1 \quad . \tag{2.10}$$

Due to the rapid decrease g possesses a Mellin transform

$$\hat{g}(s) = \int_0^\infty g(x)x^{s-1}\ dx \tag{2.11}$$

for Re s $\geq$ 1 which is analytic in Re s > 1. The integral equation (2.8) for g leads to the difference equation

$$\hat{g}(s+a) = \alpha^{-1} \frac{\Gamma(s+m-1)}{\Gamma(s+a-1)} \hat{g}(s) \tag{2.12}$$

and (2.10) is equivalent to

$$\hat{g}(1) = 1 \quad . \tag{2.13}$$

By induction we obtain from (2.12), (2.13)

$$\hat{g}(1+na) = \alpha^{-n} \prod_{k=1}^{n} \frac{\Gamma(m+(k-1)a)}{\Gamma(ka)} \quad , \qquad n \in N \quad . \tag{2.14}$$

This result may be rewritten as

$$\hat{g}(1+na) = Ab^{1+na} \frac{1}{\Gamma(na)} \prod_{j=1}^{m} \Gamma\left(\frac{1+na+j-2}{a}\right) \tag{2.15}$$

with

$$b = \left(\frac{a^m}{\alpha}\right)^{1/a} \quad , \qquad Aab \prod_{j=2}^{m} \Gamma\left(\frac{j-1}{a}\right) = 1 \quad , \tag{2.16}$$

by using repeatedly the functional equation

$$\Gamma(z+1) = z\,\Gamma(z) \tag{2.17}$$

of the Γ-function. Replacing 1+na by s in (2.15) leads to

$$\hat{g}(s) = Ab^s \frac{1}{\Gamma(s-1)} \prod_{j=1}^{m} \Gamma(\frac{s+j-2}{a}) \tag{2.18}$$

as a candidate for the solution of (2.12). This function is meromorphic with poles

$$\begin{aligned} s_{1,n} &= 1-na \quad , & & n \in N \\ s_{j,n} &= 2-j-na \quad , & j = 2,3,\ldots,m \quad , & n \in Z_+ \quad , \end{aligned} \tag{2.19}$$

(the second set is empty for m = 1).

For α irrational, all poles are simple. For α rational, coincidences

$$s_{j,n} = s_{j',n'} \quad , \quad (j,n) \neq (j',n') \tag{2.20}$$

may occur, leading to higher order poles. As all poles are lying in Re s < 1 the function $\hat{g}(s)$ is analytic in a domain containing the half-plane Re s $\geq$ 1. For s → 1 (2.18) leads to (2.13). Again using repeatedly (2.17) one verifies that (2.18) is actually a solution of (2.12). Henceforth, it will be denoted by $\hat{g}_{m,\alpha}$. The inverse Mellin transform yields the representation

$$g_{m,\alpha}(x) = A\, H^{m0}_{1m} \left(\frac{x}{b} \,\middle|\, \begin{matrix} (-1,1) \\ (\frac{k-2}{a}, \frac{1}{a})_{k=1,\ldots,m} \end{matrix} \right) \tag{2.21}$$

in terms of Fox functions (for details see [4]). It has the series expansion

$$g_{m,\alpha}(x) = Aa \sum_{k=1}^{m} \sum_{n=1}^{\infty} \frac{c_{k,n}}{\Gamma(1-k-na)} \frac{(-1)^n}{n!} (\frac{x}{b})^{k-2+na} \tag{2.22}$$

$$c_{k,n} = \prod_{j=1}^{m}{}' \; \Gamma(\frac{j-k}{a} - n) \qquad (' : \text{omit } j = k)$$

and the asymptotic behavior

$$g_{m,\alpha}(x) \sim Cx^{\sigma} \exp(-Dx^{\tau}) \tag{2.23}$$

with

$$D = \frac{1-\alpha}{a} \alpha^{1(1-\alpha)} \quad , \quad \sigma = m - 2 + \frac{m\alpha}{2(1-\alpha)} \quad , \quad \tau = \frac{a}{1-\alpha} \tag{2.24}$$

and

$$C = (2\pi)^{\frac{m-2}{2}} (1-\alpha)^{-1/2} \alpha^{\varepsilon} a^{-\delta} \prod_{k=2}^{m} \Gamma(\frac{k-1}{a})^{-1}$$

$$\varepsilon = (1+\sigma)/a \quad , \quad 2a\delta = (1-\alpha)(m-1) \quad . \tag{2.25}$$

From (2.22), (2.24) and the integral equation (2.8) one concludes (by contradiction) that $g_{m,\alpha}(x)$ is positive for x positive.

The solution $\tilde{f}_{m,\alpha}$ of the original problem (2.1)-(2.3) associated with $g_{m,\alpha}$ is given by [4]

$$\tilde{f}_{m,\alpha}(p) = Ab \; H^{m0}_{0m} (\frac{p}{b} \mid {- \atop (\frac{k-1}{a}, \frac{1}{a})_{k=1,\ldots,m}}) \tag{2.26}$$

with series expansion

$$\tilde{f}_{m,\alpha}(p) = Aab \sum_{k=1}^{m} \sum_{n=0}^{\infty} c_{k,n} \frac{(-1)^n}{n!} (\frac{p}{b})^{k-1+na} \tag{2.27}$$

with $c_{k,n}$ as in (2.22) and asymptotic expansion

$$\tilde{f}_{m,\alpha}(p) \sim Ep^{\kappa} \exp(-Bp^{a/m}) \tag{2.28}$$

with B given in (2.5),

$$2m\ \kappa = (1-\alpha)(m-1) \quad , \tag{2.29}$$

and

$$E = (2\pi)^{\frac{m-1}{2}}\ m^{-1/2}\ Aab^{1-\kappa} \quad . \tag{2.30}$$

Comparing with (2.4) we see that only $\nu = 0$ contributes to the solution.

It remains to show uniqueness. Let $\tilde{f}$ be a solution of (2.1)-(2.3). Associate with it $\hat{g}$ in the same way as $\hat{g}_{m,\alpha}$ is associated with $\tilde{f}_{m,\alpha}$. Define q by

$$q(s) = \hat{g}(s)\ [\hat{g}_{m,\alpha}(s)]^{-1} \tag{2.31}$$

for s in the half-plane Re $s \geq 1$. This function is analytic in Re $s > 1$ and satisfies there

$$q(s+a) = q(s) \quad . \tag{2.32}$$

Hence, there exists a unique entire periodic function $q_o(s)$ with period a which coincides with q in Re $s \geq 1$.

The function $\log \Gamma(s)$ has the following representation [10]

$$\log \Gamma(s) = (s-1/2)\log s - s + \frac{1}{2}\log(2\pi) + \phi(s) \tag{2.33}$$

where ϕ is given by

$$\phi(s) = 2 \int_0^\infty dt \, \frac{\arctan(t/s)}{\exp(2\pi t)-1} \quad . \tag{2.34}$$

It has the following estimate

$$|\phi(s)| \le \frac{K(s)}{12|s|} \quad , \qquad \text{Re } s \ge \delta > 0 \tag{2.35}$$

where

$$K(x+iy) = \begin{cases} \dfrac{x^2+y^2}{2|xy|} & , \quad |x| < |y| \\ 1 & , \quad |x| \ge |y| \end{cases} \quad . \tag{2.36}$$

This leads to the following estimates

$$\begin{aligned} |\Gamma(x+iy)| &\le C(a,b) e^{-\frac{\pi}{2}|y|} |y|^{b-1/2} \\ |\Gamma(x+iy)|^{-1} &\le D(a,b) e^{\frac{\pi}{2}|y|} \end{aligned} \tag{2.37}$$

in

$$\frac{1}{2} \le a \le x \le b \quad , \qquad |y| \ge b \tag{2.38}$$

with

$$\begin{aligned} C(a,b) &= (2\pi)^{1/2} \, 2^{(b-1/2)/2} \exp(c+b-a) \\ D(a,b) &= (2\pi)^{-1/2} \, a^{(1/2-a)} \exp(c+b) \\ c &= 1/(12\sqrt{2} \, a) \quad . \end{aligned} \tag{2.39}$$

Repeated application of these estimates lead to ($s = x+iy$)

$$|\hat{g}_{m,\alpha}(s)|^{-1} \le \gamma_{m,\alpha} \, |y|^{2a+1/2} \, e^{\frac{\pi}{2} |y| \frac{1-\alpha}{a}} \tag{2.40}$$

holding in

$$2+a \le x \le 2+2a \qquad , \qquad |y| \ge 2a+m \quad . \tag{2.41}$$

As

$$|\hat{g}(x+iy)| \le \hat{g}(x) \tag{2.42}$$

a similar estimate holds for $q(s)$. As a consequence of periodicity its continuation q_o satisfies

$$|q_o(s)| \le Q_o \, e^{\frac{\pi}{a}(1-\varepsilon)|y|} \le Q_o \, e^{\frac{\pi}{a}(1-\varepsilon)|s|} \tag{2.43}$$

in the whole plane for some ε with $0 < \varepsilon < 1$. By Carlson's theorem [11] such a function is constant, i.e.

$$q_o(s) = q_o(1) = 1 \tag{2.44}$$

which proves uniqueness.

3. LIMIT LAWS

The aim is to construct random variables Y_t^m, $0 \le t < T$, depending on a sequence $(X_n)_{n\geq 0}$ of i.i.d. random variables with the following property: If the distribution G of X_o belongs to the domain of attraction of the one-sided Lévy distribution F_α, $0 < \alpha < 1$, then the distribution H_t^m of Y_t^m tends to the generalized one-sided Lévy distribution $F_{m,\alpha}$ as t tends to zero.

Definition: Let $k: R_+ \to R_+$ be a bounded continuous and nondecreasing function. Define $k_n: R_+^n \to R_+$, $n \varepsilon N$, by $k_1 = k$ and by the recursion

$$k_n(x_1,x_2,\ldots,x_n) = k(x_1 + k_{n-1}(x_2,\ldots,x_n)) \quad . \tag{3.1}$$

Lemma 1: The functions $k_n: R_+^n \to R_+$ are bounded, continuous and nondecreasing satisfying

$$||k_n|| = ||k|| \quad , \quad n \varepsilon N \tag{3.2}$$

where $||\ ||$ is the supremum norm.

Proof. By induction.

Lemma 2: The functions $k_n: R_+^n \to R_+$ satisfy the relation

$$k_n(x_1,x_2,\ldots,x_{n-1},x_n) = k_{n-1}(x_1,x_2,\ldots,x_{n-1}+k(x_n)) \tag{3.3}$$

for $n \geq 2$.

Proof. For n = 2 (3.3) coincides with (3.1). Let n > 2 and assume

$$k_{n-1}(x_2,\ldots,x_n) = k_{n-2}(x_2,\ldots,x_{n-1}+k(x_n)) \quad . \tag{3.4}$$

Inserting (3.4) into (3.1) yields (3.3) after using (3.1) for n-1.

Lemma 3. Let $(x_n)_{n\geq 1}$ be an arbitrary sequence with $x_n \geq 0$, $n \geq 1$. Then the sequence $(y_n)_{n\geq 1}$ with $y_n = k_n(x_1,\ldots,x_n)$ is convergent.

Proof. Lemma 1 and 2 imply

$$0 \leq y_1 \leq y_2 \leq \ldots \leq y_{n-1} \leq y_n \leq \ldots \leq ||k|| \quad . \tag{3.5}$$

Definition: Let R_+^∞ be the set of sequences $\underline{x} = (x_n)_{n\geq 1}$, $x_n \in R_+$, $n \in N$. Define the function k_∞: $R_+^\infty \to R_+$ by

$$k_\infty(\underline{x}) = \lim_{n\to\infty} y_n \tag{3.6}$$

where

$$y_n = k_n(x_1, x_2, \ldots, x_n) \quad . \tag{3.7}$$

In view of Lemma 3 k_∞ is well-defined.

Lemma 4: The function k_∞: $R_+^\infty \to R_+$ satisfies the relation

$$k_\infty(\underline{x}) = k(x_1 + k_\infty(S\underline{x})) \tag{3.8}$$

where the "shift" S: $R_+^\infty \to R_+^\infty$ is defined by

$$\underline{x} = (x_n)_{n\geq 1} \to S\underline{x} = (x_{n+1})_{n\geq 1} \quad . \tag{3.9}$$

Proof. The sequence $(z_n)_{n\geq 1}$ with

$$z_n = k_n(x_2, \ldots, x_{n+1}) \tag{3.10}$$

converges to $k_\infty(S\underline{x})$. From (3.1) (3.7),(3.10) we obtain

$$y_n = k(x_1 + z_{n-1}) \quad . \tag{3.11}$$

As k is continuous the r.h.s. converges to $k(x_1 + k_\infty(S\underline{x}))$ whereas the l.h.s. converges to $k_\infty(\underline{x})$.

<u>Definition</u>: Let $(X_n)_{n\geq 0}$ be a sequence of i.i.d. random variables with values in R_+. For arbitrary $a: (0,T) \to R_+$ define the random variable Y_t by

$$Y_t = a(t)\, X_o + k_\infty(a(t)\underline{X}) \tag{3.12}$$

where $a(t)\underline{X} = (a(t)X_n)_{n\geq 1}$.

<u>Theorem 1</u>: Let G and H_t be the distributions of X_o and Y_t, respectively. Then H_t satisfies the integral equation

$$H_t(x) = \iint_{C_{t,x}} dG(y)dH_t(z) \tag{3.13}$$

where $C_{t,x} \subset R_+^2$ is given by

$$C_{t,x} = \{(y,z) \mid a(t)y + k(z) < x\} \tag{3.14}$$

Proof. The random variable

$$Z_t = a(t)X_1 + k_\infty(a(t)S\underline{X}) \tag{3.15}$$

is independent of X_o and has H_t as distribution. Using (3.8) yields

$$Y_t = a(t)X_o + k(Z_t) \tag{3.16}$$

which leads to (3.13) in view of the independence of the two terms of the r.h.s. of (3.16).

Corollary. The Laplace transforms $\tilde{g}$ and $\tilde{h}_t$ of G and H_t, respectively, satisfy

$$\tilde{h}_t(p) = \tilde{g}(a(t)p) \int_0^\infty \exp(-pk(x))dH_t(x) \quad . \tag{3.17}$$

Subtracting the defining relation

$$\tilde{h}_t(p) = \int_0^\infty \exp(-px)dH_t(x) \tag{3.18}$$

multiplied with $\tilde{g}(a(t))$ from (3.17) yields after division by t

$$t^{-1}[1-\tilde{g}(a(t)p)]\tilde{h}_t(p) = \int_0^\infty \phi_t(x;p)dH_t(x) \tag{3.19}$$

with

$$\phi_t(x;p) = t^{-1} [\exp(-pk(x)) - \exp(-px)] \quad . \tag{3.20}$$

Let now G be in the domain of attraction of the one-sided Lévy distribution F_α, $0 < \alpha < 1$. This is the case iff the asymptotic behaviour of G is given by [2]

$$1-G(x) \sim \frac{x^{-\alpha} L(x)}{\Gamma(1-\alpha)} \quad , \quad x \to \infty \quad . \tag{3.21}$$

Here, L is a slowly varying function [2], i.e. it satisfies

$$\lim_{y\to\infty} \frac{L(yx)}{L(y)} = 1 \quad , \quad x \in R_+ \quad . \tag{3.22}$$

Equivalently, the Laplace transform $\tilde{g}$ of G has the following asymptotic behaviour [2]

$$1-\tilde{g}(p) \sim p^{\alpha} L(p^{-1}) \quad , \quad p \to 0 \quad . \tag{3.23}$$

Define

$$b(x) = \inf \{y \in R_+ \mid x[1-G(y)] \leq \frac{1}{\Gamma(1-\alpha)}\} \quad . \tag{3.24}$$

Then [2]

$$\lim_{x\to\infty} \frac{xL(b(x))}{b(x)^{\alpha}} = 1 \tag{3.25}$$

which leads to [2]

$$\lim_{x\to\infty} x[1-\tilde{g}(\frac{p}{b(x)})] = p^{\alpha} \quad . \tag{3.26}$$

Hence, setting

$$a(t) = \alpha^{1/\alpha} b(t^{-1})^{-1} \tag{3.27}$$

we obtain

$$\lim_{t\downarrow 0} t^{-1}[1-\tilde{g}(a(t)p)] = \alpha p^{\alpha} \quad . \tag{3.28}$$

Remark. For the special case $L(x) = 1$, $x \in R_+$, we obtain explicitly

$$a(t) = (\alpha t)^{1/\alpha} \quad . \tag{3.29}$$

Definition: A sequence $(t(n))_{n\geq 1}$ is called c-sequence if it converges to zero and if the sequence $(H_n)_{n\geq 1}$ with $H_n = H_{t(n)}$ has a limit.

Recall the basic fact of probability theory [2] that there exist c-sequences. The limit H may depend on the choice of the sequence and need not be a proper probability distribution. If it is not proper it is "defective", i.e.

$$\lim_{x\to\infty} H(x) = 1-\rho \quad , \quad 0 < \rho \leq 1 \quad . \tag{3.30}$$

Taking (3.28) into account we obtain (in an obvious notation)

$$\text{c-}\lim_{t\downarrow 0} t^{-1}[1-\tilde{g}(a(t)p)]\tilde{h}_t(p) = \alpha p^{\alpha}\, \tilde{h}(p) \quad . \tag{3.31}$$

We now specify the function $k: R_+ \to R_+$ entering (3.20) by setting

$$k(x) = x[1+\psi_t^m(x)]^{-1} \tag{3.32}$$

where

$$\psi_t^m(x) = \begin{cases} tx & , \quad m = 2 \\ t^{\mu}x \tanh(t^{1-\mu}\, x^{m-2}) & , \quad m > 2 \end{cases} \tag{2.33}$$

with $0 < \mu < 1$ and $t \in (0,T(m))$ with $T(2) = \infty$ and

$$T(m) = \{\frac{1}{m-2}\, y(m)^{-\frac{m-1}{m-2}} \cosh^2 y(m)\}^{\frac{m-2}{\mu(m-1)-1}} \tag{3.34}$$

for $m \geq 3$ where $y(m)$ is the unique solution of

$$y \tanh y = \frac{m-1}{2(m-2)} \quad . \tag{3.35}$$

Lemma 5. For fixed $t \in (0,T(m))$ the function $k: R_+ \to R_+$ defined by (3.32), (3.33) is continuous, bounded and increasing if (for $m \geq 3$)

$$(m-1)^{-1} < \mu < 1 \quad . \tag{3.36}$$

The supremum norm is given by

$$||k|| = \lim_{x\to\infty} k(x) = t^{-\mu} \tag{3.37}$$

Proof. By elementary calculus.
Define $\phi(\cdot;p):R_+ \to R$ by

$$\phi(x;p) = e^{-px}\, px^m \quad , \tag{3.38}$$

Note that this is the pointwise limit of $\phi_t(x;p)$ defined in (3.20) as t tends to zero for the choice (3.32),(3.33) of k.

Lemma 6: $\phi_t(\cdot;p)$ tends in supremum norm to $\phi(\cdot;p)$ for $p > 0$.

Proof. By investigating the extrema of the difference for small t one finds for $m = 2$ and $m \geq 4$

$$\begin{aligned} \lim_{t\downarrow 0} t^{-1}\, ||\phi_t(\cdot;p) - \phi(\cdot;p)|| &= p^{2-2m}\, e^{-c}\, c^{2m-1}(\tfrac{c}{2} - 1) \\ c &= m+1 + [(m-1)^2 + 2]^{1/2} \end{aligned} \tag{3.39}$$

and for $m = 3$ ($\frac{1}{2} < \mu < 1$!)

$$\lim_{t\downarrow 0} t^{2\mu-2}\, ||\phi_t(\cdot;p) - \phi(\cdot;p)|| = p^{-4}\, e^{-5}\, 5^{5/3} \quad . \tag{3.40}$$

The l.h.s. of (3.19) has a limit given by (3.31) as t tends to zero (through a c-sequence). Hence, the r.h.s. has a limit. Anticipating the interchangeability of limit and integration (dealt with in Section 4) we obtain for this limit

$$\text{c-}\lim_{t\downarrow 0} \int \phi_t(x;p)dH_t(x) = \int_0^\infty e^{px}\, px^m\, dH_t(x) \quad . \tag{3.41}$$

Combining (3.41) with (3.31) we obtain

$$p(-\frac{d}{dp})^m \tilde{h}(p) = \alpha p^\alpha \tilde{h}(p) \quad . \tag{3.42}$$

In view of the results of Section 1 we conclude that

$$\text{c-}\lim_{t\downarrow 0} H_t = H = AF_{m,\alpha} \tag{3.43}$$

with $0 \leq A \leq 1$, possibly depending on the choice of the c-sequence.

Assuming G to have no atom at the origin it is shown in Section 4 that the limit is non-defective (i.e. $A = 1$) for all c-sequences. Together with a topological lemma (lemma 7 Section 4) this implies

$$\lim_{t\downarrow 0} H_t = F_{m,\alpha} \quad . \tag{3.44}$$

With this result we have achieved the goal set at the beginning of this section.

Remark. One may ask whether also $F_\alpha = F_{1,\alpha}$ (i.e. the ordinary one-sided Lévy distributions) occurs as limit law in an analogous way as $F_{m,\alpha}$, $m \geq 2$. In view of (3.32), (3.33) an obvious choice for k is

$$k(x) = x(1+t)^{-1} \tag{3.45}$$

leading via (3.1), (3.6), (3.7) to

$$k_\infty(\underline{x}) = \sum_{n=1}^{\infty} x_n(1+t)^{-n} \quad . \tag{3.46}$$

Note that k is continuous and increasing but not bounded, hence a proof will require some modifications. From (3.12), (3.46) we obtain

$$Y_t = a(t) \sum_{n=0}^{\infty} X_n \ (1+t)^{-n} \quad . \tag{3.47}$$

For infinite sums Kolmogorov's "three series theorem" [2] is relevant. I conjecture that Y_t tends in distribution to F_α as t tends to zero if the distribution G of X_o is in the domain of attraction of F_α and if a(t) is chosen according to (3.24), (3.27).

4. TECHNICALITIES

Some steps were left out in Section 3. These gaps will be filled in here. We start with a topological lemma of a general nature.

Lemma 7. Let X be a sequentially compact Hausdorff space and f_o a map of the interval (0,a), a > 0, into X. A sequence $(t_n)_{n\geq 1}$, $t_n \in (0,a)$, converging to zero is called c-sequence if the sequence $(x_n)_{n\geq 1}$ of the images $x_n = f_o(t_n)$ is convergent. If all c-sequences have the same limit, say x, then the extension f of f_o with f(0) = x and $f(t) = f_o(t)$, $t \in (0,a)$, is continuous at t = 0. In particular, for any sequence (t_n) in (0,a) converging to zero the sequence of images $(f(t_n))$ converges to x.

Proof. Assume f not to be continuous at t = 0. There exists a neighborhood U of x and a sequence (t'_k) converging to zero with $f(t'_k) \notin U$. As X is sequentially compact there exists a subsequence $(t'_{k(n)}) \equiv (t_n)$ which is a c-sequence with $(f(t_n))$ converging to $y \neq x$, a contradiction.

Let M be the set of all finite measures μ on R_+ with $\mu(R_+) \leq 1$ and M^1 the subset with $\mu(R_+) = 1$. Provided with the vague topology [12] M is compact and Hausdorff. We denote by J the bijection of M onto the set of distribution functions $F:R_+ \to [0,1]$ ($F(0) = 0$, F left-continuous and non-decreasing, $F(\infty) = A \leq 1$); explicitly $(J\mu)(x) = \mu([0,x))$.

If we can show that the limit H in (3.43) is independent of the chosen c-sequence we may apply Lemma 7 to the map $f_o:(0,T(m)) \to M$ with $f_o(t) = \mu_t = J^{-1} H_t \in M$.

Before doing so we have to justify (3.41) where integration and c-limit were interchanged.

Let C_f denote the set of all bounded continuous functions $f:R_+ \to R$ with a finite value c at infinity (i.e. $|f(x)-c|$ arbitrarily small for x sufficiently large). The subset C_o of C_f consists of those f with $c = 0$. Convergence in C_f is defined with respect to the supremum norm.

Lemma 8. Let $(\mu_n)_{n\geq 1}$ be a converging sequence in M with limit μ and $(f_n)_{n\geq 1}$ a sequence in C_f converging to $f \in C_o$. Then limit and integration may be interchanged:

$$\lim_{n\to\infty} \int f_n \, d\mu_n = \int f \, d\mu \quad . \tag{4.1}$$

Proof. See [13] (note that in Lemma 1 of [13] the sequence was erroneously assumed to be in C_o instead of C_f; the proof remains unaffected).

Corollary. If $(\mu_n)_{n\geq 1}$ and its limit μ are in M^1 then (4.1) remains true for $(f_n)_{n\geq 1}$ converging to $f \in C_f$.

Proof. Set $c = f(\infty)$. As $\mu_n \in M^1$ we have

$$\int f_n \, d\mu_n = \int (f_n - c) d\mu_n + c \tag{4.2}$$

The first term of the r.h.s. fulfils the requirements of Lemma 8. Taking the limit and using $\mu \in M^1$ yields the result.

Remark. The function $\phi_t(\cdot;p)$ defined in (3.20) with k given by (3.32), (3,33) is in C_f as

$$\lim_{x\to\infty} \phi_t(x;p) = t^{-1} \exp(-pt^{-\mu}) \tag{4.3}$$

(with $\mu = 1$ for $m = 2$). The function $\phi(\cdot;p)$ defined in (3.38) belongs to C_o. Hence Lemma 8 is applicable to (3.41).

To exclude the possibility of a defective limit H one could try to show directly that the family (Y_t), $0 < t < T(m)$, of random variables is stochastically bounded [2]. This approach has been unsuccessful so far. An indirect proof will be given below based on an additional assumption on the distribution G in the integral equation (3.13).

Let M^1_o be the subset of M^1 consisting of the probability measures μ without atom at the origin ($\mu(\{0\}) = 0$). The additional assumption is $J^{-1} G \in M^1_o$. From (3.13), (3.14) it follows that also $J^{-1} H_t \in M^1_o$. Defining $\theta\mu$ by

$$(\theta\mu)([0,x)) = 1-\mu((x^{-1},\infty)) \tag{4.4}$$

we obtain an involution of M^1_o. A sequence $(\mu_n)_{n\geq 1}$ in M^1_o converging in M to μ is mapped onto the sequence $(\theta\mu_n)_{n\geq 1}$ in M^1_o converging to $\mu^* \in M^1$ with an atom at the origin whose weight is given by the defect of μ.

The involution θ transforms (3.13) into the following integral equation

$$T_t(x) = \iint_{D_{t,x}} dR(y)dT_t(z) \tag{4.5}$$

with $D_{t,x} \subset R_+^2$ given by

$$D_{t,x} = \{(y,z) \mid [a(t)y^{-1} + k(z^{-1})]^{-1} < x \} \quad . \tag{4.6}$$

The distributions R and T_t are related to G and M_t by

$$R = J\,\theta\,J^{-1}\,G \quad , \qquad T_t = J\,\theta\,J^{-1}\,H_t \quad . \tag{4.7}$$

Applying the Stieltjes transform

$$(S\mu)(q) = \int_0^\infty (x+q)^{-1}\,d\mu(x) \quad , \quad q > 0 \quad , \quad \mu \in M \quad , \tag{4.8}$$

to (4.5) yields

$$\int_0^\infty K_t(x,q)dT_t(x) = 0 \tag{4.9}$$

with

$$tK_t(x,q) = \frac{\phi}{(x+q)(x+q+\phi)} - \frac{x+\phi}{(x+q+\phi)^2}\,a(t)\chi$$

$$\chi = (SR)(a(t)\,\frac{q(x+\phi)}{x+q+\phi})$$

$$\phi = \begin{cases} t & , \quad m = 2 \\ t^\mu \tanh(t^{1-\mu}\,x^{2-m}) \;, & m > 2 \end{cases} \quad . \tag{4.10}$$

Using

$$\lim_{t\downarrow 0} t^{-1}\phi = x^{2-m} \tag{4.11}$$

and

$$\lim_{t\downarrow 0} t^{-1}\, a(t)(SR)(a(t)w) = \alpha\Gamma(1+\alpha)w^{\alpha-1} \tag{4.12}$$

we obtain for x > 0 pointwise

$$\lim_{t\downarrow 0} K_t(x,q) = K(x,q) \tag{4.13}$$

with

$$K(x,q) = (x+q)^{-2}\, x^{2-m} - \alpha\Gamma(1+\alpha)x^{\alpha+1}\, q^{\alpha-1}(x+q)^{-\alpha-1} \quad . \tag{4.14}$$

In deriving (4.12) we made use of (L = Laplace transform)

$$(SR)(q) = q^{-2}\, L(1-LG)(q^{-1}) \quad . \tag{4.15}$$

Together with (3.23), (3.25) and Theorem 6 (Chap. 14, § 1) [9] this yields (4.12).

Remark. The functions $g_{m,\alpha}$ introduced in Section 1 satisfy the integral equation

$$\int_0^\infty K(x,q)g(x)dx = 0 \quad . \tag{4.16}$$

This is seen by applying the Mellin transform to (4.16) and using the explicit form (4.14) of K. The result is the difference equation (2.12).

For m = 2 the functions K_t and K are in C_f and K is the norm limit of K_t [13]. Hence, by the corollary to Lemma 8 taking the c-limit of (4.9) yields

$$\int_0^\infty K(x,q)dT(x) = 0 \tag{4.17}$$

with

$$J^{-1} T = \operatorname*{c-lim}_{t\downarrow 0} \theta J^{-1} H_t \quad . \tag{4.18}$$

If the c-limit H of H_t were defective J^{-1} T would have an atom at the origin. As H is given by (3.43) with A < 1 the remaining part of T has the density $Ag_{2,\alpha}$ which satisfies (4.16). Hence, T would satisfy

$$\int_0^\infty K(x,q)dT(x) = (1-A)K(0,q) \neq 0 \tag{4.19}$$

(note $K(0,q) = q^{-2}$ for m = 2), which contradicts (4.17).

Similarly, for m $\geq$ 3 the divergence of K(x,q) as x approaches zero has precursors which prevent T_t accumulating weight near the origin as t tends to zero. Thus the c-limit of H_t cannot be defective.

REFERENCES

[1] Lévy, P.: 'Théorie de l'additon des variables aléatoires'. Paris: Gauthier-Villars 1954.

[2] Feller, W.: 'An introduction to probability theory and its applications', Vol. II. New York: John Wiley 1971.

[3] Gnedenko, B.V., Kolmogorov, A.N. : 'Limit distributions for sums of independent random variables'. Reading: Addison Wesley 1954.

[4] Schneider, W.R. : 'Generalized one-sided stable distributions'. Proceedings of the second BiBoS-Symposium. Albeverio, S., Blanchard, Ph., Streit, L., (eds.). Lecture notes in mathematics. Berlin: Springer (1986).

[5] Schneider, W.R.: 'Stable distributions: Fox function representation and generalization'. First Ascona-Como international conference (1985): 'Stochastic processes in classical and quantum systems'. To appear in Lecture notes in physics. Berlin: Springer (1986).

[6] Fox, C.: 'The G and H Functions as symmetrical Fourier kernels'. Trans. Amer. Math. Soc. 98, 395-429 (1961).

[7] Braaksma, B.L.J.: 'Asymptotic expansions and analytic continuations for a class of Barnes-integrals'. Compos. Math. 15, 239-341 (1963).

[8] Srivastava, H.M., Gupta, K.C., Goyal, S.P.: 'The H-functions of one and two variables with applications'. New Delhi, Madras: South Asian Publishers, 1982.

[9] Doetsch, G.: 'Handbuch der Laplace-Transformation'. Basel, Stuttgart: Birkhäuser 1971.

[10] Whittaker, E.T., Watson, G.N.: 'A course of modern analysis'. Cambridge: Cambridge University Press 1980 (reprint of fourth edition 1927).

[11] Titchmarsch, E.C.: 'The theory of functions'. London: Oxford University Press 1960.

[12] Bauer, H.: 'Wahrscheinlichkeitstheorie und Grundzüge der Masstheorie'. Berlin, New York: Walter de Gruyter 1974.

[13] Schneider, W.R.: 'Rigorous scaling laws for Dyson measures'. In: Stochastic Processes - Mathematics and Physics. Proceedings of the first BiBoS-Symposium. Albeverio, S., Blanchard, Ph., Streit, L., (eds.). Lecture notes in mathematics, Vol. 1158, Berlin: Springer (1986).

THE OSCILLATOR REPRESENTATION OF THE METAPLECTIC GROUP APPLIED TO QUANTUM ELECTRONICS AND COMPUTERIZED TOMOGRAPHY

Walter Schempp
Lehrstuhl fuer Mathematik I
University of Siegen
D-5900 Siegen
Federal Republic of Germany

ABSTRACT. Starting from two basic postulates which are motivated by the radial symmetry of the transverse energy patterns of graded-index optical fibers and the quantum mechanical treatment of the laser oscillations, this paper deals with a unified approach to the quantized transverse eigenmode spectrum of circularly symmetric and rectangular multimode parabolic-index optical waveguides. The approach is based on the topologically irreducible, continuous, unitary, linear representations of the diamond solvable Lie group $D(\mathbf{R})$ which forms an extension of the real Heisenberg two-step nilpotent Lie group $\tilde{A}(\mathbf{R})$ by the one-dimensional compact torus group $\mathbf{T}$. As an application, the coupling coefficients of transverse eigenmodes in coaxial circular and rectangular laser resonators and optical waveguides are computed in terms of Krawtchouk polynomials via the smallest real three-step nilpotent Lie group $B(\mathbf{R})$. Finally, as an application to computerized axial tomography, an extension of the group theoretical method is indicated to establish a singular value decomposition for the classical Radon transform $\mathfrak{R}$ in n-dimensional Euclidean space $\mathbf{R}^n$ $(n \geq 2)$ via the discrete spectrum of the reductive dual pair $(\tilde{O}(n,\mathbf{R}), \tilde{Sp}(1,\mathbf{R}))$. As a consequence, the symbolic calculus furnishes a new group theoretical proof of the inversion formula for $\mathfrak{R}$ and the dual inversion formula for the back-projector ${}^t\mathfrak{R}$.

1. Optical Fiber Communication

Lightwave electronics, including optical communication via silica fibers and optoelectronic devices, has become one of the most promising fields of applied physics and electrical engineering since the laser first appeared.

S. Albeverio et al. (eds.), Stochastic Processes in Physics and Engineering, 305–344.

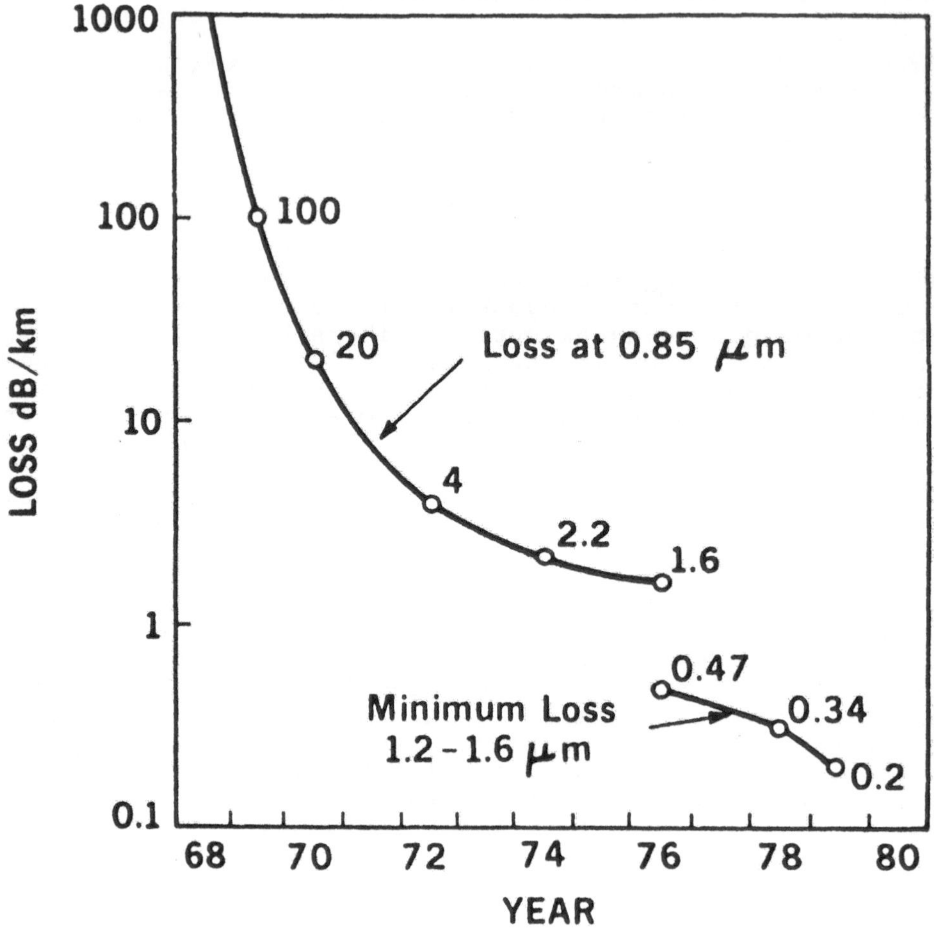

Fiber losses as reported in the past 10 years

The overwhelming advantages of optical fiber communication are

- extreme low loss of the optical signals over a wide range of wavelengths (less than 1 dB/km, corresponding to a 25% loss per km)
- immense bandwidth (1 and 100 GHz, respectively, for multimode and single-mode fibers over 1 km) that makes it possible to use extreme short pulses

As a consequence, longer repeater spacings and a higher information capacity are available. Further advantages are

- immunity to electromagnetic interference
- natural abundance of raw materials
- small size (silica fibers have a diameter of 10-100 μm and a total diameter of about 1-2 mm inclusive the coatings)
- low weight
- lower cost than the traditional media (such as coaxial cables or hollow metallic waveguides)

Characteristic of the progress in lightwave communication technology is the enormous reduction of the transmission loss of optical fibers accomplished in the last decade as illustrated by the diagram.

One of the most important factors that helped make optical fiber communication a reality is the invention of the light-emitting diode (LED) and the semiconductor injection laser. These are the only lasers pumped directly by an electric current. They are consequently easily incorporated into the electronic circuitry used in the communication technology. In addition, by cleaving and coupling the cavities of semiconductor injection lasers in an appropriate way, the newly developed device of C^3 laser produces a monochromatic output radiation of wavelengths ($0.8\ \mu m < \lambda < 1.6\ \mu m$) that fall in the optimal (low loss, low group velocity dispersion) transmission region of the silica fibers and allows therefore to exploit completely the high information capacity of modern broad-band optical fiber transmission systems. Engineers now believe that in near future the optical fiber coupled to semiconductor injection lasers and LEDs will change dramatically the entire aspect of terrestrial communication.

2. Eigenmode Patterns (Experimental Results)

The photograph below shows the intensity pattern of a higher transverse eigenmode excited on a round silica graded-index optical fiber about 85 μm in core diameter. Moreover, the output energy pattern of a transverse eigenmode of a laser resonator with quadratic cross-section is also displayed in the photograph below.

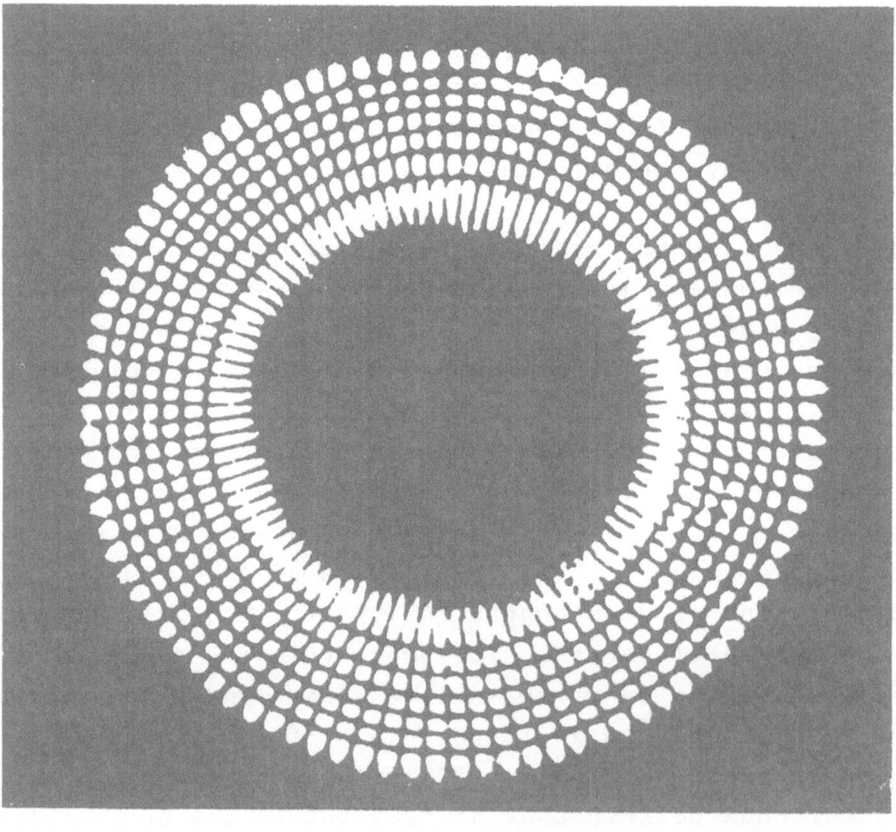

Intensity pattern of a higher transverse eigenmode in a circular waveguide

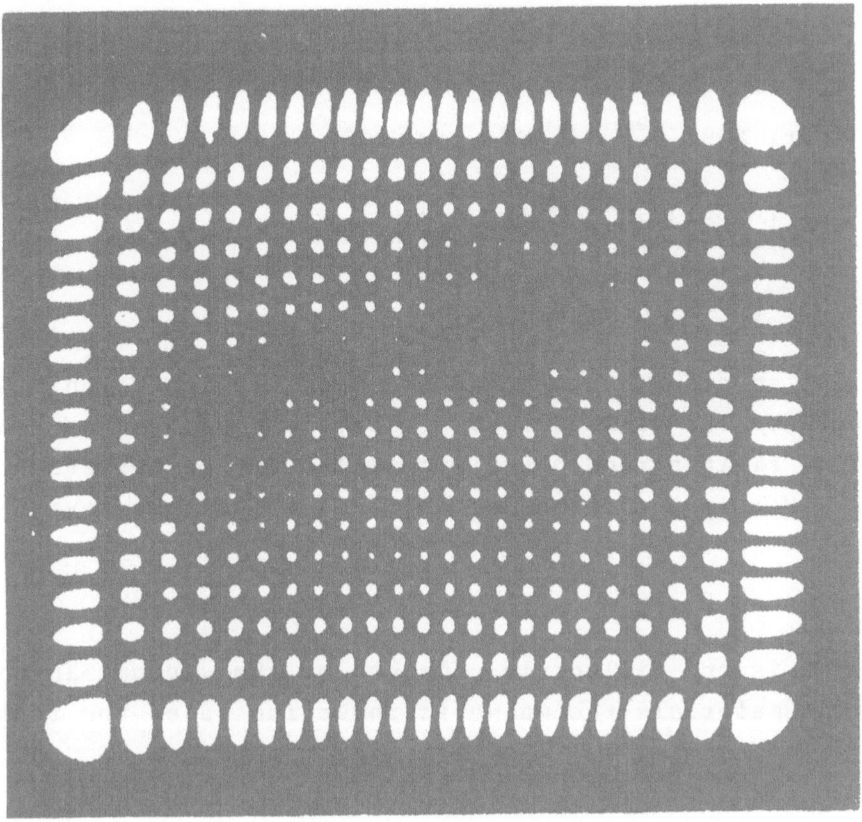

Intensity pattern of a higher transverse eigenmode
of a rectangular laser

3. The Semiclassical and Quantum Theory of the Laser

The semiclassical laser theory explains and even predicts various properties of laser light. It deals with the interaction of laser radiation with laser active atoms in gases or solids by assuming that the light field is a classical electromagnetic field obeying Maxwell's equations, whereas the motion of the electrons within the atoms is treated by means of quantum theory. Specifically, the semiclassical theory of the laser starts with the differential form of

Maxwell's equations for the electromagnetic radiation in the laser cavity

$$\mathrm{curl}\; E = -\frac{\partial B}{\partial t} \quad \text{(induction law)},$$

$$\mathrm{curl}\; H = J + \frac{\partial D}{\partial t} \quad \text{(Oersted's law)},$$

$$\mathrm{div}\; B = 0.$$

The last equation states that the magnetic induction B has no sources. The dielectric displacement D depends on the electric field strength E via the electric polarization density P (dipole moment per unit volume) of the cavity medium

$$D = \varepsilon_o E + P$$

where ε_o is the dielectric constant of the vacuum. In non-magnetic materials the magnetic induction takes the form

$$B = \mu_o H$$

where μ_o denotes the magnetic susceptibility of the vacuum. Ohm's law states that the current density J takes the form

$$J = \sigma E$$

where σ denotes the electric conductivity. Put $\varepsilon_o \mu_o = \frac{1}{c^2}$.

It is well known that the constant c has the dimension of a velocity and turns out to be numerically equal to the velocity of light in vacuum. If we differentiate Oersted's law with respect to time t and express $\frac{\partial H}{\partial t}$ by curl E, we get by the transversality condition for the electric field

$$\mathrm{div}\; E = 0$$

the linear partial differential equation

$$\Delta_3 E - \frac{1}{c^2}\frac{\partial^2 E}{\partial t^2} + \mu_0 \sigma \frac{\partial E}{\partial t} = \mu_0 \frac{\partial^2 P}{\partial t^2}$$

which reduces in the case $P = 0$ (no total dipole moment) to the telegraph equation. In the case $\sigma = 0$ (no current) the telegraph equation reduces to the homogeneous wave equation for the electric field strength

$$\Delta_3 E = \frac{1}{c^2}\frac{\partial^2 E}{\partial t^2}$$

Thus classical Maxwell's equations describe the light field as electromagnetic waves. To solve for the fields in dielectric waveguides the procedure followed in classical electromagnetics is to couple the solutions of the wave equations together by means of boundary conditions. This method is particularly well suited for describing the unguided radiation modes and the evanescent field in the cladding far from the core. There are several instances, however, for which a classical field theory fails to give experimentally observed results, whereas a quantized field succeeds. This is true, for instance, of spontaneous emission of light, a phenomenon which is a typical quantum mechanical process. Since the semiclassical theory cannot treat this process which really depends upon the photon nature of light, it becomes necessary to develop a completely quantum mechanical theory of the laser source. This is done by means of quantization of the electromagnetic field by introducing creation and annihilation operators obeying the canonical commutation relations in a purely formal analogy between the Hamiltonian operator of the standing electric wave (= mode Hamiltonian) and the standard Hamiltonian of the quantum mechanical harmonic oscillator. The quantization procedure allows to express the electric field strength, the magnetic induction, the vector potential, and the total energy of the field by the creation and annihilation operators which are, from the mathematical point of view, the essential ingredients of the Bargmann-Fock-Segal realization (and the coherent state model) of the linear Schrödinger representation U_1 of the real Heisenberg nilpotent Lie group $\tilde{A}(\mathbb{R})$. On the other hand, since classical Maxwell's equations form a macroscopic theory dealing with fields which vary in space by a discernible amount over distances large compared with atomic dimensions, i.e., a theory concerned with an average field and not with the field within the local vicinity of an atom, the procedure of quantization the Maxwell's equations cannot describe, for instance, the propagating modes in the active stripes of the very closely coupled two-cavity resonators of a C^3 laser (cleaved-coupled-cavity semiconductor injection laser) in an absolutely satisfactory manner. Therefore we avoid the classical Maxwell's equations completely and start our group representational approach to a unified treatment of the transverse eigenmodes of laser resonators and graded-index optical fibers from two simple postulates based on (i) a symmetry property expressing the geometry of the laser cavity or dielectric waveguide and (ii) the canonical commutation relations of quantum mechanics. Specifically, we will realize the canonical commutation relations by the real Heisenberg Lie algebra $\mathfrak{n}$. Thus we will have recourse to the essentially unique, topologically irreducible, unitary, linear representation U_1 of $\tilde{A}(\mathbb{R})$ which determines a

unique procedure of quantizing the quadratic polynomials and certainly not a method to quantize the polynomials of higher degree than two. In order to escape potential objections against the group theoretical method within the range of quantization just described, let us quote the preface of Eugen Wigner's classic "Gruppentheorie und ihre

Anwendung auf die Quantenmechanik der Atomspektren". Professor E.P. Wigner writes in 1931: "Man hat gegen die gruppentheoretische Behandlung der Schrödingergleichung oft den Einwand erhoben, daß sie 'nicht physikalisch' sei. Es scheint mir aber, daß die bewußte Ausnutzung elementarer Symmetrieeigenschaften dem physikalischen Gefühl eher entsprechen muß, als die mehr rechnerische Behandlung. Der erwähnte Einwand dürfte darauf zurückzuführen sein, daß die Gruppentheorie einen wesentlich anderen Charakter hat, als die dem Physiker hauptsächlich geläufigen Teile der Mathematik, so daß es immerhin einige Zeit erfordert, bis man sich mit ihr befreundet hat". - Since the representation theory of Lie groups has become a highly specialized and sophisticated discipline of modern mathematics, the aforementioned statement of Professor Wigner continues to remain valid without any reservation and modification.

4. The Aim of the Paper

The purpose of the paper is to present a unified approach to the quantized transverse eigenmodes of circularly symmetric and rectangular optical waveguides. The former are important for round optical fibers and gas lasers with cylindrical resonators, whereas the latter are relevant for integrated optics devices like planar dielectric-clad slab waveguides, LEDs and semi-conductor injection lasers. We start our investigations of the guided radiation in graded-index optical fibers by assuming the following two simple postulates:

(i) The optical fiber is formed by the circular cylinder $\mathbf{T} \times \mathbf{R}$,

(ii) the standard basis $\{X_0, Y_0\}$ of each plane $\mathbf{R} \oplus \mathbf{R}$ transverse to the optical fiber satisfies the canonical commutation relations.

The conditions (i) and (ii) suggest to look at an extension $D(\mathbb{R})$ of the real Heisenberg two-step nilpotent Lie group $\tilde{A}(\mathbb{R})$ by the one-dimensional compact torus group $\mathbb{T}$. In Section 9 infra we will give an explanation of the precise meaning of the postulates (i) and (ii) supra by identifying

(i') the circular cylinder $\mathbb{T} \times \mathbb{R}$ with the coadjoint stabilizer $\mathbb{T} \times \tilde{\mathbb{C}}$ in the <u>diamond</u> <u>solvable</u> <u>Lie</u> <u>group</u> $D(\mathbb{R})$ of the linear form

$$\lambda Z_0^*: aX_0+bY_0+cZ_0 \rightsquigarrow \lambda c \qquad (\lambda \neq 0)$$

on the three-dimensional real Heisenberg Lie algebra $\mathfrak{n}$ with center $\mathfrak{z} = \mathbb{R}Z_0$,

and by identifying

(ii') each plane $\mathbb{R} \oplus \mathbb{R}$ transverse to the optical fiber axis with the kernel $\mathfrak{n}/\mathfrak{z}$ of λZ_0^*.

The main tools of the group representational approach are the <u>Wigner</u> <u>distribution</u> in its <u>dual</u> version ("the ambiguity functions") and the harmonic analysis on $D(\mathbb{R})$ which is based on the Mackey machinery and the Kirillov coadjoint orbit method. As an application we calculate the coupling coefficients of the quantized transverse eigenmodes in coaxial (<u>circular</u> and <u>rectangular</u>) laser resonators and optical waveguides in terms of Krawtchouk polynomials by an application of a suitable topologically irreducible, continuous, unitary, linear representation of the smallest real three-step nilpotent Lie group $B(\mathbb{R})$. A particularly interesting phenomenon is the <u>parity</u> <u>preservation</u> under the coupling of transverse eigenmodes in the rectangular case. Finally, we replace the circular cylinder $\mathbb{T} \times \mathbb{R}$ which represents according to (i) the optical fiber by the spherical cylinder $S_{n-1} \times \mathbb{R}$ and compute by an application of the oscillator representation of the metaplectic group **Mp**(n,**R**) the singular value decomposition of the classical Radon transform $\mathcal{R}$ in n-dimensional Euclidean space $\mathbf{R}^n$ ($n \geq 2$). From this result we may read off the inversion formula

$$\Delta_n^{(n-1)/2}({}^t\mathcal{R}\circ\mathcal{R}) = c_n . \mathrm{id} \qquad (c_n \in \mathbf{C})$$

for $\mathcal{R}$ as well as the dual inversion formula for the back-projector ${}^t\mathcal{R}$. Since our reasoning is based on the spectral theory of the reductive dual pair ($\hat{\mathbf{O}}(n,\mathbf{R}),\widetilde{\mathbf{SL}}(2,\mathbf{R})$) in **Mp**(n,**R**), which represents the natural framework for dealing with the transforms $\mathcal{R}$ and ${}^t\mathcal{R}$, the present paper forms another step in the research program [9] proposed to the 14th ICGTMP.

Acknowledgment. The manuscript represents an expanded version of a lecture presented to the First International Conference on the 'Physics of Phase Space' in College Park, Maryland, May 1986. The author is grateful to Professor E.P. Wigner for his interest and his encouragement.

5. The Wigner Distribution and the Ambiguity Functions

Correlation is a fundamental signal processing operation which is extensively used for delay estimation, signal-to-noise enhancement, pulse compression, and spectrum analysis in radar, lidar, sonar, and communication systems. To start with, let $\mathcal{S}(\mathbf{R})$ be the complex Schwartz space of all infinitely differentiable complex-valued functions ψ on the real line **R** which, together with their derivatives of all orders, vanish at infinity faster than any inverse power of the distance from the origin. For any waveform $\psi \in \mathcal{S}(\mathbf{R})$ introduce the associated Wigner distribution function

$$P(\psi;q,p) = \int_{\mathbf{R}} \psi(q+\tfrac{1}{2}s)\overline{\psi}(q-\tfrac{1}{2}s)e^{-2\pi ips}ds$$

We have

$$P(\psi;.,.) \in \mathcal{S}(\mathbf{R} \oplus \mathbf{R}).$$

Form the Fourier cotransform

$$H(\psi;x,y) = \overline{\mathcal{F}}_{\mathbf{R}^2} P(\psi;p,q) = \int_{\mathbf{R}^2} P(\psi;q,p)e^{2\pi i(qx+py)}dqdp$$

$$= \int_{\mathbf{R}} \psi(t+\tfrac{1}{2}x)\overline{\psi}(t-\tfrac{1}{2}x)e^{2\pi iyt}dt,$$

where $(x,y) \in \mathbf{R} \oplus \mathbf{R}$ are the dual coordinates of $(p,q) \in \mathbf{R} \oplus \mathbf{R}$. In the field of radar, the auto-correlation function $H(;.,.)$ is called to be the <u>auto-ambiguity function</u> (x = time delay, y = Doppler frequency shift). If $\varphi \in \mathcal{S}(\mathbf{R})$, the <u>cross-ambiguity function</u> is defined according to the prescription

$$H(\psi,\varphi;x,y) = \int_{\mathbf{R}} \psi(t+x)\overline{\varphi}(t)e^{2\pi iyt}\, dt$$

(asymmetric form). There are ambiguity function processors available which form over the time-frequency detector plane the cross-correlation $H(\psi,\varphi;.,.)$ between a given pair $(\psi,\varphi) \in \mathcal{S}(\mathbf{R}) \times \mathcal{S}(\mathbf{R})$ of wave functions.

<u>6. The Real Heisenberg Two-Step Nilpotent Lie Group</u> $\widetilde{A}(\mathbf{R})$

Shortly after Heisenberg introduced the canonical commutation relations in quantum mechanics, which underlie the uncertainty principle, Weyl [12] showed they could be interpreted as the structure relations for the Lie algebra $\mathfrak{n}$ of the three-dimensional, real, connected, simply connected, two-step nilpotent Lie group $\tilde{A}(\mathbf{R})$ formed by the unipotent matrices

$$\begin{pmatrix} 1 & x & z \\ 0 & 1 & y \\ 0 & 0 & 1 \end{pmatrix} = (x,y,z) \in \mathbf{R}^3$$

$\tilde{A}(\mathbf{R})$ is called the real Heisenberg nilpotent Lie group. Its Lie algebra, the three-dimensional real Heisenberg Lie algebra $\mathfrak{n}$, satisfies

$$\tilde{A}(\mathbf{R}) = \exp(\mathfrak{n})$$

and is spanned over $\mathbf{R}$ by the nilpotent matrices

$$X_0 = \begin{pmatrix} 0 & 1 & 0 \\ 0 & 0 & 0 \\ 0 & 0 & 0 \end{pmatrix}, \quad Y_0 = \begin{pmatrix} 0 & 0 & 0 \\ 0 & 0 & 1 \\ 0 & 0 & 0 \end{pmatrix}, \quad Z_0 = \begin{pmatrix} 0 & 0 & 1 \\ 0 & 0 & 0 \\ 0 & 0 & 0 \end{pmatrix}.$$

The matrix Z_0 which appears on the right hand side of the commutation relations

$$[X_0, Y_0] = X_0 Y_0 - Y_0 X_0 = Z_0$$

spans the center $\mathfrak{z}$ of $\mathfrak{n}$ over $\mathbf{R}$. All the other Lie bracket products of
the standard basis $\{X_0, Y_0, Z_0\}$ of $\mathfrak{n}$ vanish. Therefore $\mathfrak{z}$ is the commutator ideal of $\mathfrak{n}$ and consequently $\tilde{A}(\mathbf{R})$ is a two-step nilpotent Lie group. It is precisely the preceding Lie algebraic form of the canonical commutation relations for the standard basis $\{X_0, Y_0\}$ of the plane $\mathfrak{n}/\mathfrak{z}$ transverse to the optical fiber axis which we referred to in the postulate (ii) of Section 4 supra.

Let $\mathfrak{n}^*$ denote the dual vector space of $\mathfrak{n}$ over $\mathbf{R}$. Then the coadjoint action of $\tilde{A}(\mathbf{R})$ in $\mathfrak{n}^*$ with respect to the dual basis $\{X_0^*, Y_0^*, Z_0^*\}$ of $\{X_0, Y_0, Z_0\}$ is given by the matrix

$$\begin{pmatrix} 1 & 0 & y \\ 0 & 1 & -x \\ 0 & 0 & 1 \end{pmatrix}.$$

It follows that the Kirillov coadjoint orbit picture of $\tilde{A}(\mathbf{R})$ is formed by the single point orbits

$$\mathcal{O}_{\mu,\nu} = \{(\mu,\nu,0)\} \quad (\mu,\nu) \in \mathbf{R} \oplus \mathbf{R}$$

and the affine planes

$$\mathcal{O}_{\lambda} = \{(\mu,\nu,\lambda) \in \mathfrak{n}^* \mid (\mu,\nu) \in \mathbf{R} \oplus \mathbf{R}\}$$

passing through the points λZ_0^* ($\lambda \in \mathbf{R}$, $\lambda \neq 0$). The Kirillov correspondence shows that there exist exactly two kinds of topologically irreducible, continuous, unitary, linear representations of $\tilde{A}(\mathbf{R})$, namely

- one-dimensional representations (characters) corresponding to $\mathcal{O}_{\mu,\nu}$,
- infinite-dimensional representations $(U_\lambda)_{\lambda \in \mathbf{R}^\times}$ corresponding to $\mathcal{O}_\lambda$.

Up to isomorphy, the family $(U_\lambda)_{\lambda \in \mathbf{R}^\times}$ acting on the complex Hilbert space $L^2(\mathbf{R})$ is given by

$$U_\lambda(x,y,z)\psi(t) = e^{2\pi i \lambda(z+yt)}\psi(t+x) \qquad (\psi \in \mathcal{S}(\mathbf{R})).$$

The restriction of U_λ to the center

$$\tilde{C} = \{(0,0,z) \mid z \in \mathbf{R}\} = \exp(\mathfrak{c})$$

is the character $e_\lambda : z \rightsquigarrow e^{2\pi i\lambda z}$ of $\tilde{C}$. Moreover, we have

$$H(\psi,\varphi;x,y) = \langle U_1(x,y,0)\psi|\varphi\rangle \qquad (\psi,\ \varphi \in \mathcal{S}(\mathbf{R}))$$

for all $(x,y) \in \mathbf{R} \oplus \mathbf{R}$. Thus the cross-ambiguity function H is given by the coefficient function of the <u>linear Schrödinger representation</u> U_1 of $\tilde{A}(\mathbf{R})$ modulo $\tilde{C}$.

According to the Stone-von Neumann-Segal theorem, U_1 is up to isomorphy the <u>unique</u> topologically irreducible, continuous, unitary, linear representation of $A(\mathbf{R})$ with the basic character $e_1 : z \rightsquigarrow e^{2\pi iz}$ of $\tilde{C}$ as <u>central character</u>, and this representation of $\tilde{A}(\mathbf{R})$ is <u>square integrable modulo</u> $\tilde{C}$.

There are many rather different looking ways of realizing the linear Schrödinger representation U_1 of $\tilde{A}(\mathbf{R})$. We should mention the following realizations:

- the Bargmann-Fock model (which is closely related to the coherent state representation)
- the isotypic component of the right regular representation of $\tilde{A}(\mathbf{R})$ acting in the representation space of $\mathrm{Ind}_{\tilde{C}}^{\tilde{A}(\mathbf{R})} e_1$
- the lattice representation

This richness of harmonic analysis on $\tilde{A}(\mathbf{R})$ adds greatly to its applicability inside and outside mathematics.

<u>7. The Oscillator Representation</u>

Let $\lambda \neq 0$ denote a fixed real number. The kernel of the linear form $\lambda Z_0^* \in \mathfrak{n}^*$ can be identified with the real plane $\mathfrak{n}/\mathfrak{z}$ carrying the symplectic form

$$B_\lambda : (X,Y) \rightsquigarrow \lambda[X,Y]$$

and with the flat coadjoint orbit $\mathcal{O}_\lambda$ of $\tilde{A}(\mathbf{R})$ through λZ_0^*. The coadjoint stabilizer of λZ_0^* in $\tilde{A}(\mathbf{R})$ is the center $\tilde{C}$ of $\tilde{A}(\mathbf{R})$ with Lie algebra $\mathfrak{z}$. Identify the group of automorphisms of $\tilde{A}(\mathbf{R})$ which leave the center $\tilde{C}$ pointwise fixed with the real symplectic group $\mathbf{Sp}(\mathfrak{n}/\mathfrak{z}, B_\lambda) = \mathbf{Sp}(1,\mathbf{R}) = \mathbf{SL}(2,\mathbf{R})$. For any element $\sigma \in \mathbf{Sp}(1,\mathbf{R})$ the topologically irreducible, unitary, linear representation

$$(x,y,z) \rightsquigarrow U_\lambda(\sigma^{-1}(x,y,z))$$

of $\tilde{A}(\mathbf{R})$ is unitarily isomorphic to U_λ. By the Stone-von Neumann-Segal theorem there exists a unitary operator $R_\lambda(\sigma)$ on the complex Hilbert space $L^2(\mathbf{R})$ which satisfies the intertwining identity

$$R_\lambda(\sigma)^{-1} \circ U_\lambda(x,y,z) \circ R_\lambda(\sigma) = U_\lambda(\sigma^{-1}(x,y,z))$$

for all elements $(x,y,z) \in \tilde{A}(\mathbf{R})$. In view of Schur's lemma, R_λ forms a <u>projective</u> linear representation of $\mathbf{Sp}(1,\mathbf{R})$. The 2-cocycle γ_{R_λ} on $\mathbf{Sp}(1,\mathbf{R})$ associated with R_λ can be explicitly calculated by means of the Leray-Maslov-Souriau index [6]. It is well-known that $H^2(\mathbf{Sp}(1,\mathbf{R}), \mathbf{Z}/2\mathbf{Z}) = \mathbf{Z}/2\mathbf{Z}$ implies the existence of a unique non-trivial group extension $\mathbf{Mp}(\mathfrak{n}/\mathfrak{z}, B_\lambda)$ of $\mathbf{Sp}(\mathfrak{n}/\mathfrak{z}, B_\lambda)$ by $\mathbf{Z}/2\mathbf{Z}$. The twofold covering $\mathbf{Mp}(1,\mathbf{R})$ of $\mathbf{Sp}(1,\mathbf{R})$ is called the real <u>metaplectic group</u>. Let $\tilde{\sigma} \rightsquigarrow \sigma$ denote the projector of $\mathbf{Mp}(1,\mathbf{R})$ onto $\mathbf{Sp}(1,\mathbf{R})$ with kernel $\{+1,-1\}$. Then R_λ gives rise via γ_{R_λ} to a continuous, unitary, linear representation S_λ of $\mathbf{Mp}(1,\mathbf{R})$ acting on $L^2(\mathbf{R})$ such that the relation

$$S_\lambda(\sigma)^{-1} \circ U_\lambda(x,y,z) \circ S_\lambda(\sigma) = U_\lambda(\sigma^{-1}(x,y,z))$$

holds for all $(x,y,z) \in \tilde{A}(\mathbf{R})$. The representation S_λ of $\mathbf{Mp}(1,\mathbf{R})$ is called the <u>oscillator representation.</u>

It is not difficult to extend the preceding reasoning to the real Heisenberg nilpotent Lie group $\tilde{A}(\mathbf{R}^n)$ in order to construct the oscillator representations S_λ ($\lambda \neq 0$) of the metaplectic group $\mathbf{Mp}(n,\mathbf{R})$, $n \geq 1$, acting on $L^2(\mathbf{R}^n)$.

8. The Real Three-Step Nilpotent Lie Group $B(\mathbf{R})$

Obviously the three-dimensional real Heisenberg two-step nilpotent Lie group $\tilde{A}(\mathbf{R})$ can be extended to a four-dimensional real Lie group $B(\mathbf{R})$ without destroying the property of being nilpotent. Indeed, let G denote the maximal unipotent subgroup of $\mathbf{Sp}(1,\mathbf{R})$ which fixes pointwise the closed subgroup

$$\exp(\mathbf{R}Y_0) = \{(0,y,0) \in \tilde{A}(\mathbf{R}) \mid y \in \mathbf{R}\}$$

of $\tilde{A}(\mathbf{R})$. Form the external semi-direct product

$$B(\mathbf{R}) = G \ltimes \tilde{A}(\mathbf{R})$$

with the convention that the open end of the symbol $\ltimes$ always points to the normal subgroup. Then $B(\mathbf{R})$ represents the smallest real three-step nilpotent Lie group (cf. Ratcliff [8]). Let $\{V_0, X_0, Y_0, Z_0\}$ denote the standard basis of its Lie algebra $\mathfrak{b}$. Then we have

$$\mathfrak{b} = \mathbf{R}V_0 \oplus \mathfrak{n},$$

all nonzero Lie bracket products of $\mathfrak{b}$ are generated by the commutation relations

$$[V_0, X_0] = Y_0, \quad [X_0, Y_0] = Z_0,$$

and the center of $\mathfrak{b}$ coincides with the center $\mathfrak{z} = \mathbf{R}Z_0$ of $\mathfrak{n}$.

The construction of a series $(U_{(a,\lambda)})_{(a,\lambda)\in \mathbf{R}\times\mathbf{R}^{\times}}$ of topologically irreducible, continuous, unitary linear representations of $B(\mathbf{R})$ which extend the family of infinite-dimensional representations $(U_\lambda)_{\lambda\in\mathbf{R}^{\times}}$ of $\tilde{A}(\mathbf{R})$ is also straightforward. Introduce coordinates in the real vector space $\mathfrak{b}^*$ dual to $\mathfrak{b}$ by considering the dual basis $\{V_0^*, X_0^*, Y_0^*, Z_0^*\}$. For all pairs $(a,\lambda)\in \mathbf{R}\times\mathbf{R}^{\times}$ the parabolic cylinder

$$\mathcal{O}_{(a,\lambda)} = \{(a-\frac{1}{2\lambda}\nu^2, \mu, \nu, \lambda)\in \mathfrak{b}^* \mid (\mu,\nu)\in \mathbf{R}\oplus\mathbf{R}\}$$

lies over the two-dimensional affine vector subspace $\mathcal{O}_\lambda$ of $\mathfrak{b}^*$ passing through the point λZ_0^*. If $U_{(a,\lambda)}$ denotes a representative of the isomorphy class which corresponds to the coadjoint orbit $\mathcal{O}_{(a,\lambda)}$ of $B(\mathbf{R})$ under the Kirillov correspondence and acts in the complex Hilbert space $L^2(\mathbf{R})$, then we have

$$U_{(a,\lambda)}(v,x,y,z) = e^{2\pi i(a-\frac{\lambda}{2}t^2-\frac{\lambda}{2}xt-\frac{1}{6}x^2)v} . U_\lambda(x,y,z)$$

for all elements $(v,x,y,z)\in B(\mathbf{R})$. Obviously $U_{(a,\lambda)}$ restricts to U_λ for all $(a,\lambda)\in\mathbf{R}\times\mathbf{R}^{\times}$ and admits therefore the central character e_λ.

9. The Diamond Solvable Lie Group D(R)

Let us combine the postulates (i) and (ii) for optical fibers as exposed in Section 4 by identifying the peripheral circle $\mathbf{T}$ of any fiber cross-section with the maximal compact subgroup $\mathbf{SO}(2,\mathbf{R})$ of $\mathbf{Sp}(1,\mathbf{R}) = \mathbf{SL}(2,\mathbf{R})$ the elements of which act as automorphisms of $\tilde{A}(\mathbf{R})$ by rotating the first two coordinates $(x,y)\in\mathbf{R}\oplus\mathbf{R}$ of the elements $(x,y,z)\in\tilde{A}(\mathbf{R})$ and leaving the third "central" coordinate $z\in\mathbf{R}=\tilde{C}$ fixed. Form the external semi-direct product

$$D(\mathbf{R}) = \mathbf{T} \ltimes \tilde{A}(\mathbf{R}).$$

Then

$$D(\mathbf{R}) = \{(e^{2\pi i\theta}, x, y, z) \mid (\theta, x, y, z) \in \mathbf{R}^4\}$$

and the group law of the extension $D(\mathbf{R})$ of $\tilde{A}(\mathbf{R})$ by $\mathbf{T}$ reads as follows:

$$(e^{2\pi i\theta_1}, x_1, y_1, z_1).(e^{2\pi i\theta_2}, x_2, y_2, z_2) =$$

$$(e^{2\pi i(\theta_1+\theta_2)}, x_1+x_2\cos 2\pi\theta_1-y_2\sin 2\pi\theta_1, y_1+x_2\sin 2\pi\theta_1+y_2\cos 2\pi\theta_1,$$

$$z_1+z_2+x_1(x_2\sin 2\pi\theta_1+y_2\cos 2\pi\theta_1))$$

The group $D(\mathbf{R})$ is called the <u>diamond group</u>. It forms a real, connected, non-exponential, solvable Lie group. The standard basis $\{H_o, X_o, Y_o, Z_o\}$ of the four dimensional Lie algebra $\mathfrak{d}$ of $D(\mathbf{R})$ over $\mathbf{R}$ satisfies the commutation relations

$$[H_o, X_o] = Y_o, [H_o, Y_o] = -X_o, [X_o, Y_o] = Z_o.$$

Let $\mathfrak{d}^*$ denote the dual vector space of the <u>diamond Lie algebra</u> $\mathfrak{d}$ over $\mathbf{R}$. Then the coadjoint action of $D(\mathbf{R})$ in $\mathfrak{d}^*$ with respect to the dual

basis $\{H_o^*, X_o^*, Y_o^*, Z_o^*\}$ of $\{H_o, X_o, Y_o, Z_o\}$ is given by the matrix

$$\begin{pmatrix} 1 & x\sin 2\pi\theta - y\cos 2\pi\theta & x\cos 2\pi\theta + y\sin 2\pi\theta & -(x^2+y^2)/2 \\ 0 & \cos 2\pi\theta & -\sin 2\pi\theta & y \\ 0 & \sin 2\pi\theta & \cos 2\pi\theta & -x \\ 0 & 0 & 0 & 1 \end{pmatrix}$$

In the Kirillov picture, the non-degenerate coadjoint orbits of D(**R**) are either <u>circular cylinders</u> or <u>paraboloids of revolution</u> in $\mathfrak{d}^*$. For each real number $\lambda \neq 0$ the coadjoint stabilizer in the diamond group D(**R**) of the linear form λZ_0^* is given by $\mathbf{T} \times \tilde{C}$, independently of the parameter value λ. Thus we may identify

(i') the optical fiber $\mathbf{T} \times \mathbf{R}$ with the coadjoint stabilizer $\mathbf{T} \times \tilde{C}$ of λZ_0^* in D(**R**),

and

(ii') each plane $\mathbf{R} \oplus \mathbf{R}$ transverse to the optical fiber axis with the kernel $\mathfrak{n}/\mathfrak{z}$ of λZ_0^*.

10. The Quantized Transverse Eigenmode Spectrum of Circular Optical Waveguides

A waveform $\psi \in L^2(\mathbf{R})$, $\psi \neq 0$, is called a <u>transverse eigenmode</u> of a circularly symmetric optical waveguide $\mathbf{T} \times \tilde{C}$ if the auto-ambiguity function $H(\psi;.,.)$ is <u>radial</u> on the plane $\mathfrak{n}/\mathfrak{z}$ transverse to the axis, i.e., invariant under the action of SO(2,**R**).

Let $\lambda \neq 0$ be a fixed real number. The projective linear representation R_λ of Sp(1,**R**) gives rise to a projective linear representation R'_λ of $\mathbf{T} \times \tilde{C}$ and the 2-cocycle γ_{R_λ} on Sp(1,**R**) associated with R_λ gives rise to the 2-cocycle $\gamma_{R'_\lambda}$ on $\mathbf{T} \times \tilde{C}$ associated with R'_λ. Indeed, let j denote the homomorphism of

$$D(\mathbf{R}) = (\mathbf{T} \times \tilde{C}).\tilde{A}(\mathbf{R})$$

into $\mathrm{Aut}(\tilde{A}(\mathbf{R}))$ which maps each element $(e^{2\pi i\theta},x,y,z) \in D(\mathbf{R})$ onto the inner automorphism of $\tilde{A}(\mathbf{R})$ associated with

$(e^{2\pi i\theta}, x, y, z)$. Then the conjugation $j(e^{2\pi i\theta}, 0, 0, z)$ by the element $(e^{2\pi i\theta}, z) \in T \times \tilde{C}$ induces an automorphism of $\tilde{A}(\mathbf{R})$ which coincides with the rotation

$$\begin{pmatrix} \cos 2\pi\theta & -\sin 2\pi\theta \\ \sin 2\pi\theta & \cos 2\pi\theta \end{pmatrix} \in \mathbf{SO}(2,\mathbf{R}) \hookrightarrow \mathbf{Sp}(1,\mathbf{R})$$

on the first two coordinates of the elements of $\tilde{A}(\mathbf{R})$ and leaving the third central coordinate fixed. Thus $j(T \times \tilde{C}) \subseteqq \mathbf{Sp}(1,\mathbf{R})$. It follows that R'_λ and $\gamma_{R'_\lambda}$ are the images of R_λ and γ_{R_λ}, respectively, under the restriction $j|(T \times \tilde{C})$. Denote by

$$\tilde{T} \times \tilde{C} = (\mathbf{R}/\tfrac{1}{2}\mathbf{Z}) \times \tilde{C} = \{(e^{\pi i\theta}, z) \mid (\theta, z) \in \mathbf{R} \times \mathbf{R}\}$$

the inverse image of $T \times \tilde{C}$ in $\mathbf{Mp}(\mathfrak{u}/\mathfrak{r}, \mathbf{R}) \times \tilde{C} = \mathbf{Mp}(1,\mathbf{R}) \times \tilde{C}$. Obviously the mapping

$$(e^{\pi i\theta}, z) \rightsquigarrow (e^{2\pi i\theta}, z)$$

projects the double covering $\tilde{T} \times \tilde{C}$ in the natural way onto the coadjoint stabilizer $T \times \tilde{C}$ in $D(\mathbf{R})$ of λZ_0^*. The projective linear representation R'_λ of $T \times \tilde{C}$ gives rise via $\gamma_{R'_\lambda}$ to a continuous, unitary, linear representation S'_λ of $\tilde{T} \times \tilde{C}$. Form the characters

$$\chi_\lambda^n(e^{\pi i\theta}, z) = e^{2\pi i((2n+1)\theta + \lambda z)} \mathrm{id}_{L^2(\mathbf{R})} \qquad (n \in \mathbf{Z})$$

of $\tilde{T} \times \tilde{C}$. Then

$$S'_\lambda \circ U_\lambda \otimes \chi_\lambda^n \qquad (n \in \mathbf{Z})$$

is a family of topologically irreducible, continuous, unitary, linear representations of D(R) having U_λ as their restrictions to $\tilde{A}(R)$. Conversely, each representation of D(R) having these properties is unitarily isomorphic to exactly one of the representations $S'_\lambda \circ U_\lambda \otimes \chi_\lambda^n$ $(n \in Z)$; see Lion [6]. It follows

Theorem 1. The waveform $\psi \in L^2(R)$, $\|\psi\| = 1$, is an eigenmode of a circular optical waveguide if and only if there is a number $n \in N$ such that the unitary linear representation

$$T \ni e^{2\pi i\theta} \rightsquigarrow \chi_1^n . S_1'(e^{\pi i\theta}, 0)$$

acts trivially on ψ.

The geometric optics reason behind the fact that the elements $e^{\pi i\theta}$ of the double covering $\tilde{T}$ of the to rus group T occur in Theorem 1 as the argument of the representations $\chi_1^n . S_1'$ and not the underlying elements $e^{2\pi i\theta}$ of the torus group T itself is in the phase shifts (which actually turn out to be phase retardations) of $\frac{1}{2}\pi$ suffered at each of the two caustics of the optical fiber. The illustration below shows the laser beam in a plastic distributed-index rod lens of 93.5 mm length and 3 mm in diameter. Moreover, the ray projection shows the inner and outer caustic in a graded-index optical fiber. Thus the spherical cylinder $\tilde{T} \times \tilde{C}$ serves to describe the transverse phase in the optical fiber $T \times \tilde{C}$.

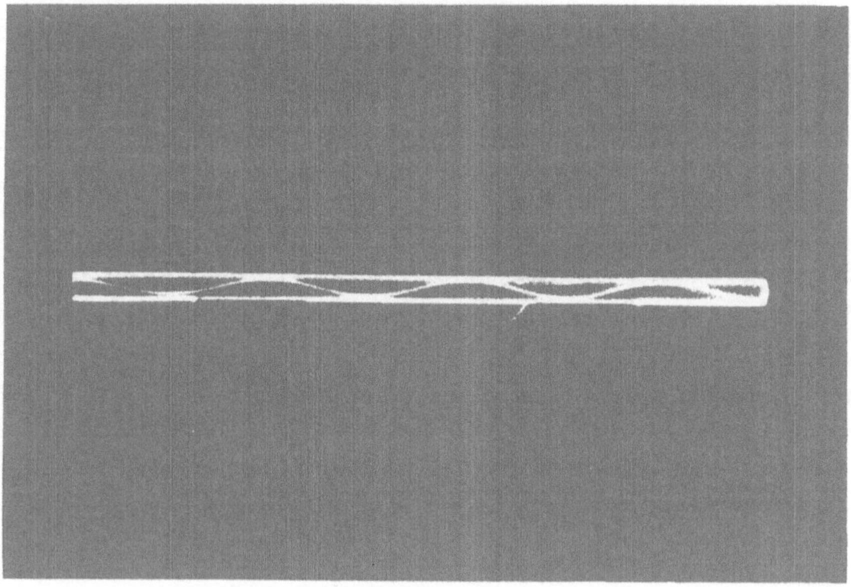

Trajectory of a YAG laser beam in an optical fiber

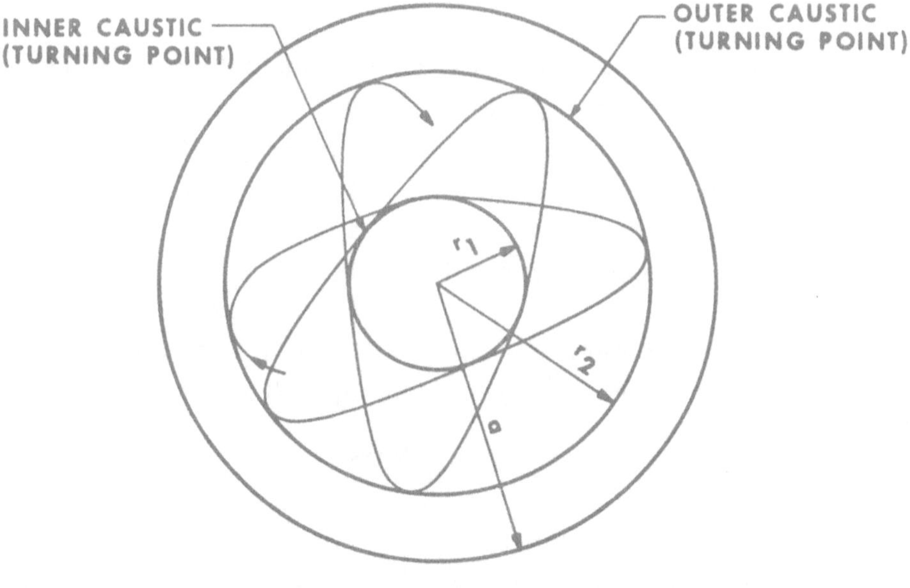

Ray projection showing caustics in a graded-index optical fiber

In particular, the projection $(e^{\pi i\theta},z) \rightsquigarrow (e^{2\pi i\theta},z)$ of $\tilde{T} \times \tilde{C}$ onto $T \times \tilde{C}$ gives a group theoretical explanation why Keller's geometric construction [4] of double-sided planes can be adapted to calculate the phase retardations at the inner and the outer caustic of a graded-index optical fiber.

To point out more clearly the connection between harmonic analysis on the diamond solvable Lie group $D(\mathbf{R})$ and quantum mechanics, let $\mathcal{H}_\lambda$ $(\lambda \neq 0)$ denote the Schrödinger Hamiltonian of the harmonic oscillator (= mode Hamiltonian) acting on the Sobolev space

$$H^2(\mathbf{R}) = \{\psi \in L^2(\mathbf{R}) \mid t^2\psi \in L^2(\mathbf{R})\}.$$

Then

$$\mathcal{H}_\lambda = \frac{1}{\lambda}\frac{d^2}{dt^2} - 4\pi^2\lambda t^2$$

is an essentially self-adjoint linear operator in $L^2(\mathbf{R})$ and forms the infinitesimal generator of $\mathbf{SO}(2,\mathbf{R})$ in $\mathbf{Sp}(1,\mathbf{R})$. The closure of $\mathcal{H}_\lambda$ has a <u>pure point spectrum</u> formed by the <u>simple</u> eigenvalues $\{-2\pi(\operatorname{sign}\lambda)(2n+1) \mid n \in \mathbf{N}\}$ and the associated eigenfunctions are given by the scaled <u>Hermite functions</u> $t \rightsquigarrow H_n(\sqrt{|\lambda|}.t)$, $n \in \mathbf{N}$.

The unitary linear representation S'_λ of $\tilde{T} \times \tilde{C}$ acts on the complex Hilbert space $L^2(\mathbf{R})$ by the operator

$$S'_\lambda(e^{\pi i\theta},z) = e^{i\theta\mathcal{H}_\lambda}$$

and the topologically irreducible, continuous, unitary, linear representations $S'_\lambda \circ U_\lambda \otimes \chi^n_\lambda$ $(n \in \mathbf{Z})$ of $D(\mathbf{R})$ act on $L^2(\mathbf{R})$ according to the prescription

$$S'_\lambda \circ U_\lambda \otimes \chi_\lambda^n(e^{2\pi i\theta},x,y,z) = e^{i(\mathcal{H}_\lambda+2\pi(2n+1))\theta} \circ U_\lambda(x\cos 2\pi\theta+y\sin 2\pi\theta, -x\sin 2\pi\theta+y\cos 2\pi\theta,z)$$

Thus Theorem 1 supra determines the quantized transverse eigenmode spectrum of circularly symmetric optical waveguides in the following way:

Corollary 1. The waveform $\psi \in L^2(\mathbf{R}), \psi \neq 0$, is an eigenmode of a circular optical waveguide if and only if

$$\psi = \zeta_n H_n \qquad (\zeta_n \in \mathbf{C})$$

for an integer $n \geq 0$. ($H_n(x)$=Hermite function of degree $n \geq 0$ $= e^{-\pi x^2} h_n(x)$). In this case

$$H(H_m,H_n;x,y) = \sqrt{\frac{n!}{m!}}\,(\sqrt{\pi}(x+iy))^{m-n} L_n^{(m-n)}(\pi(x^2+y^2)) \qquad (m \geq n \geq 0)$$

(Schwinger's formula) for all pairs $(x,y) \in \mathbf{R}^2$.
($L_n^{(\alpha)}(x)$ = Laguerre function of degree $n \geq 0$ and order α = $e^{-\frac{1}{2}x} l_n^{(\alpha)}(x)$).

The following computer plot is generated according to the preceding formulae. It is in good agreement with the experiment (cf. Section 2 supra).

Computed eigenmode in a circular waveguide

11. The Quantized Transverse Eigenmode Spectrum of Rectangular Optical Waveguides

The flatness of the non-degenerate coadjoint orbits $\mathcal{O}_\lambda (\lambda \neq 0)$ in the Kirillov picture of the real Heisenberg nilpotent Lie group $\tilde{A}(\mathbf{R})$ implies the square integrability modulo $\tilde{C}$ of the topologically irreducible, continuous, unitary, linear representations $(U_\lambda)_{\lambda \in \mathbf{R}^\times}$ of $\tilde{A}(\mathbf{R})$ in the complex

Hilbert space $L^2(\mathbf{R})$. From this fact we may infer (among other consequences) that the continuous mapping

$$\psi \rightsquigarrow H(\psi,\varphi;\cdot,\cdot) \qquad (\varphi \in L^2(\mathbf{R}))$$

defines an <u>intertwining</u> <u>operator</u> of the linear Schrödinger representation $(U_1, L^2(\mathbf{R}))$ of $\tilde{A}(\mathbf{R})$ and the right <u>regular</u> representation of $\tilde{A}(\mathbf{R})$ acting in the representation space of $\mathrm{Ind}_{\tilde{C}}^{\tilde{A}(\mathbf{R})} e_1$. Moreover, the <u>orthogonality</u> <u>relations</u>

$$\iint_{\mathbf{R}\oplus\mathbf{R}} H(\psi,\varphi;x,y).\bar{H}(\psi',\varphi';x,y)\,dx\,dy = \langle\psi|\psi'\rangle.\langle\varphi'|\varphi\rangle$$

hold for all wavefunctions $\psi, \psi', \varphi, \varphi'$ in $L^2(\mathbf{R})$. See [11]. It makes the conclusion obvious that the linear Schrödinger representation U_1 of $\tilde{A}(\mathbf{R})$ extends to an isometric isomorphism of the representation space of $\mathrm{Ind}_{\tilde{C}}^{\tilde{A}(\mathbf{R})} e_1$ onto the complex Hilbert space of all Hilbert-Schmidt operators acting on $L^2(\mathbf{R})$ as kernel operators with kernels in $L^2(\mathbf{R}^2)$. As a result we mention the following result.

<u>Corollary 2.</u> The waveform $\psi \in L^2(\mathbf{R}^2)$, $\psi \neq 0$, is an eigenmode of a rectangular optical waveguide if and only if

$$\psi = \zeta_{n,m} H_n \otimes H_m \quad (\zeta_{n,m} \in \mathbf{C})$$

for integers $n \geq 0$, $m \geq 0$.

The computer plot displayed below illustrates the preceding theorem (cf. Section 2 supra).

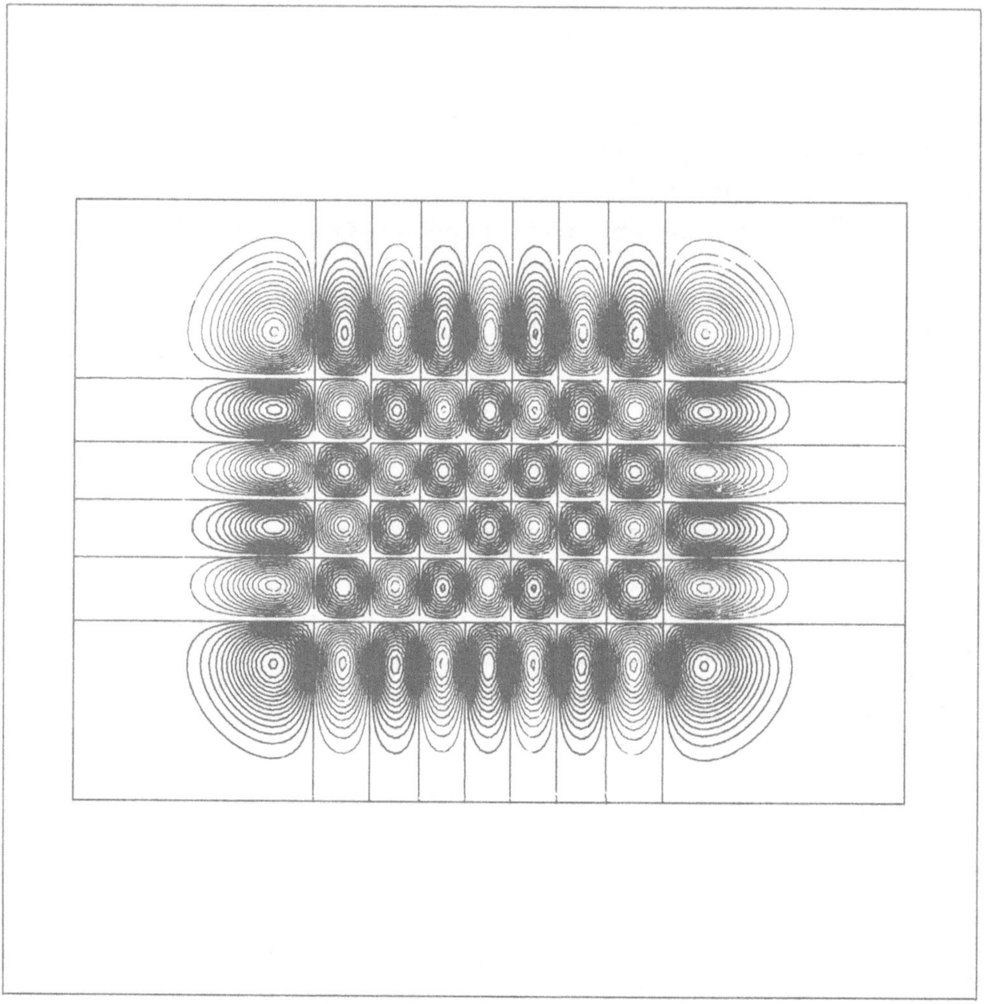

Computed eigenmode of a rectangular laser

12. The Coupling Coefficients of Coaxial Transverse Eigenmodes

If the wave functions ψ', φ' and ψ, φ belonging to the Schwartz space $\mathcal{S}(\mathbf{R})$ represent two transverse eigenmodes of two co-axial optical devices like laser resonators or dielectric waveguides, their coupling coefficient is defined according to the prescription

$$C(\psi', \varphi', \psi, \varphi) = \int_{n/t} H(\psi', \varphi'; x, y) . \bar{H}(\psi, \varphi; x, y) dxdy.$$

The integral has to be evaluated at the coupling plane n/t transverse to the common axis of the optical devices. In the present section we will calculate the coaxial coupling coefficents explicitly in the circular as well as in the rectangular case (cf. Kogelnik [5]).

Lasers emit a very narrow cone of monochromatic radiation referred to as a <u>laser beam</u>. Denote its wavelength by λ. The transverse electromagnetic eigenmode output of a laser is a beam with a Gaussian wavefront. Define the <u>Gaussian beam parameters</u> by

$$\alpha' = \frac{2}{w'^2} \quad , \quad \alpha = \frac{2}{w^2}$$

and

$$q = (\frac{1}{w'^2} + \frac{1}{w^2} + \frac{1}{2} ik (\frac{1}{r'} - \frac{1}{r}))$$

where

$$k = \frac{1}{\lambda}$$

denotes the <u>wave number</u>, w', w the <u>beam radii</u> and r', r the <u>radii of curvature</u> of the phase fronts at the coupling plane $\mathcal{M}/\mathcal{K}$ of the optical devices. Transfer the symplectic reference planes of the two beams to the coupling plane $\mathcal{M}/\mathcal{K}$ by the symplectic mappings σ', σ with matrices

$$\begin{pmatrix} \frac{1}{\sqrt{\alpha'}} & 0 \\ 0 & \sqrt{\alpha'} \end{pmatrix}, \quad \begin{pmatrix} \frac{1}{\sqrt{\alpha}} & 0 \\ 0 & \sqrt{\alpha} \end{pmatrix} \in \mathbf{Sp}(\mathcal{M}/\mathcal{K}, B_1)$$

associated with the beam radii w' and w, respectively, and by the symplectic mappings τ', τ with matrices

$$\begin{pmatrix} 1 & 0 \\ +\frac{k}{r'} & 1 \end{pmatrix}, \quad \begin{pmatrix} 1 & 0 \\ -\frac{k}{r} & 1 \end{pmatrix} \in \mathbf{Sp}(\mathcal{M}/\mathcal{K}, B_1)$$

associated with the radii of curvature r', and r, respectively, of the phase fronts. The symplectic mappings σ', τ' and σ, τ <u>preserve</u> the <u>ambiguity</u> <u>surfaces</u> over the coupling plane $\mathcal{M}/\mathcal{K}$ transverse to the common beam direction and $R_1(\tau') \circ R_1(\sigma')\psi'$ resp. $R_1(\tau) \circ R_1(\sigma)\psi$ are the waveforms to be coupled where

$$R_1(\sigma')\psi'(t) = \psi'(\sqrt{\alpha'}t), \quad R_1(\tau')\psi'(t) = e^{i\frac{k}{2r'}t^2}\psi'(t)$$

$$R_1(\sigma)\psi(t) = \psi(\sqrt{\alpha}t), \quad R_1(\tau)\psi(t) = e^{-i\frac{k}{2r}t^2}\psi(t)$$

for $t \in \mathbf{R}$. In terms of the representations $(U_{(a,\lambda)})_{(a,\lambda) \in \mathbf{R}\times\mathbf{R}^{\times}}$ of the group $B(\mathbf{R})$ we get for the ambiguity functions at the coupling plane (cf. Section 8):

$$\langle U_{(0,\sqrt{\alpha'})}\, \psi'(\tfrac{k}{r'},x,y,0) | \varphi'\rangle, \quad \langle U_{(0,\sqrt{\alpha})}\, \psi(-\tfrac{k}{r},x,y,0) | \varphi\rangle.$$

As the next step, recall the definition of the Krawtchouk polynomials $K_n(x;p,N)$. For all integers $n \geq 0$ these hypergeometric polynomials are given by

$$K_n(x;p,N) = {}_2F_1(-n,-x;-N;p^{-1})$$

where $N \geq 0$, $0 \leq x \leq N$, $p \in]0,1[$. In terms of shifted factorials, $K_n(x;p,N)$ takes the form

$$K_n(x;p,N) = \sum_{0 \leq k \leq n} \frac{(-n)_k(-x)_k}{(-N)_k k!} \left(\frac{1}{p}\right)^k \qquad (0 \leq n \leq N).$$

The Krawtchouk polynomials are orthogonal on $\{0,1,\dots,N\}$ with respect to the binomial distribution. Thus their orthogonality relations take the form

$$\sum_{0 \leq x \leq N} K_n(x;p,N) K_m(x;p,N) \binom{N}{x} p^x (1-p)^{N-x} = 0 \qquad (m \neq n)$$

for $n \leq N$, $m \leq N$ and $p \in]0,1[$. By virtue of the preceding facts and Corollary 1 and Corollary 2 of Theorem 1 it follows by taking Laplace transforms

Theorem 2. Let $m \geq n \geq 0$, $m' \geq n' \geq 0$ be integers. Keep to the preceding notations.

a) In the circular case set $m = p$, $m - n = l$, $m' = p'$, $m' - n' = l'$. Then

$$C_{p,l,p',l'} = 0 \quad \text{for} \quad l \neq l',$$

i.e., there is no coupling between transverse eigenmodes of different angular moments. Moreover,

$$C_{p,l,p',l} = \left(\frac{2}{ww'q}\right)^{l+1} \frac{(p+p'+l)!}{\sqrt{p!p'!(p+l)!(p'+l')!}}(1-\frac{\alpha}{q})^{p}(1-\frac{\alpha'}{q})^{p'}.$$

$$K_p(p';\frac{(q-\alpha)(q-\alpha')}{q(q-\alpha-\alpha')},p+p'+l)$$

b) In the <u>rectangular</u> case we have

$$C_{m,n,m',n'} = 0 \text{ for } \begin{cases} m+m' \equiv 1 \bmod 2, \\ n+n' \equiv 1 \bmod 2, \end{cases}$$

i.e., there is <u>no coupling between even and odd</u> transverse eigenmodes. In other words, the parity is preserved under the coupling of eigenmodes. In the case

$$m' = 2\mu', \quad m = 2\mu, \quad n' = 2\nu', \quad n = 2\nu$$

we get in terms of the Krawtchouk polynomials

$$C_{m,n,m',n'} = (-\frac{1}{2})^{\mu+\nu+\mu'+\nu'} \frac{2}{ww'q} \frac{(2\mu+2\mu')!(2\nu+2\nu')!}{(\mu+\mu')!(\nu+\nu')!\sqrt{2\mu!2\mu'!2\nu!2\nu'!}}$$

$$(1-\frac{\alpha}{q})^{\mu+\nu}(1-\frac{\alpha'}{q})^{\mu'+\nu'}.K_{\mu}(\mu';\frac{(q-\alpha)(q-\alpha')}{q(q-\alpha-\alpha')}, \mu+\mu'-\frac{1}{2})$$

$$K_{\nu}(\nu';\frac{(q-\alpha)(q-\alpha')}{q(q-\alpha-\alpha')}; \nu+\nu'-\frac{1}{2})$$

A similar result holds in the case

$$m' = 2\mu'+1, \quad m = 2\mu+1, \quad n' = 2\nu'+1, \quad n = 2\nu+1.$$

The <u>fundamental transverse eigenmode</u> is the same in the circular case (p=l=p'=l'=0 and in the rectangular case

$(m=n=m'=n'=0)$. The *fraction of power* κ coupled from one fundamental eigenmode to the other is given by

$$\kappa = |C_{0000}|^2 = \frac{4}{\left(\frac{w}{w'}+\frac{w'}{w}\right)^2 + \rho^2\,\frac{w'^2}{w^2}}$$

where the ratio $\frac{w'}{w}$ is the *mismatch in beam radius*, and

$$\rho^2 = \frac{w^2}{R\lambda}\left(\frac{R}{R'} - 1\right)^2$$

is the *mismatch between the curvatures* of the two phase fronts at the coupling plane.

The plot below shows the fraction κ of power coupled between the fundamental transverse eigenmodes as a function of $(\frac{w'}{w})^2$ for various values of ρ.

13. The Discrete Spectrum of the Reductive Dual Pair $(\tilde{O}(n,\mathbb{R}),\tilde{Sp}(1,\mathbb{R}))$ in $Mp(n,\mathbb{R})$

Let $n\geq 2$ be an arbitrary integer and $O(n,\mathbb{R})$ the orthogonal group of the Euclidean vector space $\mathbb{R}^n$, i.e., the isometry group with respect to the standard scalar product $\langle . | . \rangle$ of $\mathbb{R}^n$. Then $O(n,\mathbb{R})$ acts on the $(2n+1)$-dimensional real Heisenberg nilpotent Lie group $\tilde{A}(\mathbb{R}^n)$ as a subgroup of the symplectic group $Sp(n,\mathbb{R})$ which consists of all the automorphisms of $\tilde{A}(\mathbb{R}^n)$ leaving the center $\tilde{C}$ pointwise fixed. Let $\tilde{O}(n,\mathbb{R})$ denote the preimage of $O(n,\mathbb{R})$ and $\tilde{Sp}(1,\mathbb{R})$ the preimage of $Sp(1,\mathbb{R}) = SL(2,\mathbb{R})$ in the metaplectic group $Mp(n,\mathbb{R})$. Then $(\tilde{O}(n,\mathbb{R}),\tilde{Sp}(1,\mathbb{R}))$ forms a reductive *dual pair* in $Mp(n,\mathbb{R})$ in the following sense (cf. Gelbart [2]): Each of the groups $\tilde{O}(n,\mathbb{R})$ and $\tilde{Sp}(1,\mathbb{R})$ gives rise by restricting the oscillator representation S_1 of $Mp(n,\mathbb{R})$ to a family of unitary linear operators acting in $L^2(\mathbb{R}^n)$ and commuting with the other.

For each integer $k \geq 0$ let $H_k(\mathbf{R}^n)$ denote the complex vector space of solid spherical harmonic functions of degree k on $\mathbf{R}^n$. Recall that the elements of $H_k(\mathbf{R}^n)$ are the homogeneous polynomials P with complex coefficients of degree k in n real variables such that $\Delta_n P = 0$ where $\Delta_n = \langle \frac{\partial}{\partial x} | \frac{\partial}{\partial x} \rangle$ denotes the Laplace operator of $\mathbf{R}^n$. It is well-known that the natural action of $\mathbf{O}(n,\mathbf{R})$ on $H_k(\mathbf{R}^n)$ defines a finite dimensional, irreducible, linear representation $(D_k, H_k(\mathbf{R}^n))$ of $\mathbf{O}(n,\mathbf{R})$. If $N(k,n)$ denotes the complex dimension of D_k, i.e.,

$$N(k,n) = \dim_{\mathbf{C}} H_k(\mathbf{R}^n) \qquad (k \in \mathbf{N})$$

then we get

$$\sum_{k \geq 0} N(k,n) X^k = \frac{1+X}{(1-X)^{n-1}}$$

where X denotes an indeterminate. It follows

$$N(k,n) = \frac{(2k+n-2)(k+n-3)!}{k!(n-2)!} \qquad (k \geq 1)$$

and $N(0,n) = 1$. Let $\mathbf{C}_+$ denote the open upper half-plane and $(D_m, H_m(\mathbf{C}_+))$ for any positive integer or half integer m the <u>harmonic</u> representation of $\mathbf{Sp}(1,\mathbf{R})$ with lowest weight m. Notice that $(D_m, H_m(\mathbf{C}_+))$ belongs for $m>1$ to the holomorphic discrete series of $\mathbf{Sp}(1,\mathbf{R})$ and that we have the Hilbert sum decomposition (cf. Howe [3])

$$L^2(\mathbf{R}^n) = \hat{\bigoplus_{k \in \mathbf{N}}} H_k(\mathbf{R}^n) \otimes_{\mathbf{C}} H_{k+\frac{1}{2}n}(\mathbf{C}_+).$$

Consider the element

$$J_o = \begin{pmatrix} 0 & 1 \\ -1 & 0 \end{pmatrix}$$

of the Lie algebra $\mathfrak{sl}(2,\mathbf{R})$ and denote by Exp the exponential mapping of $\mathfrak{sl}(2,\mathbf{R})$ into $\mathbf{Mp}(1,\mathbf{R})$. The compact one-parameter group $\mathrm{Exp}(\pi\theta J_0)$ $(\theta \in \mathbf{R})$ calculated in the metaplectic group $\mathbf{Mp}(1,\mathbf{R})$ covers the one-parameter subgroup

$$\mathbf{SO}(2,\mathbf{R}) = \left\{\begin{pmatrix} \cos 2\pi\theta & -\sin 2\pi\theta \\ \sin 2\pi\theta & \cos 2\pi\theta \end{pmatrix} \,\Big|\, \theta \in \mathbf{R}\right\}$$

of rotations in $\mathbf{Sp}(1,\mathbf{R})$. In particular $\mathrm{Exp}(-\frac{\pi}{4}J_0)$ covers the Weyl element $\begin{pmatrix} 0 & 1 \\ -1 & 0 \end{pmatrix}$ of $\mathbf{Sp}(1,\mathbf{R})$. Since the non-degenerate symmetric bilinear form $\langle . | . \rangle$ on $\mathbf{R}^n$ has signature n, the Fourier cotransform $\overline{\mathcal{F}}_{\mathbf{R}^n}$ acting on $L^2(\mathbf{R}^n)$ can be expressed as follows

$$\mathcal{F}_{\mathbf{R}^n} = e^{-\frac{\pi i}{4}n} S_1(\mathrm{Exp}(-\tfrac{\pi}{4}J_0)).$$

Let

$$\mathcal{H} = \Delta_n - 4\pi^2\langle x|x\rangle$$

denote the Schrödinger Hamiltonian of the n-dimensional harmonic oscillator. In view of the identity

$$\frac{d}{dt}S_1(\mathrm{Exp}(tJ_0))\Big|_{t=0} = \frac{i}{2\pi}\mathcal{H}$$

we get

$$S_1(\mathrm{Exp}(\pi\theta J_0)) = e^{\frac{i}{2}\theta\mathcal{H}} \qquad (\theta \in \mathbf{R})$$

and in particular

$$S_1(\mathrm{Exp}(-\tfrac{\pi}{4}J_0)) = e^{-\frac{i}{8}\mathcal{H}}.$$

Consequently, the Fourier transform $\mathcal{F}_{\mathbf{R}^n}$ acting on $L^2(\mathbf{R}^n)$ admits the form

$$\mathcal{F}_{\mathbf{R}^n} = e^{\frac{\pi i}{4}n} e^{\frac{i}{8}\mathcal{H}}$$

Notice that the fundamental mode $x \rightsquigarrow e^{-\pi\langle x|x\rangle}$ forms the lowest weight vector of the harmonic representation $(D_{k+\frac{1}{2}n}, H_{k+\frac{1}{2}n}(C_+))$ of $Sp(1,\mathbf{R})$ and that $H_k(\mathbf{R}^n)e^{-\pi\langle x|x\rangle}$ are the eigenvector spaces of the unitary linear operator $e^{\frac{i}{8}\mathcal{H}}$ with eigenvalues $e^{-\frac{\pi}{2}i(k+\frac{n}{2})}$ $(k \in \mathbf{N})$. Since $L^2(\mathbf{R}^n)$ is the Hilbert sum of the finite dimensional vector subspaces $H_k(\mathbf{R}^n)e^{-\pi\langle x|x\rangle}$ the discrete spectrum of the reductive dual pair $(\tilde{O}(n,\mathbf{R}), \tilde{S}p(1,\mathbf{R}))$ is completely determined.

14. The Singular Value Decomposition of the Radon Transform

In Section 9 supra we identified the optical fiber with the circular cylinder $\mathbf{T} \times \mathbf{R}$ and then the Lie group $\mathbf{T} \times \mathbf{R}$ with the coadjoint stabilizer in the diamond solvable Lie group $D(\mathbf{R})$ of the linear form $\lambda Z_0^* \in \mathfrak{d}^*$. Let us replace the one-dimensional torus group $\mathbf{T} = \mathbf{S}_1$ by the compact unit sphere $\mathbf{S}_{n-1}$ of the n-dimensional Euclidean vector space $\mathbf{R}^n$ $(n \geq 2)$. It is well known that $\mathbf{S}_{n-1}$ admits the structure of a Lie group if and only if n = 2 and n = 4 (Samelson's theorem). Identify $\mathbf{S}_{n-1}$ with the compact homogeneous manifold $\mathbf{O}(n)/\mathbf{O}(n-1)$ and form the spherical cylinder $\mathbf{S}_{n-1} \times \mathbf{R}$. Moreover, let ε_x denote the Dirac measure on $\mathbf{R}^n$ located at the point $x \in \mathbf{R}^n$. The <u>Radon transform</u> $\mathcal{R}f \in \mathcal{S}(\mathbf{S}_{n-1} \times \mathbf{R})$ of $f \in \mathcal{S}(\mathbf{R}^n)$ is defined according to the prescription

$$\mathcal{R}f(\omega, r) = \int_{\mathbf{R}^n} f(x)\, \varepsilon_{(r-\langle\omega|x\rangle)}\, dx.$$

Notice that the measure on $\mathbf{R}^n$ against which the function $f \in \mathcal{S}(\mathbf{R}^n)$ has to be integrated carries mass only on a hyperplane in $\mathbf{R}^n$ and that the spherical cylinder $S_{n-1} \times \mathbf{R}$ on which $\mathcal{R}f$ is defined forms a twofold covering of the differentiable manifold $\mathbf{P}^n(\mathbf{R})$ of all hyperplanes in $\mathbf{R}^n$. The relationship between the Radon transform $\mathcal{R}$ and the Fourier transform is given by the projection theorem:

$$\mathcal{R} = \mathcal{F}_{\mathbf{R}^n} \otimes \bar{\mathcal{F}}_{\mathbf{R}^1}$$

This identity is basic for our approach to a singular value decomposition of $\mathcal{R}$. Let ${}^t\mathcal{R}: \mathcal{S}(S_{n-1} \times \mathbf{R}) \longrightarrow \mathcal{S}(\mathbf{R}^n)$ denote the back-projector. An application of the results on the discrete spectrum of the reductive dual pair $(\tilde{O}(n,\mathbf{R}), \tilde{S}p(1,\mathbf{R}))$ gathered in the preceding section combined with the Rodrigues formula for the Hermite polynomials $(h_n)_{n\geq 0}$ and the well-known relationship between the Hermite polynomials of even degree and the Laguerre polynomials allows us to generate the eigenfunctions of the self-adjoint operator ${}^t\mathcal{R}\circ\mathcal{R}$.

Theorem 3. The linear mapping

$${}^t\mathcal{R}\circ\mathcal{R}: L^2(\mathbf{R}^n, e^{\pi\langle x|x\rangle}) \longrightarrow L^2(\mathbf{R}^n, e^{\pi\langle x|x\rangle})$$

has the eigenfunctions

$$S_{n-1} \times \mathbf{R} \ni (\omega, r) \rightsquigarrow r^k L_{(m-k)/2}^{(k+\frac{1}{2}n-1)}(2\pi r^2) Y_{kl}(\omega)$$

where $m \geq 0$ is an arbitrary integer, $0 \leq k \leq m$ such that

$$m - k \equiv 0 \pmod 2,$$

and $(Y_{kl})_{0\leq l\leq N(k,n)}$ forms a Hilbert basis of $H_k(\mathbf{R}^n)$. The linear mapping

$$\mathcal{R}\circ{}^t\mathcal{R}: L^2(S_{n-1}\times \mathbf{R}, 1\otimes e^{\pi r^2})\longrightarrow L^2(S_{n-1}\times \mathbf{R}, 1\otimes e^{\pi r^2})$$

has the eigenfunctions

$$(\omega, r)\rightsquigarrow H_m(r)Y_{kl}(\omega).$$

The corresponding eigenvalues are given by

$$\sqrt{\frac{\pi^{n-1/2}\, m!}{2^m(\frac{1}{2}(m-k))!\,\Gamma(\frac{1}{2}(n+m+k))}}.$$

For another approach to a singular value decomposition of the Radon transform which avoids group theoretical arguments, see Davison [1]. Also see the recent paper by Louis [7].

Let $\Delta_n^{1/2}$ denote the pseudodifferential operator on $\mathbf{R}^n$ with symbol $y\rightsquigarrow\sqrt{\langle y|y\rangle}$. Then we may read off from the preceding results the following

<u>Corollary 1.</u> The inversion formulae

$$\Delta_n^{(n-1)/2}\,({}^t\mathcal{R}\circ\mathcal{R}) = c_n.\mathrm{id}\,,$$

$$\Delta_1^{(n-1)/2}\,(\mathcal{R}\circ{}^t\mathcal{R}) = c_n.\mathrm{id}$$

hold where $c_n\in\mathbf{C}$ is a constant.

Since the harmonic representation $D_{k+\frac{1}{2}n}$ acts on the functions $f \in H_{k+\frac{1}{2}n}(\mathbf{C}_+)$ by means of a Hankel transform of order $k+\frac{1}{2}n-1$, we get in the case $n = 2$:

Corollary 2. For all $r > 0$ the Laguerre polynomial $l_m^{(k)}$ of degree $m \geq 0$ and order $k > -1$ admits the integral representation

$$l_m^{(k)}(r) = \frac{1}{m!} r^{-k/2} e^r \int_0^{+\infty} e^{-t} t^{m+k/2} J_k(2\sqrt{tr})\,dr.$$

The reductive dual pair approach is, of course, not restricted to a treatment of the classical Radon transform but is also applicable to various kinds of Radon transforms defined on Grassmannian manifolds. It is the purpose of a forthcoming paper to work out the singular value decomposition and the inversion formulae of such transforms. For an application of the notion of dual reductive pair to zeta functions, see the paper [10].

References

1. Davison, M.E.: A singular value decomposition for the Radon transform in n-dimensional Euclidean space. Numer. Funct. Anal. and Optimiz. **3** (1981), 321-340
2. Gelbart, S.: Examples of dual reductive pairs. Proc. of Symposia in Pure Math. **33** (1979), part 1, pp. 287-296
3. Howe, R.: On some results of Strichartz and of Rallis and Schiffman. J. Functional Analysis **32** (1979), 297-303
4. Keller, J.B.: A geometrical theory of diffraction. Proc. of Symposia in Applied Math. **8** (1958), pp. 27-52
5. Kogelnik, H.: Coupling and conversion coefficients for optical modes. Proc. of the Symposium on Quasi-Optics, pp. 333-347. Brooklyn, N.Y.: Polytechnic Press 1964
6. Lion, G.: Extensions de représentations de groupes de Lie nilpotents et indices de Maslov. C.R. Acad. Sc. Paris **288** (1979), Série A, pp. 615 - 618

7. Louis, A.K.: Laguerre and computerized tomography: Consistency conditions and stability of the Radon transform. In: Polynômes orthogonaux et applications. Proceedings, Bar-le-Duc 1984, pp. 524-531. Lecture Notes in Math., Vol. **1171**. Berlin-Heidelberg-New York-Tokyo: Springer 1985

8. Ratcliff, G.: Symbols and orbits for 3-step nilpotent Lie groups. J. Functional Analysis **62** (1985), 38-64

9. Schempp, W.: Applied nilpotent harmonic analysis - a research program. Proc. of the 14th ICGTMP, pp. 59-68. Singapore: World Scientific 1986

10. Schempp, W.: On the zeta function attached to the reductive dual pair (**MO**(p,q,**R**),**Mp**(1,**R**)) in the metaplectic group **Mp**(p+q,**R**). C.R. Math. Rep. Acad. Sci. Canada **8** (1986), 161-166

11. Schempp, W.: Harmonic analysis on the Heisenberg nilpotent Lie group, with applications to signal theory. London: Longman (in print)

12. Weyl, H.: Gruppentheorie und Quantenmechanik. Darmstadt: Wissenschaftliche Buchgesellschaft 1981

STEPPING STONE MODELS IN POPULATION GENETICS AND POPULATION DYNAMICS

Tokuzo SHIGA

Department of Applied Physics, Tokyo Institute of Technology,Oh-okayama, Tokyo 152, Japan

ABSTRACT

We summarize the results of two kinds of stepping stone models arising in population genetics and population dynamics. Although these two describe different phenomena they are closely related through a duality relation. We further attempt to generalize this framework as much as possible.

1. Introduction

By "a stepping stone model" we mean a stochastic evolution of a large population system, of which individual lives in a site of some lattice space. Biologically the lattice space is interpreted as a collection of colonies, which realizes a geographical structure combining with migration rates. Such a model was first introduced by Kimura [2] into population genetics, and many works have been made concerning this model, (see Maruyama [5]). They actually treated Markov chain models and calculated some genetical quantities, but their results remain mathematically primitive. We will here treat a diffusion model derived by taking a diffusion approximation from the Markov chain model. Then we can obtain deeper results by making use of nice techniques such as duality and coupling, which have been successfully used in the theory of interacting particle systems (cf. Liggett [4]).

We will also discuss a different type of stepping stone model arising in population dynamics. The model is formulated as a continuous time Markov chain which is defined by taking account of a strong restraint of population in each colony into an ordinary branching process with a spatially movement.

These two stepping stone models apparently describe different phenomena. Nevertheless there is a closed relation between them and many results for one are translated to the other immediately.

The purpose of this report is mainly to survey results concerning stationary states and ergodic theorems for the genetical stepping stone

S. Albeverio et al. (eds.), Stochastic Processes in Physics and Engineering, 345–355.

model, and an extinction problem and limit behaviors for the population dynamical stepping stone model. Furthermore we will attempt to generalize the space of colonies to a continuum space as much as possible.

2. Genetical stepping stone model

Let us first discuss a genetical stepping stone model. Let S be a countable set. Each element i of S is interpreted as a colony. We here assume that there are two alleles A and a in each colony, and denote by x_i $(1-x_i)$ the gene frequency of the A-allele (resp. the a-allele) in the colony i. We further assume that a time evolution is caused by mutation, selection, migration and random sampling. Then the time evolution is described by the following stochastic differential equation (SDE):

$$(2.1)_a \qquad dx_i(t) = \sqrt{x_i(t)(1-x_i(t))}\, dB_i(t) + b_i(x(t))dt$$

$$(2.1)_b \qquad 0 \leqq x_i(t) \leqq 1 \quad \text{for every} \quad t \geqq 0 \quad \text{and} \quad i \in S$$

$$(2.1)_c \qquad b_i(x) = v(1-x_i) - ux_i + sx_i(1-x_i) + \sum_{j\in S} m_{ji}(x_j - x_i)$$

where $B_i(t)$, $i \in S$ is a collection of independent one-dimensional standard Brownian motions, s is a real constant, and u, v and m_{ji} $(j \neq i)$ are non-negative constants.

Under the assumption; $\sup_{i\in S} \sum_{j\neq i} m_{ji} < +\infty$, the SDE (2.1) is uniquely solvable, so that we have a diffusion process $(x(t), P_x)$ taking values in $X = [0,1]^S$ (the configulation space of gene frequencies).

Genetically, u and v are mutation rates from A to a and a to A respectively, s is a selection rate i.e. $s > 0$ $(s < 0)$ means A is superior (resp. inferior) to a in selection, and m_{ji} stands for a migration rate from j to i. Therefore our main concern is the influences of these parameters on the genetical evolution.

For each $x \in X$, δ_x stands for the unit mass measure at x. Especially δ_1 (δ_0) is a uniform state in which all individuals are of the A-allele (resp. the a-allele) in all colonies. For a probability measure μ on X, (which we call a state and denote the totality of states by $\mathcal{P}(X)$), $(x(t), P_\mu)$ stands for the diffusion process with the initial state μ, so that $P_\mu(\cdot) = \int \mu(dx) P_x(\cdot)$. Let $T_t\mu$ be the distribution of $x(t)$ under the probability law P_μ. If $T_t\mu$ is constant in $t \geqq 0$, then μ is a stationary state, of which totality is denoted by $\mathcal{S}$. To investigate the structure of $\mathcal{S}$ it suffices to see $\mathcal{S}_{ext}$, the totality of extremal element of $\mathcal{S}$, since $\mathcal{S}$ is non-empty, compact, convex and simplex.

Among the genetical factors the mutation is most influential in the evolution.

Theorem 2.1 ([7]). Suppose $u > 0$, $v > 0$ and s is arbitrary. Then there is only one stationary state, and the process $(x(t), P_x)$ is ergodic.

Thus when a both-way mutation works, the selection force and the geographical factor do not appear so explicitly, and a genetical equilibrium is realized very soon. However, if a one-way mutation is involved, then one can observe some critical phenomenon concerning the selection parameter. Let us confirm it in the following homogeneous case:

(2.2) $S = Z^d$ (d-dimensional cubic lattice), $m_{ji} = m_{i-j}$ and (m_{ji}) is irreducible.

Theorem 2.2 ([9]). Suppose $u > 0$ and $v = 0$, in addition to (2.2). Then there exists a critival value $0 < s_c < \infty$ such that if $s < s_c$, δ_0 is the unique stationary state and $\lim_{t\to\infty} T_t\mu = \delta_0$ holds for every $\mu \in \mathcal{P}(X)$, while if $s > s_c$, there is another stationary state ν such that

$$\lim_{t\to\infty} T_t\mu = \mu(\{0\})\delta_0 + (1-\mu(\{0\}))\nu$$

for every Z^d-translation invariant state μ .

If we assume $u = 0$ and $v > 0$, instead of $u > 0$ and $v = 0$, then the same statement is valid by replacing 0 by 1 . We notice that if $u > 0$ and $v = 0$, δ_0 is always a stationary state. From the theorem it follows that δ_0 is stable for small s , but when s increases across the critical value, δ_0 becomes unstable and, instead of this, another stable stationary state ν appears.

Theorem 2.3 ([9]). Suppose $u = v = 0$ and $s > 0$, in addition to (2.2). Then

$$\lim_{t\to\infty} T_t\mu = \mu(\{0\})\delta_0 + (1-\mu(\{0\}))\delta_1$$

for every Z^d-translation invariant state μ . If $s < 0$, the same result is valid by replacing 0 and 1 .

In this case both δ_0 and δ_1 are stationary states, and their stability is determined by the sign of s . In Theorem 2.2 and 2.3 it is an indispensable assumption that the set of colonies is an infinite lattice space, but even then the geographical structure does not work effectively. However, if both the mutation pressure and the selective force are neglected, then the migration factor becomes more sensitive in the genetical evolution, and this case corresponds to Kimura's original model [2], [3].

From now on, let us assume $u = v = s = 0$, (but (2.2) is not assumed). Then the stepping stone model is described by

$$dx_i(t) = \sqrt{x_i(t)(1-x_i(t))}\, dB_i(t) + \sum_{j\in S} m_{ji}(x_j(t) - x_i(t))dt \, . \tag{2.3}$$

Let $M = (m_{ij})$ with $m_{ii} = \sum_{j\neq i} m_{ji}$, and let M^* be the transposed of M. Then $P_t = \exp tM^*$ defines a transition probability of a continuous time Markov chain on the state space S. We note the P_t-Markov chain should be interpreted as a temporally backward movement of the population.

From (2.3) it follows obviously

$$EX_i(t) = \sum_{j\in S} P_t(i,j)EX_j(0) \, . \tag{2.4}$$

Thus the P_t-Markov chain (X_t, P_i) (in particular P_t-harmonic functions) plays an essential role to investigate limit behaviors of the diffusion model.

Let us introduce two classes of P_t-harmonic functions H and H^*; let $((X_t^1, X_t^2), P_{(i_1,i_2)})$ be the product process of two P_t-Markov chains (X_t, P_{i_1}) and (X_t, P_{i_2}), and let Ω_∞ be the event that X_t^1 and X_t^2 collide infinitely many often. Let

$$H = \{h : S \to [0,1],\ P_t h = h \text{ for every } t \geqq 0\} ,$$

$$H^* = \{h \in H : \lim_{t\to\infty} h(X_t^1) = \lim_{t\to\infty} h(X_t^2) = 0 \text{ or } 1 \text{ a.s. on } \Omega_\infty \text{ w.r.t. } P_{(i_1,i_2)} \text{ for all } i_1 \text{ and } i_2\} .$$

Then we have

<u>Theorem 2.4</u> ([7]). (i) For each $h \in H^*$ there exists a unique extremal stationary state ν_h such that

$$\int x_i \nu(dx) = h(i) \quad \text{for all} \quad i \in S ,$$

(ii) $$S_{ext} = \{\nu_h : h \in H^*\} .$$

(iii) Let $h \in H^*$ and $\mu \in P(X)$. Then

$$\lim_{t\to\infty} T_t\mu = \nu_h \quad \text{if and only if}$$

$$\lim_{t\to\infty} \sum_{j\in S} P_t(i,j) \int x_j \mu(dx) = h(i) \text{ , and}$$

$$\lim_{t\to\infty} \sum_{j\in S} \sum_{k\in S} P_t(i,j)P_t(i,k)\int x_j x_k \mu(dx) = h(x)^2 \quad \text{for all} \quad i \in S .$$

Let us explain Theorem 2.4 by two illustrative examples.

Example 1. Let $S = Z^d$ and assume (2.2). If $d = 1$ or 2, then $\mathcal{H}^* = \{0,1\}$ and $\mathcal{S}_{ext} = \{\delta_0, \delta_1\}$, so that only genetically uniform states appear. On the other hand if $d \geq 3$, then $\mathcal{H}^* = \mathcal{H} = \{c : 0 \leq c \leq 1\}$. Hence we see $\mathcal{S}_{ext} = \{\nu_c : 0 \leq c \leq 1\}$. Moreover ν_c is a fluctuating state for each $0 < c < 1$ which has a strong correlation such that

$$\int x_i x_j \nu_c(dx) - c^2 \sim K(c)/|i-j|^{d-2} \quad \text{as} \quad |i-j| \to \infty$$

where $0 < K(c) < \infty$ is a constant depending on $0 < c < 1$. Furthermore, it can be shown that any finite dimensional marginal distribution of ν_c $(0 < c < 1)$ has a smooth density with respect to the Lebesque measure ([9]).

Example 2. Let S be a set that consists of countably many half-lines that join together at the origin, i.e. let $L_n = \{(n,i): i = 1,2,3,\ldots\}$ $(n = 1,2,3,\ldots)$, $S = \{0\} \cup L_1 \cup L_2 \cup L_3 \cup \ldots$.

Suppose that $m_{(n,i),(n,i-1)} = p$ ($(n,0)$ is identified with 0) and $m_{(n,i),(n,i+1)} = q$ for $n \geq 1$ and $i \geq 1$, $m_{0,(n,1)} = a_n$ and $m_{j,k} = 0$ for all other $j \neq k$. We further assume that (a_n) are positive and summable. Then if $p \leq q$, then $\mathcal{H}^* = \{0,1\}$ and $\mathcal{S}_{ext} = \{\delta_0, \delta_1\}$. On the other hand, if $p > q$, then $\mathcal{H}^* = \{h_A : A \subset \{1,2,3,\ldots\}\}$ where $h_A(n,i) = P_{(n,i)}$ (X_t diverges to infinity along L_m for some $m \in A$ as $t \to \infty$). Therefore $\mathcal{S}_{ext} = \{\nu_A : A \subset \{1,2,3,\ldots\}\}$.

This example has an interesting interpretation. Imagine the origin as a big city like Tokyo in Japan. There are many railroads from the country-sides towards the center. Also they have no communication with each other except through the center. If $p < q$ we observe the tendency that much more people move towards their home towns rather than towards the center. Then the genetical character would spread from the center to the country-sides, and the uniform states would be realized in the whole country. On the other hand $p > q$ means that much more people gather in Tokyo from countrysides, which fits in with the present situation of Japan. Thus the coexistence of various cultures, dialects and genetical characters is realized.

3. A stepping stone model in population dynamics

Let us introduce a stochastic population model which is a continuous time Markov chain $(\underline{n}(t), P_{\underline{n}})$ on a space of population vectors

$N = \{\underline{n} = (n_i)_{i\in S}: n_i = 0,1,2,\cdots, |\underline{n}| = \sum_{i\in S} n_i < +\infty\}$. Now we take a countable space S as a set of colonies. The infinitesimal transition is given by

$$(3.1)\qquad \begin{array}{lll} \underline{n} \to \underline{n} + e_i & \text{with } \beta n_i & \text{(birth)} \\ \underline{n} \to \underline{n} - e_i & \text{with } \delta n_i + \gamma n_i(n_i - 1) & \text{(death)} \\ \underline{n} \to \underline{n} - e_i + e_j & \text{with } q_{ij} n_i & \text{(migration)} \end{array}$$

where e_i stands for an element of N with all components zero except 1 in the colony i, and β, γ, δ and q_{ij} $(i \neq j)$ are non-negative constants.

If the quadratic term is neglected in the death rate (i.e. $\gamma = 0$), then the model is nothing but a branching process with countably many types. However, the quadratic term is very important from the view point of population dynamics, since foods and living materials are limited in each colony. Also the quadratic term is interpreted as a competition factor. If S is a finite set and if $\beta > 0$ and $\gamma > 0$, then the population attains to an equilibrium or extincts very soon according as $\delta = 0$ or $\delta > 0$ respectively. While if S is infinite the circumstance changes drastically. To see this let us consider a spatially homogeneous situation:

$$(3.2)\qquad S = Z^d, \quad q_{ij} = q_{j-i,0} \quad \text{and} \quad (q_{ij}) \text{ is irreducible.}$$

Then one can observe a critical phenomenon as follows.

<u>Theorem 3.1</u> ([9]). Let $\gamma > 0$ and $\delta > 0$ be fixed. Then there exists a critical parameter $0 < \beta_c < \infty$ such that if $\beta < \beta_c$, $P_{\underline{n}}(\lim_{t\to\infty} |\underline{n}(t)| = 0) = 1$ for every $\underline{n}$, while if $\beta > \beta_c$,

$$P_{\underline{n}}(\lim_{t\to\infty} |\underline{n}(t)| = +\infty) > 0 \quad \text{for every } \underline{n} \neq 0 .$$

<u>Theorem 3.2</u> ([9]). Suppose that $\gamma > 0$, $\delta = 0$ and $\sum_{i\in Z^d} |i| q_i < +\infty$.

$$P_{\underline{n}}(\lim_{t\to\infty} |\underline{n}(t)| = +\infty) = 1 \quad \text{for every } \underline{n} \neq 0 .$$

We remark that Theorem 3.2 does not imply the existence of an equilibrium of population. However, under a symmetry assumption; $q_i = q_{-i}$ for all $i \in Z^d$, one can show

$$\lim_{t\to\infty} P_{\underline{n}}(n_i(t) \geq 1) > 0 \quad \text{for all } i \in Z^d \text{ and } \underline{n} \neq 0 .$$

Furthermore, we conjecture that for every $\underline{n} \neq 0$ the distribution of

$(\underline{n}(t),P_{\underline{n}})$ converges to a unique equilibrium that is an infinite product of identical Poisson distributions with parameter β/γ where it should be noted that the migration rate (q_i) does not involve the equilibrium distribution.

4. Duality between two stepping stone models

As was seen in the above, two stepping stone models describe different phenomena, but they are closely related from a mathematical point of view. Let $(x(t),P_x)$ be the stepping stone model in population genetics associated with the SDE (2.1), and let $(\underline{n}(t),P_{\underline{n}})$ be the stepping stone model in population dynamics which is governed by (3.1). For each $\underline{n} \in N$ we set $f_{\underline{n}}(x) = \prod_{i\in S} (1-x_i)^{n_i}$ and $f_{\underline{o}}(x) = 1$.

Then we have the following relation between them.

Theorem 4.1. Assume that $s \geqq 0$, $\beta = s$, $\gamma = \frac{1}{2}$, $\delta = u$ and $q_{ij} = m_{ji}$. Then for each $\underline{n} \in N$ and $x \in X$

$$E_x(f_{\underline{n}}(x(t))) = E_{\underline{n}}(f_{\underline{n}(t)}(x)\exp(-\int_o^t v|\underline{n}(s)|ds)) .$$

Accordingly, in order to investigate the population genetical model it suffices to analize the population dynamical model. We actually took this way, but even the analysis of the dynamical model is not so easy, mainly due to the presence of the quadratic term in the death rate. (For the detail see [9]).

5. General formulation of the genetical stepping stone model

In this section we would like to attempt to generalize the stepping stone model associated with (2.1) as much as possible so that one may treat a continuum space of colonies. Let S be a locally compact separabel metric space with the metric ρ and the Borel field $\mathcal{B}_S$, and let m be a σ-finite measure on $(S,\mathcal{B}_S)$. $L^2(m)$ stands for the space of all m-square integrable functions on S. $B_t = (B_t(\phi): \phi \in L^2(m))$ is an $L^2(m)$-cylindrical $(\mathcal{F}_t)$-Brownian motion if it is a Gaussian system defined on a probability space $(\Omega,\mathcal{F},P)$ with a filtration $(\mathcal{F}_t)$ such that

(i) for each $\phi \in L^2(m)$, $B_t(\phi)$ is a one-dimensional $(\mathcal{F}_t)$-Brownian motion with the variance $t\|\phi\|_m^2$, and

(ii) for every $\phi \in L^2(m)$, $\psi \in L^2(m)$, $\alpha \in R$ and $\beta \in R$ $B_t(\alpha\phi+\beta\psi) = \alpha B_t(\phi) + \beta B_t(\psi)$ holds almost surely.

Then there corresponds a Gaussian random measure $B(dt\,dx)$ satisfying that

$$B_t(\phi) = \iint I_{[o,t]}(s)\phi(x)B(ds\,dx) .$$

Furthermore, for every $(\mathcal{F}_t)$-adapted functional Φ satisfying that

$$\int_0^t\int_S \Phi(s,x,\omega)^2 ds\,dx < +\infty \quad \text{for every} \quad t \geqq 0 ,$$

one can define a stochastic integral

$$(\Phi B)_t = \int_0^t\int_S \Phi(s,x,\omega)B(ds\,dx)$$

as a continuous $(\mathcal{F}_t)$-local martingale having the following quadratic variation process

$$<\Phi B>_t = \int_0^t\int_S \Phi(s,x,\omega)^2 ds\; m(dx) .$$

Suppose that we are given the following a, b and G :

(i) $a(y): [0,1] \to R$ continuous and $a(0) = a(1) = 0$,

(ii) $b(y): [0,1] \to R$ continuous, $b(0) \geqq 0$ and $b(1) \leqq 0$,

(iii) G is a generator of a Markov process on the state space S .

Then a formal analogue of the equation (2.1) is

$$(5.1)_a \qquad X_t(x) = X_0(x) + \int_0^t a(X_s(x))dB_s(x) + \int_0^t b(X_s(x))ds + \int_0^t GX_s(x)ds$$

$$(5.1)_b \qquad X_t \in C_{[0,1]} \equiv \{\phi : S \to [0,1] \ \text{continuous}\} \quad \text{for all} \quad t \geqq 0 .$$

Here the stochastic integral of the r.h.s. of $(5.1)_a$ is just formal and one should understand like $dB_s(x) = $ "$B(ds\,dx)/m(dx)$" .

In order to make sense the equation (5.1) let us introduce the following conditions on G : let $C_0(C_\infty)$ be the totality of continuous functions with compact support (resp. vanishing at infinity) defined S .

(G.1) $(G,\mathcal{D}(G))$ generates a strong continuous Markov semigroup P_t on C_∞ , and a subset of C_0 is a core of $(G,\mathcal{D}(G))$,

(G.2) P_t admits a transition density $P_t(x,y)$ w.r.t. $m(dy)$ that are jointly continuous in $t > 0$, x and y ,

(G.3) for some $0 < \alpha < 1$, $\sup_{0\leqq t\leqq T} \sup_x P_tP_t^*(x,x)/t^\alpha < +\infty$,

(G.4) for every compact subset K, every $h_o > 0$ and every $t > 0$ there exist constants $C > 0$ and $\beta > 0$ such that

$$\int_0^t \int_S (P_{s+h}(x,y) - P_s(x',y))^2 ds\, m(dy) \leqq C(h^\beta + \rho(x,x')^\beta)$$

for every $0 < h < h_o$, $x \in K$ and $x' \in K$.

Under the condition (G) one can consider the following stochastic integral equation instead of (5.1):

$$(5.2)_a \qquad X_t(x) = P_t X_o(x) + \int_0^t \int_S P_{t-s}(x,y) a(X_s(y)) B(ds\, dy) + \int_0^t \int_S P_{t-s}(x,y) b(X_s(y)) ds\, m(dy)$$

$$(5.2)_b \qquad X_t \in C_{[0,1]} \quad \text{for all} \quad t \geqq 0 .$$

Then we have

Theorem 5.1.

(i) Suppose that $a(y)$ and $b(y)$ are Lipschitz continuous in addition to the above stated assumptions. Then for every $X_o \in C_{[0,1]}$ there is a unique (pathwise) solution of (5.2).

(ii) Let $a(y) = \sqrt{y(1-y)}$ and $b(y) = v(1-y) - uy + sy(1-y)$ as in $(2.1)_c$. Then for every $X_o \in C_{[0,1]}$ there exists a unique (in the law sense) solution of (5.2).

Example 1. Let S be a countable set and m be the uniform measure on S. Then G is an infinitesimal matrix of a continuous time Markov chain. Then (5.2) reduces to the original stepping stone model of (2.1).

Example 2. Let $S = R^d$ and $G = -(-\Delta)^{\alpha/2}$ $(0 < \alpha \leqq 2)$. Then the condition tion (G) is fulfilled if and only if $d = 1$ and $1 < \alpha \leqq 2$. Otherwise there is no solution of the equation (5.2) unless $a(y)$ is trivial.

Example 3. Let $a(y) = \varepsilon y(1-y)$ $(\varepsilon > 0)$ and $b(y) = y(1-y)$. Futher assume $S = R^1$ and $G = -(-\Delta)^{\alpha/2}$ with $1 < \alpha \leqq 2$. Then the equation (5.2) is a random K-P-P equation;

$$\frac{\partial X_t(x)}{\partial t} = GX_t(x) + (1 + \varepsilon\dot{B}_t(x)) X_t(x)(1-X_t(x))$$

where $\dot{B}_t(x)$ is a space-time white noise.

Example 4. Let $S = \{x = (x_1,\ldots,x_d) : x_1,\ldots,x_d$ are integers except at most one component$\}$, which is called "a d-dimensional cubic net". On the set S one can define a Brownian motion, of which generator would satisfy the condition (G).

In order to establish "duality" we assume $a(y) = \sqrt{y(1-y)}$ and $b(y) = v(1-y)-uy+sy(1-y)$ as in (2.1), and denote by (X_t, P_X) the associated diffusion process on the state space $C_{[0,1]}$. Let $\hat{S} = \bigcup_{n=0}^{\infty} S^n/(\text{permutation})$ be the configulation space of finitely many particles. Each element of $\hat{S}$ is denoted by $\underline{x} = (x_1, \ldots, x_n)$ which is permutation invariant, and set $|\underline{x}| = n$. For $\underline{x} = (x_1, \ldots, x_n)$ we define $\underline{x}^i = (x_1, \ldots, x_{i-1}, x_{i+1}, \ldots, x_n) \in S^{n-1}$ and $\underline{x}^{i*} = (x_1, \ldots, x_n, x_i) \in S^{n+1}$, which are obtained from $\underline{x}$ by deleting and adding x_i, respectively. Now let us consider the following (formal) generator on the state space $\hat{S}$:

$$Af(\underline{x}) = \sum_{i=1}^{n} G_i f(\underline{x}) + \frac{1}{2} \sum_{1 \leq i \neq j \leq n} \delta(x_i, x_j)(f(\underline{x}^i) - f(\underline{x})) \tag{5.3}$$

$$+ u \sum_{i=1}^{n} (f(\underline{x}^i) - f(\underline{x})) + s \sum_{i=1}^{n} (f(\underline{x}^{i*}) - f(\underline{x}))$$

for $\underline{x} = (x_1, \ldots, x_n) \in \hat{S}$ and $Af(\underline{0}) = 0$, where $\delta(x,y)$ is a "δ-function" at the diagonal set $\{(x,y) \in S \times S : x = y\}$. The condition (G) would guarantee the existence of the local time at the diagonal set for the two particle motion of independent P_t-Markovian particles. Using the local time we can construct a Markov process $(\underline{x}(t), P_{\underline{x}})$ uniquely on the state space $\hat{S}$, that is governed by A of (5.3).

For each $\underline{x} = (x_1, \ldots, x_n) \in \hat{S}$ and $X = (X(x)) \in C_{[0,1]}$ set

$$F(\underline{x}, X) = \prod_{i=1}^{n} (1 - X(x_i)) \quad \text{and} \quad F(\underline{0}, X) = 1 .$$

Then one can establish the following duality:

<u>Theorem 5.2</u>. Let $u \geqq 0$ and $s \geqq 0$. Then for $\underline{x} \in \hat{S}$ and $X \in C_{[0,1]}$

$$E_X(F(\underline{x}, X_t)) = E_{\underline{x}}(F(\underline{x}(t), X) \exp(-v \int_0^t |\underline{x}(s)| ds)) .$$

<u>Acknowledgements</u>

This report was completed while the author stayed at BiBoS, Bielefeld University and ETH, Forschungsinstitut für Mathematik, Zürich. He would like to express his gratitude to Professors S. Albeverio, Ph. Blanchard and L. Streit of BiBoS and Professor H. Föllmer of ETH for their kind hospitality.

REFERENCES

[1] R.A. Holley & T.M. Liggett; Ergodic theorems for weakly interacting infinite systems and the voter model, Ann. Prob. 3 (1975), 643-663.

[2] M. Kimura; "Stepping stone" model of population, Ann. Rept. Nat. Inst. Genetics Japan 3 (1953), 62-63.

[3] M. Kimura & G.H. Weiss; The stepping stone model of population structure and decrease of genetical correlation with distance, Genetics 49 (1964), 561-576.

[4] T.M. Liggett; Interacting particle systems, Springer-Verlag, New York, Berlin, Heidelberg, Tokyo (1985).

[5] T. Maruyama; Stochastic problems in population genetics, Lecture Notes in Biomathematics 17, Springer-Verlag (1977).

[6] S. Sawyer; Results for the stepping stone model for migration in population genetics, Ann. Prob. 4 (1976), 699-728.

[7] T. Shiga; An interacting system in population genetics, J. Math. Kyoto Univ. 20 (1980), 212-242 and 723-733.

[8] T. Shiga; Continuous time multi-allelic stepping stone models in population genetics, J. Math. Kyoto Univ. 22 (1982), 1-40.

[9] T. Shiga & K. Uchiyama; Stationary states and their stability of the stepping stone model involving mutation and selection, to appear in Z. Wahr. Verw. Geb. (1986).

POWER SPECTRUM AND FREDHOLM DETERMINANT RELATED TO INTERMITTENT CHAOS

Y. TAKAHASHI
Dept. of Pure and Applied Sciences
University of Tokyo
Komaba, Meguro, Tokyo 153, Japan

ABSTRACT. We present a mathematical aspect to the problem of intermittent chaos. The singularity of power spectra, such as $1/\omega$-spectrum, is shown for a class of correlation functions under the dynamics called semi-Bernoulli systems and is also proved to be common among this class. The singularity is described by a power series called the Fredholm determinant, which may be regarded as det(I-zL) for the transfer operator L although L has a strange spectral property: every complex number in the open unit disk is an eigenvalue. Finally examples are given among maps of intervals and among statistical mechanics of one dimensional lattice models to conclude that the $1/\omega$-spectrum is a critical phenomenon at the tri-critical point.

0. INTRODUCTION

The power spectrum is an important tool to analyze time series in various branches of sciences and engineering but it does not seem to attract much attention of mathematicians. One of the reasons consists in the following theorem of Kolmogorov and Sinai:

Every purely nondeterministic dynamical system has the common spectral type (called the countable Lebesgue spectrum).

Roughly to say, this theorem says that, if you observe two systems with mixing property of the strongest kind, then the power spectrum of a (coordinate) function in one system can be observed in another system (possibly, for a very bad function). Consequently the problem of power spectrum cannot be set up as a problem for dynamical systems but it is a problem for the pair of a dynamical system and a function (taken as the coordinate).

On the other hand, the power spectrum works well in various situations and sometimes the typical character of the power specteum seems even independent of the choice of coordinate functions. In particular, this is the case of experiments around the problem of intermittent chaos where they discuss the singularity of the power spectrum.

The intermittent chaos is a phenomenon of a time series which is characterized by consecutive bursts and laminar parts. It is believed

S. Albeverio et al. (eds.), Stochastic Processes in Physics and Engineering, 357–379.

to be one of the basic routes to the on-set of turbulence and it can also be observed among time series obtained from simple dynamical systems, such as maps of intervals. (By the way, we emphasize here that maps of intervals as dynamical systems should not be regarded as a system with one degree of freedom: for instance, every finite Markov chain can be expressed by a piecewise linear map of the unit interval.)

Since the work of Pomeau and Manneville[18] several physicists have reported its close relationship to the long time tail or to the power spectrum of inverse power type, $1/\omega^{\alpha}$(e.g.cf.[2,15,16]).

The power laws are usually convincing in physical literatures from the view point of the renormalization group by varying the control parameter and then observing the asymptotic behaviour of the power-like laws of individual spectra. Nevertheless the $1/\omega$-type singularity near $\omega=0$ is extremely difficult to verify by experimental methods since $\omega=0$ corresponds to an infinite time interval.

Now let us state our purpose of the present report.

(a) To present a mathematical mechanism in dynamical systems to exhibit the power spectra of the inverse power type at the origin for the time series obtained from the observation of coordinate functions.

Here the mechanism should be stable under small perturbations of the systems and of the choice of coordinate functions.

By the way the long time tail and the power law are mutually equivalent as is known by the Tauberian theorems and their inverse, Abelian theorems([34]).

(b) To characterize the exponents of the power laws.

(c) To clarify the meaning of the power laws in the frame work of the (mathematical theory of) statistical mechanics.

Here it is natural to expect that the inverse power law is also a kind of critical phenomena since it is the case in the theory of phase transisions in statistical physics and this theory has been extended to the maps of intervals([17,29,30]) In particular one may conjecture a universality in the power law: the exponent depends only on the rough structure of dynamical systems and does not depend upon the detailed structures.

Let us summarize the answers to these problems since they are stated below in the mathematical terminologies (cf.[26,32,8,3]).

(A1) A class of dynamical systems is introduced in 2.2. We call them semi-Bernoulli. They are the systems that admit Bernoulli systems as Poincare maps. In other words, they are special flows (or suspensions) over Bernoulli systems in continuous time cases. The typical example of a semi-Bernoulli system in differentiable dynamical systems is a suspension over a diffeomorphism with hyperbolic structure, which is structurally stable, i.e., stable under small perturbations of the system. All the systems with hyperbolic attractors or strange attractors seem semi-Bernoulli (except for a technical condition (2.10) below) so far as the proofs are given ([8,1,19,25]).

(A2) Each semi-Bernoulli system admit a special, natural coordinate (the height from the Poincare section) and it is associated with a power series d called the Fredholm determinant[27] (which is the reciprocal of a kind of Artin-Mazur-Ruelle zeta funcdtion [21]). If one takes as the coordinate any function near to the height function, then the power spectrum $S(\omega)$ has the asymptotics

$$S(\omega)\sim \text{const.} S^*(\omega) \quad \text{as} \quad \omega\to 0$$

where const. depends on the choice of the coordinate but

$$S^*(\omega) = 2.\mathrm{Re}\, d(\omega)^{-1}$$

is common. ($f(\omega)\sim g(\omega)$ means $\lim f(\omega)/g(\omega) = 1$.)

(B) The asymptotics of S^* or d at $\omega=0$ is determined by the tail distribution of the return time (or the ceiling function). The exponent α given by the formula

$$S^*(\omega)\sim \text{const.}\, \omega^{-\alpha}L(\omega) \quad \text{as} \quad \omega\to 0$$

can vary exactly from 0 to 1 , where L is a slowly varying function (such as constants, log, 1/log etc., cf.[6]). Hence the exact $1/\omega$-spectrum does exist in the mathematical sense (with the correction term L that makes S^* integrable).

(c) To see the relationship to the statistical mechanics we construct a classical statistical mechanics of one dimensional lattice system, which shows the phase transition and the critical phenomenon. It is proved to be isomorphic to a family of maps of the unit interval: each lattice gas (or spin) configuration corresponds to a point of the interval and the spatial translation on the lattice is conjugate to the action of the map. Although we present only one example among possible constructions, we may conclude from this that the power law for the maps is exactly the same as that associated with the critical phenomenon for the lattice systems. The dependence of the exponent on the system parameters is given in Section 5 for this example.

Finally the author would like to express his sincere thanks to BiBoS and its members for the hospitality during his stay of half a year.

1. TRANSFER OPERATOR

1.1. Definition

Let (Z,λ) be a probability measure space and consider a nonsingular transformation g on (Z,λ) . Since the measure λg^{-1} is absolutely continuous w.r.t. λ by the definition, g induces an action on the space of the measures which are absolutely continuous w.r.t. λ . In other words, one can associate an operator $L=L_g$ on $L_1(Z,\lambda)$, called the <u>transfer operator</u> such that

$$\int_Z Lu(z)v(z)\lambda(dz) = \int_Z u(z)v(gz)\lambda(dz) \tag{1.1}$$

$$(u\in L_1(Z,\lambda)\ ,\ v\in L_\infty(Z,\lambda))\ .$$

For instance, if $Z = A^{\mathbf{N}}$ with A finite and g is the shift on Z , i.e.,

$$(gz)_n = z_{n+1} \quad (n \in \mathbf{N}) \quad \text{for} \quad z = (z_n) \in Z ,$$

then the transfer operator can be given as

$$(1.2) \qquad Lu(z) = \sum_{z' \in g^{-1}z} j(z')u(z')$$

where j is a function (called the jacobian) defined as

$$j(a,z_0,z_1,\ldots) = L\, 1_{[a]}(z) \qquad z=(z_n)$$

with

$$1_{[a]}(z) = \begin{cases} 1 & \text{if } z_0=a \\ 0 & \text{otherwise} . \end{cases}$$

In fact, using the general formula

$$(1.3) \qquad L(u(w\, g)) = w(L_g u) ,$$

one obtains

$$\begin{aligned} Lu(z) &= \sum_{a \in A} L(1_{[a]}u)(z) \\ &= \sum_{a \in A} L(1_{[a]}(u_a\, g))(z) \qquad (u_a(z) = u(a,z_0,z_1,\ldots)) \\ &= \sum_{a \in A} L(1_{[a]})(z)u_a(z) = \sum_{z' \in g^{-1}z} j(z')u(z') . \end{aligned}$$

Note that the transfer operator becomes trivial if g is invertible since then $L_g u = u\, g^{-1}$. But it has been a very useful tool for the study of non-invertible transformations, such as, maps of intervals. For example, a function u on Z is an invariant density function if

$$L_g u = u$$

and the mixing properties follow from the study of the iterates

$$L^n = L_g^n = L_{g^n} \qquad (g^n = \text{the n-fold iterate of } g) .$$

Furthermore, given a non-invertible (Z,λ,g) one can always construct its extension to invertible transformation (Y,μ,f), called the natural extension of (Z,λ,g), by taking the inverse limit (of the system $\ldots \xrightarrow{g} (Z,\lambda) \xrightarrow{g} (Z,\lambda)$). For instance in the case of the shift,

$$Y = A^{\mathbf{Z}} = \text{the space of two-sided sequences}$$

$$f \text{ is the shift on } Y; \quad (fy)_n = y_{n+1} .$$

Furthermore, a function u on Z can be identified with a function w on Y which is given in the shift case by

$$w(y) = u(y_0,y_1,\ldots) \qquad y = (y_n)_{n \in \mathbf{Z}} \in Y .$$

Conversely, every function w on Y can be approximated by function $u_n \circ f^n$ with functions u_n on Z. Hence, for example for the mixing properties of (Y, μ, f) it is enough to study the operator L.

On the other hand, given an invertible bi-measurable transformation (Y, μ, f) there are many choices of non-invertible transformation (Z, λ, g) whose natural extension is (Y, μ, f).

1.2. Spectrum of transfer operator (I): On L_1-space

The following is a version of a result in [27] for piecewise smooth maps of intervals.

Theorem 1. Let A be a finite set with more than one points, $Z = A^{\mathbf{N}}$, g shift and λ a nonsingular measure with nonvanishing jacobian. Then every complex number α with modulus $|\alpha|<1$ is the eigenvalue of L on $L_1(Z, \lambda)$ with infinite multiplicity and with infinite Jordan block size.

Proof. The essence of the proof in [27] is the existence of the following sets Z_0, B', B'' and B:

(a) a measurable set Z_0 such that $\lambda(Z_0)>0$, $gZ_0 = Z$ and $g = Z_0 \to Z$ is bi-measurable bijection.

(b) measurable sets B', B'' and B with positive λ-measures such that B_1 and B_2 are disjoint and $g: B' \to B$ and $g: B'' \to B$ are bi-measurable bijection.

In the present case, taking an element $a \in A$ one finds that $Z_0 = [a]$ gives (a). For (b) one can take $B=Z$, $B'=[a_1]$ and $B''=[a_2]$ for any distinct $a_1, a_2 \in A$.

It follows from (a) that there is a right inverse K of L, $LK = I$. Indeed,

$$Ku(z) = u(gz)\frac{d\lambda(gz)}{d\lambda(z)}1_{Z_0}(z) .$$

It is obvious that K is an isometry on $L_1(Z, \lambda)$.

From (a) one obtains a function

$$k(z) = 1_{B'}(z)\frac{d\lambda(gz)}{d\lambda(z)} - 1_{B''}(z)\frac{d\lambda(gz)}{d\lambda(z)}$$

such that $L\,k = 0$. Hence,

$$\text{Ker } L \supset \{(u \circ g)k;\ u \in L_\infty(Z, \lambda)\} .$$

In particular, $\dim(\text{Ker } L_g) = \infty$.

Now for a given $\alpha \in \mathbf{C}$ such that $|\alpha| < 1$, take any $v \in \text{Ker } L$

and put

$$u = \sum_{n=0}^{\infty} \alpha^n K^n v = (I-\alpha K)^{-1} v .$$

Then u converges since K is an isometry on $L_1(Z,\lambda)$ and

$$Lu = \sum_{n=1}^{\infty} \alpha^n LK^n v = \alpha \sum_{n=0}^{\infty} \alpha^n K^n v = \alpha u .$$

Hence u is an eigenfunction.

Next for any given $u \in L(Z,\lambda)$ put

$$w = \sum_{n=0}^{\infty} \alpha^n K^{n+1} u = (I-\alpha K)^{-1} Ku .$$

Then,

$$Lw = LKu + \sum_{n=1}^{\infty} \alpha^n LK^{n+1} u = u+\alpha w .$$

Hence one finds that α ia an eigenvalue of L on $L(Z,\lambda)$ with infinite multiplicity and with infinite Jordan block size.

Remark. In Theorem 1 we assumed a very strong condition that the jacobian is positive. But the reader will find that the conditions (a) and (b) hold under much milder assumptions. For example, they holds if for almost every z there are two distinct elements a_1, a_2 of A such that both of $j(a_i z)$, $i=1,2$, are positive.

1.3. Spectrum of the transfer operator (II): Fredholm determinant

As we have seen it in the previous section the transfer operator on L_1-space has the strange property that the open unit disk consists of eigenvalues. Nevertheless the transfer operator often admits a natural invariant subspace and its restriction to this subspace has discrete point spectrum, and which is given by zeros of a certain power series called the Fredholm determinant.

The simplest example is as follows. Let A be finite, $Z = A^{\mathbf{N}}$ and g the shift on Z . Assume that λ is a Markov measure on Z , such that

$$\lambda\{z = (z_1);\ z_0 = a_0 , \ldots, z_n = a_n\}$$

$$= \pi(a_0)p(a_0,a_1) \ldots p(a_{n-1}, a_n) \quad n \geq 0 , a_0, \ldots, a_n \in A$$

where $\pi(a) > 0$, $p(a,b) \geq 0$, $\sum_a \pi(a) = 1$, $\sum_b p(a,b) = 1$.

Then the jacobian is given as

$$(1.3) \qquad j(z) = \pi(z_0)p(z_0,z_1)/\pi(z_1) , \quad z = (z_n) \in Z .$$

Hence the finite dimensional supspaces

$$C_n = \{u:Z \to \mathbf{C} , u(z) = u(z_0, \ldots, z_n)\}$$

are invariant under the transfer operator L for $n \geq 1$. In fact, if $u(z) = u(z_0, \ldots, z_n)$, then

$$Lu(z) = \sum_{a\in A} \pi(a)p(a,z_0)u(a,z_0, \ldots, z_{n-1})/\pi(z_1)$$

$$\in C_{\max\{1,n-1\}} \, .$$

Consequently, the spectrum of L restricted on C_n with $n\geq 1$ (except for 0) is the eigenvalue set of the matrix $(L_{a,b})_{a,b\in A}$ where

$$L_{a,b} = \pi(b)p(b,a)/\pi(a) \, .$$

This fact can be stated in terms of the original system using the power series called the <u>Fredholm determinant</u>

$$(1.4) \qquad D(t) = \exp\{-\sum_{n=1}^{\infty} \frac{t^n}{n} \sum_{z\in Fix(g^n)} \prod_{i=0}^{n-1} j(g^i z)\}$$

where Fix(•) denotes the fixed point set. Then by (2.1)

$$D(t) = \exp\{-\sum_{n=1}^{\infty} \frac{t^n}{n} \sum_{z\in Fix(g^n)} \prod_{i=0}^{n-1} \pi(z_i)p(z_i,z_{i+1})/\pi(z_{i+1}))\}$$

$$= \exp\{-\sum_{n=1}^{\infty} \frac{t^n}{n} \text{trace } [(L_{a,b})_{a,b\in A}^n]\}$$

$$= \det (I-(L_{a,b})_{a,b\in A}) \, .$$

Hence we have the following74,78

<u>Theorem 2</u>. There is an invariant subspace $\underline{\underline{D}}$ of $L_1(Z,\lambda)$ under the transfer operator L such that L is an compact operator on $\underline{\underline{D}}$ and the spectrum of L on $\underline{\underline{D}}$ is given by the formula

$$(1.5) \qquad \text{Spec}(L_{|\underline{\underline{D}}}) = \{\alpha\in \mathbf{C};\ D(1/\alpha) = 0\} \quad \{0\}$$

where D(t) is the power series given by (2.2).

Proof. Take C_n, $n\geq 2$, as the invariant supspace $\underline{\underline{D}}$. Then it is immediate to see

$$(1.6) \qquad \det(I-tL_{|\underline{\underline{D}}}) = D(t) \, .$$

It is observed in [17] and proved in [27] for the case where Z is a bounded closed interval, λ is the Lebesgue measure and g is a piecewise linear map of Z to itself that this idea works well under suitable modifications even when the transformation g has no Markov property. There are several papers which give the adequate assumptions under which Theorem 2 is tuue, (cf., e.g., [9]) . Note that Axiom A diffeomorphisms admit Markov partitions so that Theorem 2 is rather easy to be proved and the Fredholm determinant $D(t)$ is a rational function. It should be noticed here that $1/D(t)$ is an Artin-Mazur-Ruelle zeta function if $U=-\log j$ is regarded as its weight function although U may not be continuous in our set-up.

Finally we note the following fact. Let us introduce the operator

$$(1.7)\qquad Mu(z) = \sum_{z' \in g^{-1}z} u(z') \qquad z \in Z = A^{\mathbf{N}}$$

associated with the shift g on Z and the <u>weighted transfer operator</u>

$$(1.8)\qquad M(U)u(z) = M(e^{-U}u)(z) = \sum_{z' \in g^{-1}z} e^{-U(z')}u(z')$$

for a continuous function U on Z.

If λ is a nonsingular probability measure for (Z,g) with positive continuous function, then

$$(1.9)\qquad L = M(-\log j) .$$

By the way if λ is an invariant measure for (Z,g), then,

$$h_\lambda(Z,g) = \inf \{\textstyle\int_Z \log[Mu(z)/u(z)]\lambda(dz)\}$$

is the metrical entropy of (Z,g,λ) where the infimum is taken over all positive continuous functions u on Z([28]) .

2. Semi-Bernoulli flows and power spectra

2.1. Power spectrum and weighted transfer operators for special flows

Let us consider a special flow (X,F^t,ν) over an invertible measure preserving transformation (Y,f,μ) with ceiling function $T(y)$: $T(y)$ is a positive measurable function on (Y,μ) ,

$$X = \{(y,s) \in Y \times \mathbf{R};\ 0 \le s < T(y)\} \quad \text{and}$$

$$(2.1)\qquad F^t x = F^t(y,s) = (f^k y,\ t+s-T_k(y))$$

$$\text{if} \quad T_k(y) \le t+s < T_{k+1}(y) \qquad (k \in \mathbf{Z},\ t \in \mathbf{R})$$

where $T_0(y) = 0$, $T_k(y) = T_{k-1}(y)+T(f^{k-1}y)$, $T_{-k}(y) = T_{-(k-1)}(y) - T(f^{-k}y)$ $(k \ge 1)$. The invariant measure ν is a constant multiple of the measure μ_T given by the formula

$$(2.2)\qquad \int_X \mu_T(dx)u(x) = \int_Y \mu(dy)\int_0^{T(y)} ds\, u(y,s) \qquad (u \in L_1(x,\nu)) .$$

In particular, ν is a finite measure if and only if

$$(2.3)\qquad C = \int_Y \mu(dy)T(y)$$

is finite. We shall pretend $\nu = \mu_T$ from now on.

Take $u \in L_2(X,\nu)$. The power spectrum $S(\omega;u)$ is defined as

(2.4) $$S(\omega;u) = \int_{\mathbf{R}} dt\, e^{2\pi i\omega t}\{\int_X u(F^t x)u(x)^* \nu(dx) - |\int_X u(x)\nu(dx)|^2\} .$$

(* denotes the complex conjugate.)
In the sense of distribution it coincides with the Fourier transform

(2.5) $$\tilde{S}(\omega;u) = \int_X \int_{\mathbf{R}} \nu(dx)dt\, e^{2\pi i\omega t}\, u(F^t x)u(x)^*$$

up to an additional constant proportional to δ_0 .

Since F^t is given by (2.1), one finds

$$\tilde{S}(\omega;u) = \int_Y \mu(dy) \int_0^{T(y)} ds \int_{\mathbf{R}} dt\, e^{2\pi i\omega t}\, u(F^t(y,s)u(y,s)^*$$

$$= \sum_{k\in\mathbf{Z}} \int_Y \mu(dy) \iint_{0\le s<T(y),\ T_n(y)\le s+t<T_{n+1}(y)} dsdt\, e^{2\pi i\omega t}\, u(t^k y, t+s-T_k(y))u(y,s)^*$$

$$= \sum_{k\in\mathbf{Z}} \int_Y \mu(dy) \int_0^{T(y)} ds\, u(y,s)^*\, e^{2\pi i\omega(T_k(y)-s)}$$

$$\times \int_0^{T(f^k y)} dt\, e^{2\pi i\omega t}\, u(f^k y,t) .$$

Consequently, putting

(2.6) $$\bar{u}(\omega,y) = \int_0^{T(y)} dt\, e^{2\pi i\omega t} u(y,t) \qquad (\omega\in\mathbf{R},\ y\in Y) ,$$

one obtains

(2.7) $$\tilde{S}(\omega;u) = \sum_{k\in\mathbf{Z}} \int_Y \mu(dy)\, e^{2\pi i\omega T_k(y)} \bar{u}(\omega,f^k y)\bar{u}(\omega,y)^* .$$

From now on let us assume that (Y,f,μ) is the natural extension of a (non-invertible) system (Z,g,λ) and that T is (identified with) a function on Z . (In other words, (Y,f,μ) is the natural extension of the special semiflow over (Z,g,λ) with ceiling function T on Z .)

Let L be the transfer operator associated with (Z,g,λ) and define a weighted transfer operator $L(\omega)$ by

(2.8) $$L(\omega)v(z) = L(e^{2\pi i\omega T}v)(z) ,\ v \in L_1(Z,\lambda) .$$

If u is a function on $\{(z,s)\in Z\times\mathbf{R};\ 0\le s<T(z)\}$, then

(2.9) $$\tilde{S}(\omega;u) = \int_Z \nu(dz)|\bar{u}(\omega,z)|^2 + 2\,\mathrm{Re} \sum_{k=1}^{\infty} \int_Z \lambda(dz)\, e^{2\pi i\omega t_k(z)} \bar{u}(\omega,z)^*$$

$$= \int_Z \nu(dz)|\bar{u}(\omega,z)|^2$$

$$+ 2 \operatorname{Re} \sum_{k=1}^{\infty} \int_Z \lambda(dz)\overline{u}(\omega,z)L(\omega)^k[\overline{u}(\omega,\cdot)](z) .$$

The relation (2.9) suggests that the singularity of the power spectrum $S(\omega;u)$ are described by the zeros of the Fredholm determinant $D(t;\omega)$ associated with the weighted transfer operator $L(\omega)$ provided that there is a good subspace $\underline{\underline{D}}_\omega$ as in Theorem 2 and the function $\overline{u}(\omega,\cdot)$ belongs to $\underline{\underline{D}}_\omega$. But here appears the difficulty that both $D(t;\omega)$ and $\underline{\underline{D}}_\omega$ depend on ω .

2.2. Semi-Bernoulli flows

By a semi-Bernoulli flow we shall mean a special flow (X,F^t,ν) over a Bersulli system (Y,f,μ) with ceiling function T which is measurable w.r.t. a Bernoulli partion α of (Y,f,μ) . Thus one may regard Y,f,μ,α and T as follows:

$Y = A^{\mathbf{Z}}$, f the shift on Y ,

$\alpha = \{[a]; a\in A\}$, $[a] = \{y=(y_n)\in Y; y_0=a\}$,

(2.10) $T(y) = T(y_0)$ for $y = (y_N)\in Y$,

μ is the product measure of infinitely many copies of a probability measure on A .

Example. Let F^t be a smooth flow on a Riemannian manifold and assume that there is a Poincare section where the derived map f admit a hyperbolic set, say Y . Denote the rerurn time to the Poincare section by $T(y)$ for $y\in Y$ and take the natural invariant measure (the Bowen-Ruelle measure) μ on the hyperbolic set Y . Then it is known that (Y,f,μ) is Bernoulli. Let X be the totality of points x such that $F^t x$ crosses Y infinitely many times both as $t \to \infty$ and $t \to -\infty$. Finally define a measure ν by the formula (2.2) . Then (X,F^t,ν) is a semi-Bernoulli flow provided that T is measurable w.r.t. a Bernoulli partition of (Y,f,μ) .

As the example above shows, the assumption on the measurability of T w.r.t. a Bernoulli partition is a technical one, which should be weakened. But we assume it in the present paper because the computation is then simple on one hand and the explicit formula for the Fredholm determinant can be given on the other hand.

By the way one can define semi-Markov flows by replacing the term Bernoulli by Markov in the definition and then find the same definition given by P. Lévy[10].

2.3. Renewal equation

One of the merits of consider semi-Bernoulli systems is that the autocorrelation function is governed by a renewal equation.

As before, let us denote

(2.11) $$C_p = \{u:X\to\mathbf{R},\ u(y,s) = u(y_{-p+1}, \ldots, y_{p-1},s)\} \quad (p\geq 1) .$$

Fix $p,q\geq 0$ and take $u\in C_p$ and $v\in C_q$ and put for $t\geq 0$

(2.12) $$R(t;u,v) = \int_Y \mu(dy) \int_0^{T(y)} ds\, u(F^t(y,s))v(y,s)^*$$
$$= G(t;u,v) + H(t;u,v)$$

where

(2.13) $$G(t;u,v) = \int_Y \mu(dy) \int_{0\leq s<T(y),\ s+t<T_{p'}(y)} ds\ u(F^t(y,s))v(y,s)^*$$

(2.14) $$H(t;u,v) = \int_Y \mu(dy) \int_{0\leq s<T(y),\ s+t\geq T_{p'}(y)} ds\ u(F^t(y,s))v(y,s)^*$$

with $p' = p+q-1$.

Note that for $t+s\geq T_{p'}(y)$

$$u(F^t(y,s)) = u(F^{t+s-T_{p+1}(y)}(f^{p+1}y,0)) .$$

It follows from the Bernoulli property of μ that

(2.15) $$H(t;u,v) = \int_Y \mu(dy) \int_Y \mu(dy') \int_{0\leq s<T(y),\ s+t\geq T_p(y)+T_{p'-p}(y')} ds\ u(F^{t+s-T_p(y)}(y',0))v(y,s)^*$$
$$= \iint_{0\leq s<T(y)} \mu(dy)ds\ v(y,s)^* Q(t+s-T_p(y);u,v)$$

where we put

(2.16) $$Q(t;u) = \int_{T_{p'-p}(y)\leq t} \mu(dy)u(F^t(y,0)) .$$

<u>Lemma 1</u>. Let us define a probability measure m on $[0,\infty)$ by

(2.17) $$m[0,t] = \mu\{y;\ T(y)\leq t\} .$$

Then the following renewal equation holds.

(2.18) $$Q(t;u) = \int_0^t Q(t-s;u)m(ds) + b(t;u)$$

where

(2.19) $$b(t;u) = \int_{T_{p'-p}(y)\leq t<T_{p'-p+1}(y)} \mu(dy)u(F^t(y,0))$$

is an absolutely integrable function in t with

$$\int_0^\infty |b(t;u)|dt \leq \int \mu(dy) \int_0^{T(y)} ds|u(y,s)| \leq |u|_{L_2(x,\nu)} \quad .$$

Proof. It is obvious that

$$Q(t;u) - b(t;u) = \int_{T_{p'-p+1}(y) \leq t} \mu(dy)u(F^t(y,0))$$

$$= \int \mu(dy')\int_{T_{p'-p}(y) \leq t-T(y')} \mu(dy)u(F^{t-T(y')}(y,0))$$

$$= \int \mu(dy')Q(t-T(y');u) = \int_0^\infty m(ds)Q(t-s;u) \ .$$

Furthermore,

$$\int_0^\infty |b(t;u)|dt = \int \mu(dy) \int_{T_{p'-p}(y)}^{T_{p'-p+1}(y)} |u(F^t(y,0))|dt$$

$$= \int \mu(dy) \int_0^{T(y)} |u(F^s(y,0))|ds \ .$$

Consequently, one obtains the estimate (2.19).

Lemma 2. Let $b(t)$ be an absolutely integrable bounded function and $q(t)$ satisfy a renewal equation

$$q(t) = \int_0^t m(ds)q(t-s)+b(t)$$

for a probability measure m on $(0,\infty)$ such that

$$\int_0^\infty sm(ds) = \int_0^\infty m(t,\infty)dt < \infty \quad .$$

Then for $\omega \neq 0$ the improper integral

$$\hat{q}(\omega) = \int_0^\infty e^{2\pi i\omega t}q(t)dt$$

converges and satisfies

$$\hat{q}(\omega) = \frac{\hat{b}(\omega)}{1-\hat{m}(\omega)} \tag{2.20}$$

provided that $\hat{m}(\omega) \neq 1$, where $\hat{m}$ and $\hat{b}$ are Fourier transform on $[0,\infty)$ of m and b , respectively

Proof.

$$(1-\hat{a}(\omega)) \int_0^R e^{2\pi i\omega t}q(t)dt - \int_0^R e^{2\pi i\omega t}b(t)dt$$

$$= \int_0^R dt \int_0^t e^{2\pi i\omega s}m(ds)e^{2\pi i\omega(t-s)}q(t-s) - \hat{m}(\omega)\int_0^R e^{2\pi i\omega t}q(t)dt$$

$$= \int_0^R e^{2\pi i\omega t} q(t) \left(\int_0^{R-t} e^{2\pi i\omega s} m(ds) - \hat{m}(\omega)\right\}dt$$

$$= \int_0^R e^{2\pi i\omega t} q(t) \left\{\int_{R-t}^{\infty} e^{2\pi i\omega} m(ds)\right\}dt$$

It follows from the renewal theorem ([6] that $q(t)$ converges as $t\to\infty$. In particular, it is bounded. On the other hand, $m(t,\infty)$ is integrable in t since m has finite mean. Consequently,

$$|(1-\hat{a}(\omega)) \int_0^R e^{2\pi i\omega t} q(t)dt - \hat{b}(\omega)|$$

$$\leq \text{const.} \int_0^R m(t,\infty)dt + \int_R^{\infty} |b(t)dt \to 0 \quad \text{as} \quad R \to \infty .$$

From Lemma 1 and 2 the existence of the Fourier transform of $R(t;u,v)$ is guaranteed and is given as the sum of Fourier transforms of $G(t;u,v)$ and $H(t;u,v)$.

<u>Theorem 3</u>. Let $u \in C_p$ and $v \in C_q$ for some $p,q \geq 1$. Then the Fourier transform

(2.21) $$S(\omega;u,v) = 2 \operatorname{Re} \int_0^{\infty} e^{2\pi i\omega t} R(t;u,v)dt$$

($R(t;u,v)$ being defined by (2.12)) exists for such ω that

(2.22) $$d(\omega): = 1 - \int_Y e^{2\pi i\omega T(y)} \mu(dy) \neq 0$$

(in particular, for every ω with sufficiently small modulus) provided that

(2.23) $$C = \int_Y \mu(dy)T(y) < \infty .$$

Furthermore, the following sesimates are valid

(2.24) $$|S(\omega;u,v)-2 \operatorname{Re} \frac{S_0(\omega;u,v)}{1-\hat{m}(\omega)}| \leq 2(p+q-1)|Tv|_2|u|_2$$

(2.25) $$|S_0(\omega;u,v)| \leq |u|_2|v|_2$$

where

(2.26) $$S_0(\omega;u,v) = \iint_0^{T(y)} \mu(dy)ds\, e^{2\pi i\omega(s+T_p(f^{-p}y))} u(y,s)$$
$$\times \left(\iint_0^{T(y)} \mu(dy)ds\, e^{2\pi i\omega(s-T_q(y))} v(y,s)\right)^* .$$

Proof. From (2.13) it follows that

$$\hat{G}(\omega;u,v) = \int_0^{\infty} e^{2\pi i\omega t} G(t;u,v)dt$$

is dominated in modulus by

$$\int_0^\infty |G(t;u,v)|dt \le \int\mu(dy) \int_0^{T(y)} ds \int_0^{T_{p'}(y)-s} dt\ |u(F^t(y,s)||v(y,s)|$$

$$\le \int\mu(dy) \int_0^{T_{p'}(y)} ds\ |T(y)||u(F^t(y,s))||v(y,s)|$$

$$\le p'|u|_2\ |Tv|_2 \qquad (p'=p+q-1)$$

On the other hand, the integral

$$\hat{H}(\omega;u,v):= \int_0^\infty e^{2\pi i\omega t}H(t;u,v)dt$$

$$(2.27) \qquad = \int\mu(dy) \int_0^{T(y)} ds\ v(y,s)^* \int_0^\infty e^{2\pi i\omega}\ Q(t+s-T_q(y);u)$$

$$= \int\mu(dy) \int_0^{T(y)} ds\ v(y,s)^*\ e^{2\pi i\omega(T_q(y)-s)}\ \hat{Q}(\omega;u)$$

exists by Lemma 2 if $d(\omega)\neq 0$ and

$$\hat{Q}(\omega;u) = \frac{1}{1-\hat{m}(\omega)} \int_0^\infty e^{2\pi i\omega t}dt \int_{T_{p'-q}(y)\le t<T_{p'-q+1}(y)} \mu(dy)u(F^t(y,0))$$

$$(2.28) \qquad = \frac{1}{1-\hat{m}(\omega)} \int\mu(dy) \int_0^{T(f^{p'-q}y)} ds\ e^{2\pi i\omega(s+T_{p'-q}(y))}$$

$$\times\ u(F^s(f^{p'-q}y,0)) .$$

Hence

$$(2.29) \qquad S_0(\omega;u,v) = (1-\hat{m}(\omega))\ \hat{H}(\omega;u,v)$$

and

$$(2.30) \qquad |S_0(\omega;u,v)| \le |u|_2\ |v|_2 .$$

Consequently one obtains (2.24)-(2.26) since $S = \hat{H} + \hat{G}$.

3. Main theorem

Let us consider a semi-Bernoulli flow (X,F^t,ν) and denote

$$C_p = \{u:X\to\mathbf{C}\ ;\ u(y,s) = u(y_{-p+1},\ldots, y_{p-1},s)\} \qquad (p\ge 1)$$

and C_0 = {constant functions}, as before. For $u \in L_2(x,\nu)$ we shall write the orthogonal decomposition of u as

$$(3.1) \qquad u = \sum_{p=0}^{\infty} u_p\ , \quad u_p \in C_p \quad (p\ge 0) .$$

The following conditions are a little simpler (and a little weaker) than the conditions in [30,31] .

Let M be the subspace of $L_2(X,\nu)$ consisting of $u \in L_2(X,\nu)$ such that the following norm is convergent:

$$|u|_M = \sum_{p=0}^{\infty} (1+p)|(1+T)u_p|_2 \quad . \tag{3.2}$$

Theorem 4. Let (X,F^t,ν) be a semi-Bernoulli flow over Bernoulli shift (Y,f,μ) with ceiling function T and S the supspace defined above. If $u \in M$, then as $\omega \to 0$

$$S((\omega;u) \sim S^*(\omega) \; |\int_X \nu(dx)u(x)|^2 + \text{const.} \tag{3.3}$$

where

$$S^*(\omega) = 2 \text{ Re } \frac{1}{d(\omega)} \tag{3.4}$$

$$d(\omega) = 1 - \int_Y \mu(dy)e^{2\pi i\omega t(y)}$$

is the Fredholm determinant of the weighted transfer operator $L(\omega)$ defined by (2.8)

Remark. The expression (3.3) may seem queer. It is clarified by Claim: If a constant function belongs to M , then T must be square integrable and vice versa. Consequently, (3.3) means that there is no singularity of $S(\omega;u)$ at $\omega=0$ if T is square integrable and that $S(\omega;u-\int u(x)\nu(dx)) = S(\omega;u)$ has a singularity at $\omega=0$ for $u\in M$ otherwise.

An abelian theorem shows the following

Proposition. Assume the same conditions as in Theorem,

(i) If T is square integrable, then

$$\lim S^*(\omega) = -d''(0)/d(0)^2 = \frac{1}{2\pi} \int T^2 d\mu/(\int T d\mu)^2 \tag{3.5}$$

(ii) Otherwise, assume that $M(t) = \mu\{y;T(y)>t\}$ varies regulary with exponent γ:

$$M(t) = t^{-\gamma}L(t) \tag{3.6}$$

where $L(t)$ is a slowly varing function at $t=\infty$

If $\gamma>1$, then, as $\omega\to 0$,

$$S^*(\omega) \sim \frac{\pi}{\Gamma(\gamma)\sin(\pi\gamma/2)} L(\omega^{-1})(2\pi\omega)^{-(2-\gamma)} \tag{3.7}$$

If $\gamma=2$, then, as $\omega\to 0$,

$$S^*(\omega) \sim \omega^{-1}\tilde{L}(\omega) \tag{3.8}$$

where $\tilde{L}$ is a slowly varying function at $\omega=0$.

Remarks. Here the exponent γ must be not less than one since we assume the integrability of T (and, hence of $M(t)$). An example of the case $\gamma=1$ is

$$M(t) \sim c/t(\log t)^2 \quad \text{as} \quad t\to\infty .$$

4. Example: piecewise linear maps

First of all let us define a class of piecewise linear continuous maps of the unit closed interval and the circle to themselves. Let

$$c_0 = 1 > c_1 > c_2 > \dots > c_N = 0 , \quad N \leq \infty ,$$

and denote

$$a_0 = c_0 - c_1 , \quad \text{and} \quad a = \frac{c_n - c_{n+1}}{c_{n-1} - c_n} \; (1 \leq n < N) .$$

These are the control parameters of maps. Define

$$f(y) = \begin{cases} a_n^{-1}(y-c_{n+1})+c_n & \text{if } y\in I_n=[c_{n+1},c_n] \text{ with } n\geq 1 \\ a_0^{-1}(c_0-y) & \text{if } y\in I_0=[c_1,c_0] \end{cases}$$

Then f is a piecewise linear continuous map of the unit closed interval I onto itself. If we define

$$f_c(y) = \begin{cases} F(y) & \text{if } y\in I_n \text{ with } n\geq 1 \\ 1+a_0^{-1}(y-c_0) & \text{if } y\in I_0 \end{cases}$$

and identify the endpoints 0 and 1 of I, then we obtain a piecewise linear continuous map of the circle.

In this case the Fredholm determinant takes the following form

$$D(t) = \exp\{-\sum_{n=1}^{\infty} \frac{1}{n} Q_n(f) t^n\} , \tag{4.2}$$

where

$$Q_n(f) = \sum_{y\in \mathrm{Fix}(f^n)} \frac{1}{|(F^n)'(y)|} .$$

Here Fix (f^n) is the set of all fixed points under the iterate f^n, as usual, although it should be modified in order to be defined for all piecewise smooth maps. (See[27] for the detail and also for a justification of being the Fredholm determinant det (I-tL).) A direct computation shows the following fact:

$$(4.3)\quad D(t) = \begin{cases} 1 - \sum_{n=1}^{N} (c_n - c_{n+1}) t^{n+1} & \text{if } N \text{ is finite,} \\ (1-a_\infty)(1 - \sum_{n=0}^{\infty} (c_n - c_{n+1}) t^{n+1}) & \text{if } N=\infty \text{ and } a_\infty = \lim a_n . \end{cases}$$

The applicability of Theorem 1 comes from the following two lemmas.

Lemma 1. The induced map on the set I_0, denoted by f_{I_0} (or f_{c,I_0}), is defined except for $y=1$ and it is conjugate to the following map g (or g_c) of I :

$$g(x) = \frac{(1-z)-c_{n+1}}{c_n - c_{n+1}} \quad \text{if} \quad 1-z\in[c_{n+1}, c_n]$$

$$[\text{or} \quad g_c(x) = \frac{z-c_{n+1}}{c_n - c_{n+1}} \quad \text{if} \quad z\in[c_{n+1}, c_n]] .$$

Lemma 2. Let us denote the indicator function of the subinterval I_n, again by I_n. Then

(i) The map f and f_c have the following common invariant density function

$$u(y) = \sum_{0\leq n<N} \frac{c_n}{c_n - c_{n+1}} I_n(y) ,$$

which is unique up to a multiplicative constant almost everywhere.

(ii) The function h is normalizable if and only if $D(t)$ is differentiable in the closed unit disc $|t|\leq 1$, or, equivalently, if and only if

$$C = \sum_{n=0}^{N} c_n < \infty .$$

Remark. Let μ be the Lebesgue measure on the unit interval I. Then the metrical entropy of the endomorphism (I,g,μ) or (I,g_c,μ) is given as

$$h_\mu(g) = h_\mu(g_c) = - \sum_{0\leq n<N} (c_n - c_{n+1}) \log(c_n - c_{n+1}) .$$

It follows from the formula for the metrical entropy of the induced systems ([3,32]) that

Example 1. Let

$$c_n = n^{-\gamma}[\log(n+1)]^{-2} \quad \gamma \geq 1 .$$

Then c_n is summable and so there is an absolutely continuous invariant probability measure. The asymptotics (21) is given as

$$S^*(\omega) \sim \text{const.}\ \omega^{-(2-\gamma)} (\log \omega)^{-2} .$$

Consequently, the exponent α is exactly 1 when $\gamma=1$.

The long time tail occurs by the following mechanism. If one observes a sample trajectory, then the longer it says near $y=0$ (laminar phase) with intermittent short time excursion to the large values of y (bursts) as $\gamma \geq 1$ becomes smaller.

When γ is less than 1, c_n is not summable and there is no finite absolutely continuous invariant measure. The sojourn time near $y=0$ becomes much longer and increases in average with the times of excursions. Thus the time average along almost all trajectories is degenerated to the Dirac mass at $y=0$.

The notion of the power spectrum can be extended to include this case by considering the time correlation with respect to the uniform measure on the interval (which is generally not invariant but nonsingular). If the system has an absolutely continuous invariant probability measure, then it can be proved directly that the new power spectrum has the same singularity.

It can also be proved by the renewal theorem and an Abelian theorem that Theorem 2 still remains valid for $\gamma<1$ if one replace (21) by

$$S^*(\omega) = \omega^{-\gamma} K(\omega)^{-1}$$

with slowly varying K .

Example 2. Using the extension stated above, we can compute the explicit exponent α of the singularity

$$S^*(\omega) \sim \text{const.}\ \omega^{-\alpha} \quad \text{as} \quad \omega \to 0 \tag{4.4}$$

in terms of the degree of the tangency

$$f(y) = y + \text{const.}\ y^{\beta}(1+o(\omega)) \quad \text{as} \quad y \to 0 \tag{4.5}$$

in our model for all $\beta > 0$. Note that $y=0$ is a fixed point of f which is "homoclinic". The dependence of α on β is as follows.

$$a = \begin{cases} 0 & \text{if} \quad 0 < \beta \leq 3/2 \\ \dfrac{2\beta-3}{\beta-1} & \text{if} \quad 3/2 < \beta \leq 2 , \\ \dfrac{1}{\beta-1} & \text{if} \ \beta > 2 \end{cases} \tag{4.6}$$

5. Lattice gas model with phase transion

It is well known that the phase transition occurs in one dimensional statistical mechanics only under the long range potentials which carry rather artificial character. Let us construct a lattice gas model which agrees with the family of maps in III. Then we may say that the inverse power law for the maps takes place just as in the critical phenomenon in statistical physics or that the statistical physical interpretation can be applied to the maps.

1. Let us denote $\sigma_i = 1$ or 0 according as the site $i \in \mathbf{Z}$ is occupied by a particle or not. Assign the potential energy $U(n)$ to each consecutive run of n particles in the configuration

$$\sigma_i = \sigma_{i+n+1} = 0 \text{ and } \sigma_{i+1} = \dots = \sigma_{i+n} = 1 .$$

Then the partition function of the N-periodic lattice system is

$$(5.1) \qquad Z_N(\beta,\zeta) = \Sigma\zeta^{n_0+\dots+n_{p+1}}\exp[-\beta(U(n_0+n_{p+1})+U(n_1)+\dots+U(n_p))]$$

where the summation is taken over $p \geq 0$, $n_0, \dots, n_{p+1} \geq 0$ such that

$$n_0 + n_1 + \dots + n_{p+1} + p = N .$$

(β and ζ are thermodynamical parameters.)
Let us introduce a power series

$$(5.2) \qquad D(t;\beta,\zeta) = \exp \sum_{n=1}^{\infty} - \frac{t^n}{n} Z_N(\beta,\zeta) .$$

Then it follows from an elementary computation

$$(5.3) \qquad D(t;\beta,\zeta) = 1 - \sum_{n=0}^{\infty} \zeta^n e^{-\beta U(n)} t^{n+1} ,$$

which is very similar to the expression (12) for the Fredholm determinant and coincides with $D(t)$ if one puts

$$\zeta^n e^{-\beta U(n)} = c_n - c_{n+1} .$$

It is not a formal resemblence. In fact, take a point $y \in Y = [0,1]$ and define a lattice configuration by

$$(5.4) \qquad \sigma_i = 0 \text{ or } 1 \text{ according as } f^i y \in (c_1,c_0] \text{ or not .}$$

Then the unit interval Y can be identified with a subset of configuration space. Conversely, it is not difficult to see that, if $\sigma = (\sigma_i)$ is give, then there is a unique point such that

$$(5.5) \qquad f^t y \in \begin{cases} [c_1,c_0] & \text{if } \sigma_i = 0 , \\ [0,c_1] & \text{if } \sigma_i = 1 . \end{cases}$$

Here the expansiveness ($|f'|>1$) is essential. Denote this point y by $\rho(\sigma)$. Then it is obvious that

$$\rho(S\sigma) = f(\rho(\sigma)) ,$$

where S denotes the spatial translation on the lattice: $(S\sigma)_i = \sigma_{i+1}$. This conjugacy ρ is often called the realization of the map F.

Under the conjugacy ρ, the properties of F and S are the same as dynamical systems. Furthermore the following theorem holds:

Theorem 5. (i) Let $P = P(\beta,\zeta)$ be the thermodynamical limit

$$P = \overline{\lim} \frac{1}{N} \log Z_N(\beta,\zeta) .$$

Then

$$e^{-P} = \sup\{t\geq 0;\ D(t) \text{ converges and is positive}\} .$$

(ii) If $D'(e^{-P})$ diverges, then the equilibrium measure is unique and is concentrated at the single configuration $\sigma_i \equiv 1$.

(iii) If $D'(e^{-P})$ converges, then there is an equilibrium measure which is mixing Markov under S. Furthermore, the clusters of consecutive $\sigma_i = 1$ are mutually independent and identically distributed.

(iv) If $D(z)$ converges in a neighborhood of $z = e^{-P}$, then the equilibrium measure is unique and is given by (iii). (Here the equilibrium measure means the state that minimizes the free energy.)

The readers will easily find that Theorem 4 and 5 state essentially the same assertions when $P=0$. The condition $P=0$ means $D(1)=0$ and, hence, $c_0=1$. But it was imposed merely because we took the unit interval for the definition of the maps. Thus $P=0$ is not any constraint. Similarly, the proof of Theorem 5 is essentially the same as that of Theorem 2 although a direct proof is done in terms of the transfer operator.

Theorem 5 asserts that the phase coexistence (the nonuniqueness of the states minimizing the free energy) is possible only when e^{-P} lies on the convergence circle of $D(z)$. It agrees with the usual criterion that the phase transition occurs when the free energy or the thermodynamical limit loses the analyticity.

In fact, if e^{-P} is less than the convergence radius of $D(z)$, then it follows from (i) that

$$D(e^{-P}) = 0 .$$

If $D'(e^{-P})$ is finite, the implicit function theorem then implies that e^{-P} is an analytic function of β and ζ. Hence the phase transition does not occur.

Consequently, the phase separation curve lies in (β,ζ)-plane where D or D' diverges. But the power spectrum exhibits the power law when D and D' converge but D" diverges. In this sense it corresponds to the tricitical point.

REFERENCES

[1] Afraimovic, V.S., Bykov, V.V. and Sil'nikov, L.P., Attracting non-rough limiting sets of Lorenz attractor type, Trudy Moskov. Obsc. 44(1982), 150-212.

Afraimovic, V.S. and Shil'nikov, L.P., Strange attractors and quasi-strange attractors, Nonlinear dynamics and Turbulence, ed.Barenblatt et al., Pitman (1984), 1-34.

[2] Aizawa, Y. and Kohyama, T., Symbolic dynamics approach to intermittent chaos-towards the comprehension of large scale self-similarity and asymptotic non-stataionarity. Chaos and Statistical Methods ed. Kuramoto, Springer (1984), 109-116.

[3] Cornfeld, I.P., Sinai, Ya. G. and Fomin, S.V., Ergodic Theory, Nauka (1980); English transiation, Springer (1981).

[4] Donsker, M.D. and Varadhan, S.R.S., Aysmptotic evaluation of certain Markov process expectations for large time I, III, IV, Comm. Pure Appl. Math. 28(1975), 1-45; 29(1976), 389-461; 36(1983), 183-212.

Varadhan, S.R.S., Large Deviation and Applications. CBMS-NSF Regional Conference Series in Applied Mathematics 46, Soc. Industrial and Applied Math. (1984).

[5] Feigenbaum, M., Quantitative universality for a class of nonlinear transformations, J. Stat. Phys. 19(1978), 25-52; The universal metric properties of nonlinear transformations, J. Stat. Phys. 21(1979), 669-706.

[6] Feller, W., An Introduction to Probability Theory and its Applications, Wiley (1950).

Hardy, G.H. and Riesz, M., The General Theory of Dirichlet Series, Cambridge Tracts in Math. and Phys, No.18, Stecher-Hafner (1964) [originally, Cambridge Univ. Press (1915)].

[7] Gollub, J.P. and Swinney, H.L., Onset of turbulence in a rotating fluid, Phys. Rev. Lett. 35(1975), 927-930.

[8] Guckenheimer, J. and Holmes, P., Nonlinear Oscillations, Dynamical Systems and Bifurcations of Vector Fields, Springer (1983).

[9] Hofbauer, F. and Keller, G., Zeta functions and transfer operators for piecewise linear transformations J.f.reine u. angewandte Math. 352(1984), 101-113.

[10] Levy, P., Oeuvres Vol.IV, Gauthier-Villars (1980).

[11] Li, T. and Yorke, J., Period three implies chaos, Amer. Math. Monthly 82(1975), 985-992.

[12] Libchaber, A. and Maurer, J., A Rayleigh-Bernard experiment: helium in a small box. Nonlinear Phenomena at Phase Transitions and Instabilities, ed. Riste (1982), 259-286.

[13] Lorenz, E.N., Deterministic nonperiodic flow, J. Atmos. Sci. 20(1963), 130-141.
[14] May, R., Biological populations obeying difference equations, stable cycles and chaos, J. Theor. Biol. 51(1975), 511-524.
[15] Mori, H., Evolution of chaos and power spectra in one-dimensional maps, Nonlinear Phenomenon in Chemical Dynamics, Springer (1981).
[16] Mori, H., So, B.C. and Okamoto, H., Spectral structure and universality of intermittent chaos, Proc. Nobel Symp. on the Physics of Chaos and Related Problems.
[17] Oono, Y. and Takahashi, Y., Chaos, external noise and Fredholm theory, Prog. Theor. Phys. 63(1980), 1804-1807.
Takahashi, Y. and Oono, Y., Towards statistical mechanics of chaos, Prog. Theor. Phys. 71(1984), 851-854.
[18] Pomeau, Y. and Manneville, P., Intermittent transition to turbulence in dissipative dynamical systems, Comm. Math. Phys. 74(1980), 189-197.
[19] Rossler, C.E., An equation for continuous chaos, Phys. Lett. 57A(1976), 397-398.
[20] Ruelle, S., Statistical Mechanics. Rigorous Results, Benjamin (1969).
[21] Ruelle, D., Thermodynamical Formalism, Addison-Wesley (1978).
[22] Ruelle, D. and Takens, F., On the nature of turbulence, Comm. Math. Phys. 20(1971), 167-192; 23(1971), 343-344.
[23] Sarkovskii, Coexistence of cycles of a continuous map of a real line to itself, Ukrain. Mat. Z. 16(1964), 61-71.
[24] Stefan, P., A theorem of Sarkovskii on the existence of periodic orbits of continuous endomorphisms of the real line, Comm. Math. Phys. 54(1977), 237-248.
[25] Sil'nikov, L.P., A contribution to the problem of the structure of an extended neighbourhood of a rough equilibrium state of saddle-focus type, Mat. Sbornik 10(1970), 91-102.
[26] Sinai, Y., G., Theory of Phase Transition, Nauka (1980).
[27] Takahashi, Y., Fredholm determinant of unimodal linear maps, Sci. Papers College Gen. Educ. Univ. Tokyo 31(1981), 61-87; An ergodic-theoretical approach to the chaotic behaviour of dynamical systems, Publ. RIMS Kyoto Univ. 19(1983), 1265-1282.
[28] Takahashi, Y., Entropy functional (free energy) for dynamical systems and their random perturbations, Proc. Taniguchi Symp. on Stochastic Analysis, Kinokuniya/North-Holland (1984), 437-467.
[29] Takahashi, Y., Observable chaos and variational principle formalism, for one-dimensional maps, Proc. 4th Japan-USSR Symp. on Probability Theory, Springer Lecture Notes in Math. Vol.1021 (1983), 676-686.
[30] Takahashi, Y., Gibbs variational principle and Fredholm theory for one-dimensional maps. Chaos and Statistical Methods, ed. Kuramoto, Springer (1984), 14-22; Phase transition which exhibits 1/f-spactrum: a rigorous result. Statistical Plasma Physics, Proc. US-Japan Workshop, Research Report IPP Nagoya, July 1984, 96-103.
[31] Takahashi, Y., One dimensional maps and power spectrum, Recent

Studies on TURBULENT PHENOMENA, 1985 ed. by T. Tatsumi et al., pp.99-116, Association for Science Documents Information, Tokyo 1985.
[32] Totoki, H., Introduction to Ergodic Theory, Kyoritsu (1971) (in Japanese).
[33] Ueda, Y., New Approach to Nonlinear Problems in Dynamics, ed. Holmes, SIAM (1980), 311-; Explosion of strange attractors exhibited by Duffing equation. Nonlinear Dynamics, ed. Helleman, New York Acad. Sci. (1981), 422-434.
[34] Zygmund, A., Trigonometric Series, Cambridge Univ. Press (1968).

ON THE QUALITATIVE BEHAVIOUR OF INTERACTING BIOLOGICAL CELL SYSTEMS

Petre Tautu

German Cancer Research Center
Dept.Mathematical Models
P.O.Box 101949
D-6900 Heidelberg 1

ABSTRACT. A conspectus is given on interacting biological cell systems (BCS) as a class of stochastic models describing the (permanent) spatial growth of cell populations. Three families of models are taken into account : (1)Eden models (E1 and E2), (2)Williams-Bjerknes models (WB1 and WB2),and (3)the k-type configuration model for carcinogenesis (C1). Given there is an explicit mathematical connection between some infinite systems with locally interacting components (shortly:interacting particle systems) and percolation models,some results about the qualitative behaviour of interacting BCS will be given in terms of first-passage percolation theory (as a generalization of renewal theory).

0.INTRODUCTION

The aim of the present paper is to introduce the interacting biological cell systems (BCS) as a special class of models created within the framework of the stochastic theory of interacting particle systems (see Liggett,1985). The most important models examined here are (1)the Eden models (Eden, 1958,1962), (2)the Williams-Bjerknes models (Williams and Bjerknes,1972), and (3)the k-type configuration model for carcinogenesis (Schürger and Tautu,1976a).

Special attention has recently been devoted to the Eden models,because they represent a simpe example of "space-filling" processes. Indeed,they can desqibe the expansion of a unique family of malignant cells as well as of a family of

S. Albeverio et al. (eds.), Stochastic Processes in Physics and Engineering, 381–402.

embryonic cells (as originally conceived by Murray Eden) provided these cell populations survive for ever. Eden models are also called "growing animals" (Peters et al.,1979) or "cancer growth models" (Stanley et al.,1985). In the same line, the Williams-Bjerknes models -primarily created to represent the spread of a unique malignant clone- can actually be models of any "invasion" process. They are the finite counterpart of the "voter models" (see Liggett,1985,Chap.V,and references therein),particularly of the "biased" voter model (Schwartz,1977). Other models,e.g.,the variants and generalizations of Eden models (Schürger,1980,1981;Rikvold,1982; Sawada et al.,1982;Meakin,1983a,etc.),the viral infection model (Schwöbel et al.,1966),genetic models as Malécot-Kimura "stepping-stone" model (Sawyer,1977;Shiga,1982),the interacting Lindenmayer stochastic systems (Tautu,1976),and many others,can be treated within the framework of the stochastic theory of interacting BCS.

The study of the asymptotic behaviour of such models begun with the seminal paper by D.Richardson (1973) ; his main result (the asymptotic shape of the set of sites occupied by a cell population growing in space eventually contains a ball of linearly expanding radius provided the growth process is permanent) is proved by using the arguments of the theory of subadditive stochastic processes. Actually,the geometric structure of a spatially developed particle system has already been predicted by M.Eden who stated (1958,p.369) that "large configurations will be essentially circular in outline...they will have a high density,i.e.,they will contain very few 'holes' and short 'tentacles'...(moreover) these configurations appear to resemble one another very closely even though they exhibit no correspondence in detail". Similar observations have been made by T.Williams (1971) for one of his models where "the configurations are as compact as they ever get". Identical remarks have recently been made by P.Meakin

(1983a,b) about the Eden and Williams-Bjerknes models, as well as about some related models (Rikvold, Sawada et al., Vold-Sutherland) ; he emphasized the importance of dimensionality: if the Hausdorff dimensionality is equal to the Euclidean dimensionality, then as the cluster grows larger and larger it will approach a constant limiting density or "porosity".

Since the theory of subadditive stochastic processes is an essential tool for treating percolation processes, it appeared obvious that under appropriate conditions a similar result to that obtained by D.Richardson should hold for percolation processes. After the first approach by D.Mollison (1974), this has been proved by J.T.Cox and R.Durrett (1981, Th.3.3; see also Schürger, 1980a). Percolation problems relative to the shapes and sizes of clusters (and especially to the possibility to obtain large or infinite clusters) are frequently encountered in statistical mechanics. Moreover, percolation methods provide a testing ground for theories of phase transitions and critical phenomena (see Stauffer, 1979). Also, an interesting parallelism between the voter model on the $d \geq 3$-dimensional lattice and measure-valued branching processes (Iscoe, 1986) must be mentioned.

The evolution of interacting BCS raises many questions particularly related to the spatial behaviour, as for example, (i) the growth velocity of a configuration, (ii) the geometric structure of this configuration (as $t \to \infty$), (iii) the asymptotic dimension, (iv) the first occupation time of a given point far from the origin, and so on. Computer simulations of the multitype configuration model for carcinogenesis (Schürger and Tautu, 1978) showed that (a) during the interaction between the first appeared malignant cells and other cells, these malignant cells may vanish; (b) many malignant foci may appear in different areas and will at late stage flow together; (c) some intermediate cells are persistent so that the malignant configuration remains "heterogeneous". These observations give occasion to new (and still unsolved) questions about (i)the local persistence of small malignant foci in a "shell" of intermediate cells,(ii)the geometric structure of the first "islands of malignancy",(iii)the growth velocity of these "black" islands as well as of the established tumour,etc.

The organization of this article is as follows. In §1 there are introduced the biological postulates necessary for the construction of interacting BCS;some examples are presented in §2. A brief review of the relevant concepts and notation pertinent to the definition and analysis of interacting BCS as Markov processes is given in §3,and some specific results are presented in §4. In the last paragraph a theorem of law of large numbers type is used for the asymptotic growth of these cell systems.

1.THEORETICAL-BIOLOGICAL POSTULATES

The spatial growth of interacting biological cell systems can be modelled mathematically by considering the following components:

(A).Space. It can be

(1)A graph G or the infinite d-dimensional integer lattice Z^d,$d\geq 1$. There are other possible lattices (see,e.g., Hammersley and Mazzarino,1983) but the Bethe tree is of theoretical interest:it is the equivalent of a conventional lattice in a space of infinite dimensionality and is adequate for the description of cascade processes (Zallen,1983,p.172: the percolation properties of such a lattice constitute the equivalent of mean-field behaviour.) Living cells are placed at each site on the lattice (as on a "matrix");in order to preserve the geometric regularity of such an idealized "tissue" the following rule is of importance:

(O)Exclusion of multiple occupancy:a site can be occupied by only one cell at a time.

(2)The Euclidean space R^d,$d\geq 1$.

(B).Time. It can be either (1)continuous or (2)discrete.

The interacting particle systems are continuous time stochastic processes.

(C).Cell population. It may be homogeneous(with indistinguishable cells) or heterogeneous(with cells of different types or properties:k-type BCS,$k\geq 2$).

(1)If the interacting BCS is homogeneous,the dynamics of the process is given by the occupancy of vacant sites on Z^d or R^d. The growth is perceived only at the periphery of the system.

(2)If k=2(two-type,"black-white"BCS),the dynamics in the system is represented by an interaction process specified in two ways:

(2.1)The interaction results with a conversion from white into black($W\to B$) that occurs with a positive probability $p\in(0,1)$;the reverse transition $B\to W$ will occur with $1-p=q$.

(2.2)The interaction takes place with some intensity, say β. It is assumed that β quantifies some dominant property (phenotypic "advantage") of one cell type (frequently the black cells) so that

(2.2.1)if $\beta=1$,there is no influence exerted of one cell type on the other one;

(2.2.2)if $\beta=\infty$,for B-cells,then a site occupied by a B-cell remains black for ever;

(2.2.3)if $1<\beta<\infty$, the influence is reversible($W\rightleftarrows B$) and a site may often change its colour.

Assumptions C.2.1,C.2.2.1 and C.2.2.2 are formally similar to the situation that a site is either vacant(White)or occupied(Black):the occupancy process is indistinguishable from the colouring process,because the state space is identical for both processes.

It is commonly assumed that the stochastic process starts with one black cell(a "germ" or a "seed")that appears is a population of white("normal",healthy")cells. For example, this is the invariable hypothesis of an infection model:the process starts with one infected cell in a homogeneous population of healthy cells. The continuous multiplication of this B-cell leads to the formation of a black "clone" (or of a "dynasty").

(3)If k≥2(multitype BCS),we have to assume a colouring process("metamorphosis"):

(3.1)The cells change their colours(properties)following a directed chain of colours,e.g.,white→...→black,that is, every new type is generated only by the preceding type.

(3.2)The alternative assumption is the reversibility of colouring,e.g.,white⇄...⇄black.

One can introduce colour dependent intensities β_i,$1\leq i\leq k$,but the use of transition probabilities is recommended.

The most complete k-type interacting BCS is the stem cell system as it is described in morphogenesis(e.g.,Hydra)or haematopoiesis. Carcinogenesis is a pathological process generated by a specific stem cell system(or an equivalent)with possibly reversible metamorphosis (see Sachs,1986:the development and reversal of malignancy).

(D)Growth and local interaction. The classical stochastic approach to biological growth is represented by the creation and application of branching processes and birth-and-death processes. If spatial motion is additionally assumed,one deals with branching Brownian motion,birth-death-and-diffusion processes,etc. For interacting BCS the following hypotheses are substantial:

(1)Cell multiplication:it is usually specified that a living cell divides into two daughter cells. The growth process is frequently assumed "permanent"(or "irreversible"),i.e., conditioned not to die. In the case of multitype BCS (assumption C3),cells of initial types may disappear as a consequence of their metamorphoses.

(2)The lifespan of a cell:it can be represented by (1) an interval of random length,which can be thought of as the waiting time or the "clock" of a cell,(2)a sequence of m 1 discrete intervals(e.g.,cell cycle "phases"),(3)a continuum of states(e.g.,the chain of reactions for DNA synthesis),etc.

(3)Local interaction:it may be of (1)short range,(2)long range,and (3) zero range. The exclusion rule (0) can be correspondingly modified:(02)a cell continues to search for a vacant site until it finds one; (03)at any given time,a cell interacts only with the cells which occupy its site at that time. Rule (02) suggests the existence of "wandering" cells on the lattice,nevertheless different from migrating cells coming from other tissues.

It is frequently assumed that the interaction in BCS should be of short range:after the division of a cell located at site x,say,on the lattice,one daughter cell must search for a place (according to rule 0),and this place must be in the "neighbourhood" of its "parent" cell.

(3')Local interaction by cell-cell contact(e.g.,transmission or exchange of ions,metabolites,genetic material, infective particles,electric signals,etc.). This kind of interaction induces the spread of characteristics,with neglect of multiplication hypothesis D1. If we call this interaction "transmission of opinions",then the laber of "voter model" is justified for this interaction process.

(4)The choice of a site can happen

(4.1)with equal probability;

(4.2)with a probability depending on some "demographic" parameter(e.g.,the number of neighbour cells not in the mitotic phase,the number of neighbour cells of different colours,etc.).

(5)The effects of local interactions are:

(5.1)the competition loser vanishes:the new daughter cell can "kill" the old occupant. This effect might be interpreted as one of the mechanisms of the "cell loss" phenomenon.

(5.2)the competition loser, is shifted to another site (e.g.,intestinal crypts) or on another superposed lattice (e.g.,skin layers).

(5.3)both concurrents "coalesce":the new daughter cell and the old occupant form a "hybrid" cell.

(5.3')the chosen neighbour changes its opinion(s),that is,it will be infected,excited,etc.

If we take into account only the situation at a site x,we deal with (a)occupancy ("mass creation","birth"),(b)coalescence(state change),independent of multiplication,or (c)vacancy(annihilation,"mass destruction","death"). The phenomenon of annihilation will not be taken into consideration:it implies a system of mobile cells or living particels(e.g., immune cells,viruses) which produce the lysis of some cells belonging to the considered BCS. One of the suggestive examples is tumour rejection (induced by cytotoxic T-lymphocytes, natural killer cells,antibody-dependent killer cells,macrophages,etc.).

2.EXAMPLES

In 1973,D.Richardson first collected nine examples of spatial ("tessellation") random growth processes G,and investigated the asymptotic behaviour of some of them within the framework of the theory of subadditive stochastic processes. All his examples are discrete time processes taking place on the square lattice Z^2. Let us consider Example 4,G^*pq:

Let $p>0$ be the probability of change $W \to B$ (if W has one or more B-neighbours),and $q>0$ the probability of change $B \to W$ (if B has one W-neighbour). Let H be the set of histories in which the black clone survives,and m the measure on the set of histories corresponding to a growth process Gpq. Then G^*pq is the growth process which has the measure $\mu(A)=m(A \cap H)/m(H)$, for each measurable set A. It is obtained from Gpq by assuming that the black clone survives.

It is easy to check up the assumptions available for G^*pq:

C.2.1,D1(permanent growth),D.3.1;assumptions of D4 and D5 type are not explicit.

Example 5 in Richardson's work deserves a special attention : "Divide the plane into squares and start the process at time 0 with one particle in each square. The particle at the origin is unwell and all others are healthy. As time passes, the particles follow independent Brownian motion trajectories. A square become black at the first instant it contains an unwell particle. A healthy particle becomes unwell as soon as it shares a square with an unwell particle." This example suggests a kind of aggregation process resembling the diffusion-limited aggregation (DLA) considered by T.A.Witten and L.M.Sander(1983). The growth process in DLA (a "kinetic particle growth") is characterized by the fact that only an "active" outside region of the aggregate contacts new particles. This would also happen when the healthy Brownian traveller meets an infected particle. Computer simulations show the formation of dendritic clusters with deep "fjords" : newly arriving particles almost always stick on a peninsula rather than penetrating a fjord (Stanley et al.,1985). It is known that DLA models possess properties associated with scale invariance and generally have a fractal dimension D smaller than the Euclidean dimension d (e.g.,Meakin,1983a,1986;Rácz and Plischke,1985,etc.). D.Richardson interpreted his example an an epidemic model ; it may be thought of as an example of a rather special interacting BCS in immunology.

R.T.Smythe and J.C.Wierman(1978,§10.6) called "Richardson model" a Gpq model with the following assumptions:A1,B1,C.2.1, D1,D.3.1,and D.4.2 (assumption D5 not explicit)(See Durrett and Liggett,1981;Durrett and Griffeath,1982).

In this paper the following five interacting BCS will be mentioned:

(1)The Eden model 1 (E1-model),based on the following assumptions:A1,B1,C1,D1,D.2.1,D.3.1,D.4.1,and D.5.1.

(2)The Eden model 2 (E2-model) in which D.4.1 is replaced by assumption D.4.2.

(3)The Williams-Bjerknes model 1 (WB1-model),based on A1, B1,C.2.2.2,D1,D.2.1,D.3.1,D.4.1,and D.5.1. Although D.Richardson identifies it with the E1-model (there is no example for E2-model),we consider the E1-model as a homogeneous interacting BCS (with indistinguishable cells).

(4)The Williams-Bjerknes model 2 (WB2-model),distinct from WB1-model by assumption C.2.2.3.

(5)The k-type configuration model for carcinogenesis(C1-model),with assumptions A1,B1,C.3.2,D1,D.2.2,D.3.1,D.4.1,and D.5.1. (A C2-model assuming a constant number of cells on a fixed subset of Z^d,$d\geq 2$,has been constructed (Schürger and Tautu,1976b) with the intention to describe carcinogenesis in vitro or special experiments.)

3.THE STOCHASTIC THEORY OF INTERACTING BCS

In order to construct interacting BCS as Markov processes, some definitions and notation are required. For details the reader is referred to the excellent book by T.M.Liggett(1985).

(E1).The countable set of sites is generally denoted by S. In this paper S will be identified by the infinite d-dimensional lattice Z^d,$d\geq 1$. A site $x\in Z^d$ is defined by the d-tuple $(x^1,\ldots,x^d)$ of integers $x^i\in Z^d$,$1\leq i\leq d$. Sites x and y in Z^d are called neighbours iff $|x-y|=1$,where $|\cdot|$ denotes the Euclidean norm. The set of all neighbours of $x\in Z^d$ is N_x. (The example of the Bethe tree -assumption A1- illustrate the fact that the word "neighbours" do not necessarily imply closeness in terms of Euclidean distance : Hammersley and Mazzarino,1983. For instance,it is pointed out (Holley and Stroock,1978) that in the nearest neighbour models (see Liggett,1985,Chap.VII) the interaction is of indeterminant range because one does not know how far away one's "nearest neighbours" reside. Also,the set of all subsets of Z^d will be denoted by L,and

L_o will represent the set of all finite subsets Λ of Z^d. [Remark:This means that the biological cell systems presented in this paper are viewed as infinite variants of the original Eden and Williams-Bjerknes models.]

(E2).The phase space for each site is denoted by W : in the simplest case,W={vacant,occupied} or W={white,black}, equivalent to W={0,1}. It is assumed that W is a (compact) metric space. The situation on the whole lattice is specified by the set $\Xi=W^{Z^d}$,which is the state space of the interacting stochastic process. We think of Ξ as the compact multiplicative group obtained by raising the multiplicative group W to the the power S.Ξ is a metric space. In the simplest case,$\Xi = \{0,1\}^{Z^d}$. Other cases of interest are W={1,...,m} (assumption D.2.2) or W=K×M,where K={1,...,k},the set of colours and M= ={1,...,m},the set of cell cycle phases (assumptions C3 and D.2.2. For other examples,see Tautu,1986).

(E3). $\xi(x)\in W$ is a W-valued random function representing the value of configuration $\xi\in\Xi$ at site $x\in Z^d$. It is the x-coordinate of configuration ξ:its value at time t is a Markov process with states $a\in W$. For example,$\xi(x)=0$ if $x\in Z^d$ is vacant (or white),and $\xi(x)=1$ if site $x\in Z^d$ is occupied (or black). Also,$\xi(x)=(i,j)$ means that site $x\in Z^d$ is occupied by a cell of colour $i\in K$ and in phase $j\in M$. Occupancy rule (O) allows us to consider Ξ the configuration space of the system ; it is a metric space provided with the product topology.

(E4).The evolution of interacting BCS is determined by a collection of continuous nonnegative transition rates ("speed functions") on Ξ ,$\{c(x,\xi,a):x\in Z^d,\xi\in\Xi,a\in W\}$,satisfying $0\le c(x,\cdot,\cdot)\in C(\Xi)$,for each $x\in Z^d$. [By $C\equiv C(\Xi)$ we denote as usual the algebra of real-valued continuous functions $f:\Xi\to R^1_+$,i.e. a Banach space with respect to the uniform(supremum)norm.] Literally,$c(x,\xi,a)$ is the rate at which a living cell located at site $x\in Z^d$ changes its state from $\xi(x)\in W$ to $a\in W$,when the entire cellular configuration is $\xi\in\Xi$. Clearly,the behaviour of the system strongly depends on the precise nature of the interaction. Three properties are essential : finite range (R), translation invariance (spatial stationarity, S), and attractiveness (A) - see the definitions in Durrett and Griffeath (1982, §2) and Holley (1985).

Example 1: Consider the E1-model. A cell located at site x on Z^d, $d \geq 1$, waits an exponential time with parameter one (assumption D.2.1), then it divides, and one daughter cell chooses a neighbour site, say $y \in N_x$, with probability $p(x,y)=(2d)^{-1}$ (assumption D.4.1). The transition rate of the process is

$$c(x,\xi)=\begin{cases}\sum_y p(x,y)\xi(y), & \text{if } \xi(x)=0\\ \sum_y p(x,y)[1-\xi(y)], & \text{if } \xi(x)=1,\end{cases}$$

where $p(x,y) \geq 0$, $\sum p(x,y)=1$, $x,y \in Z^d$, $y \in N_x$, for $|x-y|=1$. This can be written as

$$(4.1) \qquad c(x,\xi) = (2d)^{-1}\sum_y 1_{\{\xi(y)\neq\xi(x)\}}, \; |y-x|=1.$$

(E5). The construction of the stochastic interacting BCS is realized in the following three steps (Liggett, 1985, Chap. I) : (i) the given infinitesimal rates are used to write down an operator (resp. a Markov pregenerator), whose closure will ultimately be the generator of the process, (ii) construct the corresponding semigroup of contraction operators on $C(\Xi)$ by using the 1-1 correspondence between the generators and semigroups, and then (iii) define the unique Markov process corresponding to the semigroup.

Example 2: Consider the C1-model. A cell of colour $i \in K$, located at $x \in Z^d$, $d \geq 1$, runs through its cycle phyases with intensity $a_j > 0$, $j \in M$ (assumption D.2.2, not colour-dependent) and divides (when in phase $m \in M$) with intensity $d_i > 0$, $i \in K$. Both daughter cells take simultaneously the same colour $i' \in K$ with probability $d_{ii'}$, such that $d_{ii'}=0$ if $i' \notin \{i-1,i,i+1\} \cap K$, and $d_{i-1,i}+d_{ii}+d_{i+1,i}=1$. One of these daughter cells chooses site

$y \in N_x$ with probability $p(x,y)=(2d)^{-1}$, for $|x-y|=1$. The infinitesimal generator is

$$(4.2)\quad Af(\xi) = \sum_{j=1}^{m-1} \sum_{i=1}^{k} \sum_{i' \in K(i)} (2d)^{-1} c(j;i,i') \times \sum_{y} \sum_{z} [f(\xi^{(j;i,i')}(y,z) - f(\xi)],$$

$$y \in Z^d, z \in N_y, K(i) = \{i-1, i, i+1\} \cap K ,$$

where

$$(4.3)\quad c(j,i,i') = a_j d_i d_{ii'} .$$

The semigroup T_t will be defined (Step ii) by

$$T_t f = \sum_{n=0}^{\infty} \frac{t^n}{n!} A^n f , \quad 0 \le t < c(d), f \in C_o$$

where

$$c(d) = [2\sum_{j=1}^{m-1} a_j + \sum_{i=1}^{k} d_i)3^d \exp\{3^d\}]^{-1} , \quad d \ge 1$$

(Schürger and Tautu,1976a,Th.2.1). The corresponding stochastic process (Step iii) is a Hunt process $\{\xi_t\}_{t\ge 0}$ with state space (Ξ, X), that is, a time homogeneous quasi-left-continuous strong Markov process with all paths right-continuous and with left-hand limits (Schürger and Tautu,1976a,Th.2.2). [Note:For WB-models consult M.Bramson and D.Griffeath,1980a, b,1981.]

The analogy between the El-model and the invasion model (Clifford and Sudbury,1973) or the basic voter model (Holley and Liggett,1975;see Liggett,1985,Chap.V) is evidenced by (4.1). It follows that the considered class of interacting BCS, as strong Markov processes, is steadily connected with the class of Markov spin-flip systems. Actually, the latter systems have been defined by T.M.Liggett(1980,§2) as "birth-death" systems with

$$P\{\xi_t(x) \ne \xi(x)\} = c(x,\xi)h + o(h),$$

and

$$P\{\xi_t(x) \ne \xi(x), \xi_t(y) \ne \xi(y)\} = o(h), \text{for } x \ne y \in Z^d, d \ge 1.$$

4. QUALITATIVE BEHAVIOUR OF BASIC INTERACTING SYSTEMS

Let us consider the invasion (basic voter) model. It has at least two equilibrium states, namely μ_1 and μ_0, concentrated on $\xi(x)\equiv 1$ and $\xi(x)\equiv 0$. For the existence of other equilibrium states we first need that its dual (auxiliary) process (the annihilating random walk) fulfils Spitzr's condition (C) (Spitzer, 1981). Then,

(i) If (C) holds, the equilibrium states are all of the form

$$\theta\mu_1+[1-\theta]\mu_0 \quad , \quad 0\leq\theta\leq 1.$$

(ii) If (C) fails, there is an additional 1-parameter family $\{\mu_\alpha, 0\leq\alpha\leq 1\}$ of extremal invariant measures which are the weak limits, as $t\to\infty$, of $\nu_\alpha T_t$, where ν_α is the Bernoulli product measure on Ξ with density α and marginals

$$\nu_\alpha\{\xi\,|\,\xi(x)=1\}\equiv\alpha \quad , \quad \alpha\in(0,1), \text{for all } x\in Z^d$$

(Spitzer, 1981, Th. 1.6). R. Holley and T. Liggett (1975) have shown that if (ξ_t) has initial distribution ν_α, and if $\nu_\alpha P_t$ denotes the measure governing the state of the process at time t, then $\nu_\alpha P_t \Longrightarrow \mu_\alpha$ ($\Longrightarrow$ weak convergence) as $t\to\infty$. μ_α is an extreme equilibrium measure for (ξ_t), is translation invariant and mixing.

Let denote by ξ_t^α the voter model started in Bernoulli product measure ν_α. The fundamental behaviour of such an interaction system is

(I) *Clustering* : $P\{\xi_\infty^\alpha\in\cdot\}=\theta\mu_1+[1-\theta]\mu_0$, if $d=1,2$

(II) *Stability* : $P\{\xi_\infty^\alpha\in\cdot\}=\mu_\alpha$, if $d\geq 3$

(see, e.g., Cox and Griffeath, 1986, Th. 1). It is pointed out that μ_α exhibit unusual long range correlations. Surprisingly, it appears that the dichotomy I-II was anticipated by the French geneticist G. Malécot.

This system is not ergodic. The behaviour is an example of critical dimensionality:

(I) If $d=1,2$, the distribution of ξ_t^α converges weakly to a mixture of μ_0 (:all sites white) and μ_1 (:all sites black), as $t \to \infty$;

(II) If $d \geq 3$, the distribution of ξ_t^α converges to a nondegenerate steady state.

This dichotomy can be related to the propertis of the dual processes, so that we have

(I) *Clustering* system/*recurrent* random walk, for $d=1,2$

(II) *Stable* system/*transient* random walk, for $d \geq 3$.

(see Schwartz, 1977, Prop. 5.5; Donnely and Welsh, 1983; Liggett, 1985) We can explain that a black clone tends to be invasive in a 3-dimensional space (Tautu, 1978, 1980) or tends to form a "stable" pattern in the tissue (Pilz, 1986) if the system is "unbiased".

Another dimension dependence can be observed when we consider the behaviour of the occupation time process

$$T_t(A) = \sum_{x \in A} \int_0^t \xi_s^\alpha(x)\,ds \quad , \quad A \subset L_0 ,$$

where $\xi_s(x)$ is the configuration at time s of the voter model and is density preserving (Cox and Griffeath, 1983). There is no law of large numbers for T_t if $d=1$: T_t/t converges in distribution to a nonconstant limiting random variable which can be represented by means of annihilating Brownian motions (Cox and Griffeath, 1986). If $d \geq 2$, $T_t/t \to \alpha$ a.s. (Cox and Griffeath, 1983, Th. 2; see also Liggett, 1985, p. 262).

A very interesting result has recently been obtained by J.T. Cox and D. Griffeath (1986) with regard to the density of black points in a 2-dimensional voter model on a box of side $2t^{\alpha/2}$. As $t \to \infty$, this density converges as a process in $\alpha \in (0,1]$ to a time change of the Fisher-Wright diffusion process. (The reader is referred to T.G. Kurtz (1980, §5) for the random time change problem for interacting particle systems.) This convergence and a generalization of k-type invasion model will be discussed in a separate paper.

5. THE ASYMPTOTIC SHAPE OF PERMANENT GROWTH MODELS

The main theorem given by D.Richardson(1973) states -in simple words- that if A_n is the set of black sites on Z^d at time n, then the set of points $n^{-1/d}A_n$ will approach a circle. The problem can be analyzed in terms of first-passage percolation. The reader is referred for details to R.T.Smythe and J.C.Wierman(1978), R.T.Smythe(1980), and H.Kesten(1985,1986).

Let us consider a graph G with V the set of vertices (syn.: points, nodes, atoms, sites) and E the set of edges (syn.: lines, loops, bonds). With the set $E=\{e \mid e=(v_{i-1},v_i), v_k \in V, 0\le k\le\infty\}$ is associated a sequence $\{X_j\}, 1\le j\le\infty$, of positive iid random variables with finite mean, representing the time coordinates of a bond, i.e., $\{X(e_j), e_j\in E, 1\le j\le\infty\}$. The common distribution of X's is denoted by F $(F(0\_)=0)$ and often assumed exponential. As usual, a path π is defined as a sequence of sites with a sequence of bonds, say $(v_o,e_1,v_1,\ldots,e_n,v_n)$. The passage time of π is defined as

$$T(\pi) = \sum_{j=1}^{n} X(e_j) ,$$

the random time needed to traverse π. If x is a "source" vertex and y is any other vertex on G, one defines the first passage time ("travel time") from x to y to be the infimum of $T(\pi)$, taken over all paths from x to y :

$$T(x,y)=\inf\{T(\pi):\pi \text{ a path from x to y}\}.$$

If G is a lattice embedded in Euclidean space, we can divide T(x,y) by the Euclidean distance between x and y and study the limit μ of this quotient as y recedes to a great distance from x. This limit exists w.p.1 and is called time constant:

$$\mu=\mu(F,d) ,$$

and $\mu(F,d)=0$ iff $F(0)\ge p_T(d)$, where $p_T(d)$ is the critical probability of Bernoulli percolation (Kesten,1986,Th.6.1). This critical probability is decreasing in d : $p_T(d)<1/2$ for $d\ge3$, and $p_T(d)\ge 1/(2d-1)$.

Other assumptions may be suggested :

(F1) $\{X_j\}$ are not iid random variables ; their distributions may be dimension-dependent and/or colour-dependent.

(F2) $T(x,y)$ is referred to vertices instead of bonds, or between two sets of vertices, V_1 and V_2:

$$T(V_1,V_2)=\inf\{T(x,y), x\in V_1, y\in V_2\},$$

assumption which might be of interest in polychromatic percolation.

Let denote by $\underline{0}$ the origin of a percolation process and by $A(t)$ all sites that can be reached (coloured) in time t, starting from $\underline{0}$:

$$A(t)=\{v\in Z^d \,|\, T(\underline{0},v)\le t\}$$

Flattening up $A(t)$ by filling up a unit cube around each vertex, we replace it by a set $B(t)$,

$$B(t)=\{y+I \ : \ y\in A(t)\},$$

where I is the closed d-dimensional cube $I=\{z=(z^1,\dots,z^d): -1/2\le z^i\le 1/2, 1\le i\le d\}$.

The following theorem gives us the correct answer to the problem of the asymptotic shape of a simple permanent growth process (e.g., E1, E2, WB1) :

<u>Theorem 1</u> (Kesten, 1985, Th. 1.1; 1986, Th. 1.6). Assume that

(F3) $E[\min\{t_1^d,\dots,t_{2d}^d\}]<\infty$

where $t_1,\dots,t_{2d}$ are iid random variables with distribution F. Then there exists a nonrandom convex set $B_o\subset R^d$ with the following properties :

(i) B_o is invariant under permutations of coordinates and under reflections in the coordinate hyperplanes;

(ii) has nonempty interior;

(iii) is either compact or equalls all of R^d. In this respect,

(*) If B_o is compact, then for all $\varepsilon>0$,

$$(1-\varepsilon)B_o\subset 1/tB(t)\subset(1+\varepsilon)B_o \quad \text{eventually w.p.1.}$$

(**) If $B_o\in R^d$ (iff $\mu=0$), then for all $\varepsilon>0$,

$$\{x : |x| \ \varepsilon^{-1}\}\subset 1/t\ B(t) \quad \text{eventually w.p.1.}$$

If (F3) fails, then

$$\lim_{v\to\infty} \sup 1/|v|\cdot T(\underline{0},v) = \infty, \text{ w.p.1.}$$

Thus, (F3) is a necessary condition for the a.s. linear growth of B(t). The distinction (iii) between compact or noncompact B_o (or $\mu>0$ and $\mu=0$) depends only -as H.Kesten(1986,p.129) pointed out- on the atom of F at the origin.

Theorem 1 can be called the law of large configurations (following Schürger,1980). Both Eden models can be analyzed in this way : E1-model with T(x,y) attached to vertices (assumption F2)[see also Durrett and Liggett,1981,for the Richardson model]. If F(u) is exponential,i.e.,

$$F(u)=\begin{cases} 0, \text{if } u<0 \\ 1-e^{-u}, \text{if } u\geq 0 \end{cases}$$

then the shape of B(·) at certain random times has the same distribution as in E2-model (see arguments in Kesten,1986,p. 131).

B(t) can be approximated by

$$\hat{B}(t)=\{y+I : y\in Z^d,\ \hat{T}(\underline{0},y)\leq t\}$$

where $\hat{T}(\underline{0},y)$ is a modified passage time having all moments finite, such that

$$\{\hat{T}(x,y)-T(x,y), x,y\in Z^d\} \text{ is tight.}$$

Then (*) and (**) hold for $\hat{B}(t)$ without condition F3 (Kesten, 1986,Th.3.1).

The d-dimensional permanent WB2-model exhibits the same asymptotic linear behaviour as the WB1-model. If the malignant clone represents a large enough ball,then the black cell population will with overwhelming probability not become extinct and will grow at (at least) a radially linear rate (see Bramson and Griffeath,1980b). This is stated in

Theorem 2(Bramson and Griffeath,1981,Th.2). Let$\{\xi_t^A, A\in L\}$ be the d-dimensional WB2-model (with $\beta>1$). For any $A\in L, A\neq\phi$, there is a constant $c=c(d,\beta)>0$ such that

$$P\{D_{ct}\subset\xi_t^A, t\leq t'<\infty \mid \tau_\phi=\infty\} = 1,$$

where $D_r=\{x\ Z^d: \|x\|\leq r\}$, and $P\{\tau_\phi^o=\infty\}=1$ (permanent growth). Then

$$P\{\lim_{t\to\infty} \xi_t^A = Z^d\} = 1 ,$$

i.e., any finite subset A of Z^d will eventually be permanently occupied.

A similar asymptotic behaviour has been claimed for the Cl-model (Schürger and Tautu, 1976a, Conjecture 3.2).

REFERENCES

Bramson,M.,Griffeath,D.(1980a).The asymptotic behaviour of a probabilistic model for tumor growth. In:Biological Growth and Spread (W.Jäger,H.Rost,P.Tautu eds.)[Lecture Notes in Biomath.,Vol.38],pp.165-172. Berlin-Heidelberg-New York: Springer

Bramson,M.,Griffeath,D.(1980b,1981).On the Williams-Bjerknes tumour growth model.I,II. Ann.Probab.,9,173-185;Math.Proc. Cambridge Philos.Soc.,88,339-357

Clifford,P.,Sudbury,A.(1973).A model for spatial conflict. Biometrika,60,581-588

Cox,J.T.,Griffeath,D.(1983).Occupation time limit theorems for the voter model. Ann.Probab.,11,876-893

Cox,J.T.,Griffeath,D.(1986).Diffusive clustering in the two dimensional voter model. Ann.Probab.,14,347-370

Donnelly,P.,Welsh,D.(1983).Finite particle systems and infection models. Math.Proc.Cambridge Philos.Soc.,94,167-182

Durrett,R.(1981).An introduction to infinite particle systems. Stoch.Proc.Appl.,11,109-150

Durrett,R.,Griffeath,D.(1982).Contact processes in several dimensions. Z.Wahrscheinlichkeitstheorie verw.Gebiete,59, 535-552

Durrett,R.,Liggett,T.M.(1981).The shape of the limit set in Richardson's growth model. Ann.Probab.,9,186-193

Eden,M.(1958).A probabilistic model for morphogenesis. In: Symp.on Information Theory in Biology (H.P.Yockey ed.), pp.359-370. New York:Pergamon Press

Eden,M.(1961).A two-dimensional growth process. Proc.4th Berkeley Symp.Math.Statist.Probability,Vol.IV,pp.223-239. Berkeley:Univ.California Press

Gray,L.(1985).The critical behavior of a class of simple interacting systems-A few answers and a lot of questions. In:Particle Systems,Random Media,and Large Deviations (R. Durrett ed.)[Contemporary Mathematics,Vol.41],pp.149-160.

Providence:Amer.Math.Soc.

Hammersley,J.M.,Mazzariono,G.(1983).Markov fields,correlated percolation,and the Ising model. In:The Mathematics and Physics of Disordered Media (B.D.Hughes,B.W.Ninham eds.) [Lect.Notes Math.,Vol.1035],pp.201-245. Berlin-Heidelberg-New York-Tokyo:Springer

Hammersley,J.M.,Welsh,D.J.A.(1965).First-passage percolation, subadditive processes,stochastic networks,and generalized renewal theory. In:Bernoulli,Bayes,Laplace.Anniversary Volume (J.Neyman,L.M.LeCam eds.),pp.61-110. Berlin-Heidelberg-New York:Springer

Holley,R.(1985).Possible rates of convergence in finite range, attractive spin systems. In:Particle Systems,Random Media, and Large Deviations (R.Durrett ed.)[Contemporary Mathematics,Vol.41],pp.215-234. Providence:Amer.Math.Soc.

Holley,R.,Liggett,T.M.(1975).Ergodic theorems for weakly interacting infinite systems and the voter model. Ann. Probab.,3,643-663

Holley,R.,Stroock,D.(1978).Invariance principles for some infinite particle systems. In:Stochastic Analysis (A. Friedman,M.Pinsky eds.),pp.153-173. New York:Academic Press

Iscoe,I.(1986).A weighted occupation time for a class of measure-valued branching processes. Probab.Theory Rel. Fields,71,85-116

Kesten,H.(1985).First-passage percolation and a higher dimensional generalization. In:Particle Systems,Random Media, and Large Deviations (R.Durrett ed.)[Contemporary Mathemathics,Vol.41],pp.235-251. Providence:Amer.Math.Soc.

Kesten,H.(1986).Aspects of first passage percolation. In: École d'Été de Probabilités de Saint-Flour XIV-1984 (P.L. Hennequin ed.)[Lect.Notes Math.,Vol.1180],pp.125-264. Berlin-Heidelberg-New York-Tokyo:Springer

Kurtz,T.G.(1980).Representations of Markov processes as multiparameter time changes. Ann.Probab.,8,682-715

Liggett,T.M.(1980).Interacting Markov processes. In:Biological Growth and Spread (W.Jäger,H.Rost,P.Tautu eds.)[Lect.Notes in Biomath.,Vol.38],pp.145-156. Berlin-Heidelberg-New York:Springer

Liggett,T.M.(1985).Interacting Particle Systems. New York-Berlin-Heidelberg-Tokyo:Springer

Meakin,P.(1983a).Cluster-growth processes on a two-dimensional lattice. Phys.Rev.B,28,6718-6732

Meakin,P.(1983b).The Vold-Sutherland and Eden models of cluster formation. J.Colloid Interface Sci.,96,415-424

Meakin,P.(1986).A new model for biologica pattern formation. J.Theor.Biol.,118,101-113

Mollison,D.(1974).Percolation processes and tumour growth (Abstr.). Adv.Appl.Probab.,6,233-235

Peters,H.P.,Stauffer,D.,Hölters,H.P.,Loewenich,K.(1979). Radius,perimeter,and density profile for percolation clusters and lattice animals. Z.Physik B,34,399-408

Pilz,L.(1986).Occupation time of biological cell systems. In: Modelling,Identification,and Control. Proc.IASTED Intern. Symp. (M.H.Hamza ed.),pp.128-131. Anaheim-Calgary-Zürich: Acta Press

Rácz,Z.,Plischke,M.(1985).Active zone of growing clusters: Diffusion-limited aggregation and the Eden model in two and three dimensions. Phys.Rev.A,31,985-994

Richardson,D.(1973).Random growth in a tessellation. Proc. Cambridge Philos.Soc.,74,515-528

Rikvold,P.A.(1982).Simulations of a stochastic model for cluster growth on a square lattice. Phys.Rev.A,26,647-650

Sachs,L.(1986).The development and reversal of malignancy. Cancer Rev.,2,48-64

Sander,L.M.(1984).Theory of fractal growth processes. In: Kinetics of Aggregation and Gelation (F.Family,D.P.Landau eds.),pp 13-17. Amsterdam:North Holland

Sawada,Y.,Ohta,S.,Yamazaki,M.,Honjo,H.(1982).Self-similarity and a phase-transition-like behavior of a random growing structure governed by a nonequilibrium parameter. Phys. Rev.A,26,3557-3563

Sawyer,S.(1976).Results for the stepping-stone model for migration in population genetics. Ann.Probab.,4,699-728

Schürger,K.(1980a).On the asymptotic geometrical behaviour of percolation processes. J.Appl.Probab.,17,385-402

Schürger,K.(1980b).On a class of branching processes on a lattice with interactions. In:Biological Growth and Spread (W.Jäger,H.Rost,P.Tautu eds.)[Lect.Notes Biomath.,Vol.38], pp.157-164. Berlin-Heidelberg-New York:Springer

Schürger,K.(1981).A class of branching processes on a lattice with interactions. Adv.Appl.Probab.,13,14-39

Schürger,K.,Tautu,P.(1976a).A Markovian configuration model for carcinogenesis. In:Mathematical Models in Medicine (J.Berger,W.Bühler,R.Repges,P.Tautu eds.)[Lect.Notes Biomath.,Vol.11],pp.92-108. Berlin-Heidelberg-New York: Springer

Schürger,K.,Tautu,P.(1976b).Markov configuration processes on a lattice. Rev.Roumaine Math.Pures Appl.,21,233-244

Schürger,K.,Tautu,P.(1978).Die Simulation eines mathematischen Modells der Krebsentstehung. IBM Nachrichten,Heft 242, 265-273

Schwartz,D.(1977).Applications of duality to a class of Markov processes. Ann.Probab.,5,522-532

Schwöbel,W.,Geidel,H.,Lorenz,R.J.(1966).Ein Modell der Plaquebildung. Z.Naturforsch.,21,953-959

Shiga,T.(1982).Continuous time multi-allelic stepping stone models in population genetics. J.Math.Kyoto Univ.,22,1-40

Smythe,R.T.(1980).Percolation models in two and three dimensions. In:Biological Growth and Spread (W.Jäger,H.Rost, P.Tautu eds.) Lect.Notes Biomath.,Vol.38 ,pp.504-511. Berlin-Heidelberg-New York:Springer

Smythe,R.T.,Wierman,J.C.(1978).First-Passage Percolation on the Square Lattice. Lect.Notes Math.,Vol.671. Berlin-Heidelberg-New York:Springer

Spitzer,F.(1981).Infinite systems with locally interacting components. Ann.Probab.,9,349-364

Stanley,H.E.,Family,F.,Gould,H.(1985).Kinetics of aggregation and gelation. J.Polymer Sci.,73,19-37

Stauffer,D.(1979).Scaling theory of percolation clusters. Phys.Rep.,54,1-74

Tautu,P.(1976).On random systems of biological interacting objects. Rev.Roumaine Math.Pures Appl.,21,795-802

Tautu,P.(1978).Blackening a d-dimensional lattice. Rev. Roumaine Math.Pures Appl.,23,141-152

Tautu,P.(1980).Biological interpretation of a random configuration model for carcinogenesis. In:Biological Growth and Spread (W.Jäger,H.Rost,P.Tautu eds.)[Lect.Notes Biomath., Vol.38],pp.196-220. Berlin-Heidelberg-New York:Springer

Tautu,P.(1986).Random fields:Applications in cell biology. In:Stochastic Spatial Processes[Lect.Notes Math.,Vol. 1212,in press].Berlin-Heidelberg-New York-Tokyo:Springer

Williams,T.(1971).Unpubl.manuscript. Symp.Tobacco Res.Council

Williams,T.,Bjerknes,R.(1972).Stochastic model for abnormal clone spread through epithelial basal layer. Nature,236, 19-21

Zallen,R.(1983).The Physics of Amorphous Solids. New York: Wiley

S. Albeverio
Ruhr-Universität Bochum
Fachbereich Mathematik
Postfach 10 21 48
4630 Bochum
FRG

G.B. Arous
ENS
Centre de Math. Appl.
45, Rue d'Ulm
75005 Paris
France

R. Bachmann
Bayer AG
5090 leverkusen
FRG

J. Bernasconi
Brown-Boveri Research Center
5405 Baden
Switzerland

Ph. Blanchard
Universität Bielefeld
Fakultät für Physik
Postfach 86 40
4800 Bielefeld 1
FRG

G. Blankenship
University of Maryland
Electrical Engineering Department
College Park, Maryland 20110
USA

G.F. Bolz
Universität Bielefeld
Fakultät für Physik
Postfach 86 40
4800 Bielefeld 1
FRG

A. Brasche
Universität Bielefeld
Fakultät für Mathematik
Postfach 86 40
4800 Bielefeld 1
FRG

P.-L. Chow
Wayne State University
Department of Mathematics
Detroit, Michigan 48202
USA

Chr. Cocoza
Université Pierre et Marie Curie
Laboratoire de Probabilités
Tour 56
4, Place Jussieu
75230 Paris Cedex 05
France

N.J. Cutland
University of Hull
Department of Pure Mathematics
Cottingham Road
Hull, HU6 7RX
Great Britain

P. Drücke
Krupp Atlas Electronik GmbH
Gelbaldsbrücker Heerstr. 235
2800 Bremen 44
FRG

D. Dürr
present address:
BiBoS Forschungszentrum
Universität Bielefeld
Postfach 86 40
4800 Bielefeld 1

home address:
Ruhr-Universität Bochum
Fakultät fü Mathematik
Postfach 10 21 48
4630 Bochum
FRG

W. Ernst
Krupp Atlas Elektronic GmbH
Sebaldsbrücker Heerstr. 235
2800 Bremen 44
FRG

R. Figari
present address:
BiBoS Forschungszentrum
Universität Bielefeld
Postfach 86 40
4800 Bielefeld 1
FRG

home address:
Università di Napoli
Instituto di Fisica Teorica
Mostra d'Oltremare
Pad. 19
80125 Napoli
Italy

T. Fresewinkel
Ruhr-Universität Bochum
Lehrstuhl für Elektrische
Steuerung und Regelung
Postfach 10 21 48
4630 Bochum 1
FRG

S. Golin
Fakultät für Physik
Universität Bielefeld
Postfach 86 40
4800 Bielefeld 1
FRG

G. Gamez
Mathematisches Institut
Universität Erlangen
Bismarckstraße 1 1/2
8520 Erlangen
FRG

A. Grossmann
CNRS
Department of Theoretical Physics
Luminy - Case 907
13288 Marseille Cedex 09
France

F. Guerra
Instituto di Matematica
Università di Roma
G. Castelnuova
00158 Roma
Italy

Z. Haba
Institute of Theoretical Physics
University of Wroclaw
ul. Lybulskiego 36
50-205 Wroclaw
Poland

M. Hazewinkel
Centrum voor Wiskunde en Informatica
Kruislaan 413
Postbus 4079
1098 SJ Amsterdam
The Netherlands

T. Hida
University of Nagoya
Department of Mathematics
Faculty of Sciences
Nagoya 464
Japan

A. Hilbert
Universität Bielefeld
Fakultät für Physik
Postfach 8640
4800 Bielefeld 1
FRG

H. Hochadel
BASF AG
Dep. ZLI/NE
Gebäude Z 34
6700 Ludwigshafen
FRG

R. Høegh-Krohn
Universitet i Oslo
Matematisk Institutt
Postboks 10 53
Blindern Oslo 3
Norway

Ch. Kessler
Department of Pure Mathematics
University of Hull
Hull HU6 7RX
Great Britain

A. Kohl
Institut für Nachrichtengeräte
und Datenverarbeitung
RWTH Aachen
Templergraben 55
5100 Aachen
FRG

T. Koski
Abo Akademi
Matematisk Institutt
Fänriksgatan 3
205 00 Abo
Finland

P. Kotelenez
Universität Bremen
Forschungsschwerpunkt
Dynamische Systeme
FB Mathematik/Informatik
2800 Bremen 33
FRG

P. Kree
Université Paris VI
Institut de Mathematique
32, Rue Miollis
75015 Paris
France

T. Krüger
BiBoS Forschungszentrum
Universität Bielefeld
Postfach 86 40
4800 Bielefeld 1
FRG

G.J. Olsder
Delft University of Technology
Department of Mathematics
P.O. Box 356
2600 AJ Delft
The Netherlands

G. Papavassilopoulos
Delft University of Technology
Department of Mathematics
P.O. Box 356
2600 AJ Delft
The Netherlands

J. Picard
INRIA
Domaine de Voluceau - Roquencourt
P.O. Box 105
78-153 Le Chesnay
France

L. Pietronero
University of Groningen
Solid State Physics Lab.
Melkweg 1
9718 EP Groningen
The Netherlands

J. Potthoff
TU Berlin
Fachbereich Mathematik
(MA7-1)
Straße des 17 Juni 136
1000 Berlin 12
FRG

S. Rettig
Universität Darmstadt
Fachbereich Mathematik
Arb. Gr. 9
Schloßgartenstr. 7
6100 Darmstadt
FRG

M. Röckner
Universität Bielefeld
Fakultät für Mathematik
Postfach 86 40
4800 Bielefeld 1
FRG

M. Roussignol
Laboratoire de Probabilités
Tour 56
Université Pierre et Marie Curie
4, Place Jussieu
75230 Paris cedex 05
France

W. Schempp
Universität Siegen
Lehrstuhl für Mathematik I
Fachbereich 6
Hölderlingstr. 3
5900 Siegen 21
FRG

W.R. Schneider
Brown-Boveri Forschungszentrum
5405 Baden
Switzerland

M. Serva
present address:
BiBoS Forschungszentrum
Universität Bielefeld
Postfach 86 40
4800 Bielefeld 1
FRG

home address:
Università di Roma
"La Sapienza"
Instituto di Fisica
P.A. Moro
00185 Roma
Italy

T. Shiga
present address:
BiBoS Forschungszentrum
Universität Bielefeld
Postfach 86 40
4800 Bielefeld 1
FRG

home address:
Tokyo Institute of Technology
Department of Applied Physics
Oh-Okayama, Meguro
Tokyo 152
Japan

W. Splettstößer
Siemens AG
Unternehmensbereich Bauelemente
B WIS EZM D
Postfach 34 01 29
4000 Düsseldorf 31
FRG

P. Stichel
Fakultät für Physik
Universität Bielefeld
Postfach 8640
4800 Bielefeld 1
FRG

L. Streit
Fakultät für Physik
Universität Bielefeld
Postfach 86 40
4800 Bielefeld 1
FRG

Y. Takahashi
present address:
BiBoS Forschungszentrum
Universität Bielefeld
Postfach 86 40
4800 Bielefeld 1
FRG

home address:
University of Tokyo
College of Arts and Sciences
Komaba, Megruno
Tokyo 152
Japan

P. Tautu
Deutsches Krebsforschungszentrum
Abt. Mathematische Modelle
Im Neuenheimer Feld 280
6900 Heidelberg
FRG

A.C.F. Vorst
Econometric Institute
University of Rotterdam
P.O. Box 17 38
3000 DR Rotterdam
The Netherlands

W. Wedig
Universität Karlsruhe
Insitut für Technische Mechanik
Kaiserstraße 12
7500 Karlsruhe
FRG

J.C. Zambrini
present address:
BiBoS Forschungszentrum
Universität Bielefeld
Postfach 86 40
4800 Bielefeld 1
FRG

N. Zanghi
present address:
BiBoS Forschungszentrum
Universität Bielefeld
Postfach 86 40
4800 Bielefeld 1
FRG

Index